AI and Machine Learning for Mechanical and Electrical Engineering

Practical and informative, *AI and Machine Learning for Mechanical and Electrical Engineering* examines how artificial intelligence (AI) is changing the status quo in mechanical engineering, electrical systems, and management. Real-world examples and case studies demonstrate the application of AI in such diverse settings as industry and policymaking. This book illustrates how AI is playing a crucial role in enhancing productivity and innovation in various industries. It discusses transition methods and the ethical implications of using AI in mechanical engineering. Chapter highlights include the following:

- Developing a smart algorithm to integrate fault detection and classification
- Algorithms to investigate different testing scenarios for various anomalies in electric motors
- Data fusion to detect and assess electromechanical damage
- Neural networks for rolling bearing fault diagnosis
- Evolutionary algorithms to optimize deep learning models for water industry forecasts
- AI-based anomaly detection and root-cause analysis

An overarching theme is the transition from traditional mechanical, electrical, and management systems to AI-enabled smart systems. The book helps readers make sense of the challenges of integrating smart systems. It equips engineers with theoretical understanding as well as insight based on hands-on expertise. It shows how to better link and automate systems and improve productivity. This book not only shows how to implement smart solutions now but also shows the way to a more intelligent, productive, and interconnected future.

Dr. T. Rajasanthosh Kumar is an associate professor of the Department of Mechanical Engineering at Oriental Institute of Science and Technology, Bhopal, India.

Dr. Surendra Reddy Vinta is an associate professor of the School of Computer Science and Engineering at VIT-AP University, Amaravati, India.

Dr. Sagar Dhanraj Pande is head of the School of Engineering and Technology at Pimpri Chinchwad University, Pune, Maharashtra.

Dr. Aditya Khamparia is an assistant professor and coordinator of the Department of Computer Science at Babasaheb Bhimrao Ambedkar University, Satellite Centre, Amethi, India.

Innovations in Intelligent Internet of Everything (IoE)
Series Editor: Fadi Al-Turjman

Advances in Technological Innovations in Higher Education: Theory and Practices
Edited by Adarsh Garg, B.V. Babu, and Valentina E Balas

Computational Intelligence in Healthcare: Applications, Challenges, and Management
Edited by Meenu Gupta, Shakeel Ahmed, Rakesh Kumar, and Chadi Altrjman

Advances in SIoT (Social Internet of Things)
Edited by Gururaj H L, Pramod H B, and Gowtham M

Blockchain, IOT and AI Technologies for Supply Chain Management
Edited by Priyanka Chawla, Adarsh Kumar, Anand Nayyar, and Mohd Naved

Renewable Energy and AI for Sustainable Development
Edited by Sailesh Iyer, Anand Nayyar, Mohd Naved, and Fadi Al-Turjman

AIoT: Artificial Intelligence of Things
Edited by Arun Sekar Rajasekaran, Fadi Al-Turjman, and Suganyadevi S

AI and Machine Learning for Mechanical and Electrical Engineering
Edited by T. Rajasanthosh Kumar, Surendra Reddy Vinta, Sagar Dhanraj Pande, and Aditya Khamparia

For more information about the series, please visit: https://www.routledge.com/Innovations-in-Intelligent-Internet-of-Everything-IoE/book-series/IOE

AI and Machine Learning for Mechanical and Electrical Engineering

Edited by

T. Rajasanthosh Kumar, Surendra Reddy Vinta, Sagar Dhanraj Pande, and Aditya Khamparia

CRC Press
Taylor & Francis Group
Boca Raton London New York

CRC Press is an imprint of the
Taylor & Francis Group, an **informa** business
AN AUERBACH BOOK

Designed cover image: Shutterstock

First edition published 2026
2385 NW Executive Center Drive, Suite 320, Boca Raton FL 33431

and by CRC Press
4 Park Square, Milton Park, Abingdon, Oxon, OX14 4RN

CRC Press is an imprint of Taylor & Francis Group, LLC

ISBN: 9781032759487 (hbk)
ISBN: 9781032765624 (pbk)
ISBN: 9781003479062 (ebk)

DOI: 10.1201/9781003479062

Typeset in Garamond
by Newgen Publishing UK

Contents

1 **Development of a Smart Algorithm to Integrate Fault Detection and Classification of End-to-End Monitoring of Autonomous Transfer Vehicles** 1
P. PRABHAKARAN, MUDIT JOSHI, PRAMODA PATRO, ABHISHEK CHOUBEY, AND SANTOSH REDDY ADDULA

2 **Data Science and ML Algorithms to Investigate Different Testing Scenarios for Various Anomalies in Driven Electric Motor** 17
ARPIT JAIN, P. R. SIVARAMAN, ABHISHEK CHOUBEY, AND GUNA SEKHAR SAJJA

3 **A Data Fusion Technique to Detect and Assess Electromechanical Damage** 33
ARPIT JAIN, BALA GANESH N., ABHISHEK CHOUBEY, PRAMODA PATRO, AND LAKSHMAN KUMAR KANULLA

4 **AI: Classifications and Protection of the Smart Grid Systems** 52
AKSHAYA KUBBA, D. PRABAKAR, N. M. G. KUMAR, AND SHRUTI BHARGAVA CHOUBEY

5 **An Artificial Intelligence-Based Solar Radiation Prophesy Model for Green Energy Utilisation in the Energy Management System** 68
PRADEEP KUMAR MISHRA, AMIT KUMAR PATHAK, ABHISHEK CHOUBEY, AND SESIKALA BAPATLA

6 **Two-Channel Convolutional Neural Networks for Rolling Bearing Fault Diagnosis in Unbalanced Datasets** 85
RAMESH BABU PITTALA, BHUVANESWARI M. S., AND PRAMODA PATRO

7 The Implementation of Artificial Intelligence for Auto Gearbox Failure Detection 103
ASHISH A. MULAJKAR, G. LAXMAIAH, RAJU KUMAR, AND SHRUTI BHARGAVA CHOUBEY

8 Evolutionary Algorithms to Optimise Deep Learning Model for Water Industry Forecasts 117
RAVI H. GEDAM, SONIA H. BAJAJ, HEMLATA RAHUL KOSRAE, VISHAKHA AKHARA, AND GIRISH A. GARG HATE

9 Artificial Intelligence Anomaly Detection and Root Cause Analysis ...138
VISHAL GANGADHAR PURANIK, T. CHITHRAKUMAR, E. NARESH, AND SHRUTI BHARGAVA CHOUBEY

10 Artificial Intelligence and Internet of Things-Based Intelligent Scheduling for Load Distribution in Power Grids 151
SELVAM SUGANYA, MUKESH SONI, SUSHILKUMAR DANGIYA, MAHER ALI RUSHO, AND SAGAR DHANRAJ PANDE

11 Coordinated Response Strategies: Swarm Robotics for Crisis Management 182
RENU R. DANDGE, TUSHAR R. MAHORE, ANKUSH VASANT DAHAT, PALLAVI H. DHOLE, NIKHILESH P. MANKAR, AND SAGAR DHANRAJ PANDE

12 Smart Farming and Human Bioinformatics Systems Based on IoT and Sensor Devices....................198
G. BHAVYA, N. ASHWINI, K. SUMANA, MUKESH SONI, AND SAGAR DHANRAJ PANDE

13 Machine Learning Techniques Applied in Predictive Maintenance: A Review 221
P. CHENGA REDDY, KARAMALA NAVEEN, NAVEEN KILARI, AND NAGENDRA PANINI CHALLA

14 Optimization of Parameters During Tribological Investigations on *Azadirachta indica*-Based Bio-Composites 241
G. VENKATAKOTESWARARAO AND R. VINAYAGAMOORTHY

15 ANFIS Modelling Study on Surface Water Analysis 255
P. MEENAKSHI

16 **WSN-Based Optimal Crude Oil Storage Health Monitoring Framework** 276
ISMAIL KESHTA, MUKESH SONI, SAGAR DHANRAJ PANDE, ROSHAN RAJENDRAKUMAR KARWA, AND HAEWON BYEON

17 **Cybersecurity Education Gamification: A Current Review and Research Agenda** 296
RAMIZ SALAMA AND FADI AL-TURJMAN

18 **Artificial Intelligence and Cybersecurity in 6G Wireless Networks** 304
RAMIZ SALAMA AND FADI AL-TURJMAN

Index 320

Chapter 1

Development of a Smart Algorithm to Integrate Fault Detection and Classification of End-to-End Monitoring of Autonomous Transfer Vehicles

P. Prabhakaran, Mudit Joshi, Pramoda Patro, Abhishek Choubey, and Santosh Reddy Addula

1.1 Introduction

The notion of Industry 4.0 has transformed numerous industries, particularly manufacturing. Industrialisation, big data, IoT, smart measurement, and robots have transformed traditional production procedures into smarter ones. Manufacturing process integration and automation via technology have led to more sustainable and controlled production settings. Robots and other intelligent machinery

DOI: 10.1201/9781003479062-1

operate in dynamic environments with varying levels of autonomy in the industrial sector. Because of their erratic behaviour, autonomous robots are susceptible to operational and mechanical problems. To ensure minimal downtime and uninterrupted operations, it is crucial to promptly identify, diagnose, and monitor issues in operational devices. With advances in smart measuring technologies, computer hardware, and deep learning, data-based fault diagnosis in the industry has significant potential. Data-based failure diagnostics assess equipment operations by analysing retrieved sensor signal properties. A diagnosis is based on a healthy baseline. Advanced sensors can accurately monitor the environment in which machines operate. Effective data analysis enables real-time monitoring and device status tracking for human operators.

Communication between equipment and other production systems inside a network enables effective diagnostics and monitoring. An Internet of Things (IoT) device should be embedded in every aspect of the production environment and connected to all others. Through cloud computing and other virtual networks, these gadgets are interconnected and share data in real time. Data intensity might lead to bottlenecks in immediate data transfer and analysis due to multi-device bandwidth utilisation and the processing load of the centralised cloud. This makes it difficult to process all raw data on a centralised cloud server. Edge artificial intelligence (AI), which integrates AI with edge computing, is one possible solution to bottlenecks. Data collected by smart sensors on IoT devices is first processed and inferred at the node's edge close to the source, before being sent to the cloud according to edge AI's algorithms. Reduced data transfer volume, reduced dormancy, quicker solutions, scalability, and data security are all benefits of this approach. Using edge AI, autonomous transfer vehicles (ATVs) and other failure-prone industrial IoT devices may have their status monitored in real time and data sent at fast speeds.

Monitoring robots using software systems is crucial for efficient industrial processes, comparable to other equipment used in processes. They provide a real-time condition monitoring design and system implementation for industrial ATV fleets, addressing the previously noted issue. Innovative edge processing techniques using a fusion of sensors and deep learning technologies provide real-time benefits for sensor pre-processing and edge AI. An ATV for industrial use, an AI accelerator for the edge, storage for data, a tool for data visualisation, and a custom-built, open-source information distributor middleware platform are all part of the test bed. This generic solution may be used for any production equipment having sensors and IoT connection protocols.

The ATV robotics fleet condition monitoring platform is developed for this use case. Major contributions of this chapter: an architecture framework for real-time defect detection and condition monitoring using edge AI is shown and evaluated on application scenarios including industrial ATVs. On the hardware of the edge AI accelerator, an inference-driven DL algorithm is used, which takes advantage of sensor fusion. A tailored open-source middleware transfers the inference findings to data storage in real time.

This is the remaining structure of the chapter: Section 1.2 outlines the technique, components, and specifics of the proposed design (see Section 1.2 for details), while Section 1.3 shows the experimental environment and the outcomes of our strategy. Finally, Section 1.4 discusses the key findings and Section 1.5 describes the conclusions with future study.

1.2 The Proposed Method

The components of the suggested generic architecture are detailed in this section. A quick overview of the Robot Operating System (ROS) platform is provided first, which is widely used on robotic hardware and other Internet of Thing's devices. Subsequently, the advantages of cutting-edge AI and its ideas are presented. Following this, a brief explanation of the FIWARE system, its primary features, and its operation is provided. Finally, the method that has been suggested for monitoring conditions and diagnosing faults in real time is laid forth.

1.2.1 ROS

Mobile robots, like ATVs as well as other IoT devices, often include motors, wheels, cameras, and detectors as their physical components. The diversity of these components makes it more difficult to design user-friendly software interfaces for controlling and maintaining robots. The ROS middleware platform addresses these concerns by offering a standardised development environment with standard protocols for sensing drivers and interfaces. To facilitate the sharing and reuse of code blocks between robots that share similar components, ROS provides a framework. The ROS is a collection of applications that facilitate the management and operation of robots. The four main components of ROS are messages and nodes, topics, and services. Software components called "nodes" carry out calculations and exchange messages with one another. Subscribing to a topic allows any node to receive messages that are published to it. Messages may be published across a topic by any node. Similar to conventional web services, the service paradigm relies on client nodes to make requests and server nodes to react. The master node on a ROS platform is in charge of organising and controlling all communication, and all the nodes on the platform must connect to it.

1.2.2 Edge AI

In computing, an "edge" is any computational or resource network that sits between a system's data sources and its cloud servers. Small, distributed, power-efficient, and situated near data sources are typical characteristics of edge devices. When AI and computing edge come together, the result is edge AI, often called edge intelligence. A simple explanation for this setup is that data collected from hardware devices is processed locally using machine learning or DL methods by taking advantage of the computing capabilities of the edge nodes. Consequently, AI edge places AI and processing power near data sources. Edge nodes run inference models and only transmit the findings to centralised cloud servers, rather than the whole raw data acquired from IoT devices' hardware. Decentralised edge nodes process collected data, which reduces the need for network bandwidth and processing resources from centralised servers. As a result, edge AI solutions improve scalability, privacy, security, and reliability while decreasing latency and communication costs.

1.2.3 FIWARE Platform

The FIWARE platform is a middleware that, via its generic enablers (GEs), makes it easier to create smart manufacturing systems. Commonly used in a wide range of architectures and applications, GEs are software components. The platform's underlying principle is to integrate suitable GEs according to a system's needs. All FIWARE solutions revolve around a context broker (CB). Tasks like processing, analysing, and visualising context information, managing context data/API, and interacting with the IoT and robots are handled by other application-specific GEs. A central processing unit (CPU) plus a database for storing context data make up the most basic FIWARE platform. The information produced by IoT devices is handled by CB in context entities in IoT situations. Also, another GE called FIROS connects the robotic domain to the cloud so that FIWARE CBs may use and publish NGSI v2 entities instead of ROS messages.

1.2.4 Proposed System Architecture

This research presents a generic framework for monitoring conditions and detecting defects in real time using edge AI. There are five layers to the architecture: the first is the IoT, the second is the edge, the third is FIWARE, the fourth is data storage, and the fifth is visualisation. (1) Multiple devices containing ROS are comprised the IoT layer. The sensors on these gadgets are what gather data. By way of the master node, the ROS topic publication system gathers data across predefined subjects using appropriate messages. (2) AI devices in the layer's periphery that can execute DL inference algorithms make up the edge layer. They are deployed near an IoT device at the IoT layer because of their tiny size and lack of centralisation. Further, every edge device establishes an Ethernet connection to the IoT device and uses ROS to sign up for the topics offered by the allocated IoT device. The DL-based fault diagnostic method uses real-time data from sensors gathered from topics to make inferences and then publishes the findings via another subject on the master node. (3) FIROS, GEs, and a database acting as CBs create the FIWARE layer. IoT devices transmit messages to the context broker (CB) using FIROS, which converts them from master node-published topics to NGSI v2 entities. The CB then updates the database with

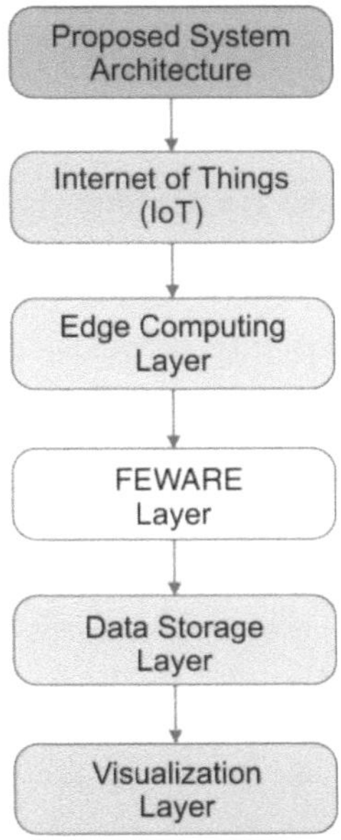

Figure 1.1 Proposed system architecture.

the new entities and attributes. (4) CB data transmission may be converted into a database-friendly format by an analysis engine that is part of the storage of the data layer, which is mainly concerned with data persistence. Information is retrieved and stored by this layer from the FIWARE software platform. (5) The methods for creating dashboards from the information kept in the server's database are part of the visualisation layer, which allows for real-time monitoring of the state of IoT devices. This layer facilitates communication between computers and end users and contributes to system-wide and monitored entity decision-making. Figure 1.1 shows the suggested architecture.

1.2.5 The Theoretical Framework

The suggested approach benefits from edge AI technology by allowing AI models to be executed in a unit closer to the data source. This work implements the short-time Fourier transform and LeNet-5 CNN models on the edge AI units for the ATV context. STFT translates non-stationary signals into the temporal-frequency domain, preserving all frequency and time information. STFT uses a windowing function to divide the waveform into tiny time segments, which are then transformed into spectrograms using a standard Fourier transform (FT). The fundamental operation formula is defined as

$$STFT_x(\tau, f) = \int_{-\infty}^{+\infty} x(t)h(t-\tau)exp^{-j2\pi ft}dt \tag{1.1}$$

A segment of the observed signal is x (t), and the STFT function is based on h (t–τ) $\exp^{-j2\pi ft}$. However, LeNet-5 is chosen for real-time defect detection because of its simplicity, computational efficiency, and inference performance. There are seven layers in the LeNet-5 method: pooling, convolutional, and fully connected, as well as classification. The formula for convolution is

$$y_j = f\left(\sum_i X_i \times W_{ij} + b_j\right) \tag{1.2}$$

The kernel for convolution (W_{ij}) uses the convolutional operator (×), the bias parameter (b_j), and the non-linear function of activation (f) to transform the input feature map (X_i) into the output feature map (y_j). Each convolution operation uses the non-linear activation function given in Equation (1.3), the ReLU function.

$$f(x) = \{x \quad x > 0 0 \quad x \leq 0 \tag{1.3}$$

Layers that pool together reduce feature map dimensionality and enable quadratic feature extraction. The ATV application uses a LeNet-5 model that performs maximum pooling operations with a 2 × 2 kernel, which permits feature maps of half size. The last output of the convolution layer establishes a dense operation connection between every neuron and the fully connected layer. Final decision-making in the fully linked layer's dense neurons is handled by the Softmax classifier in the output layer. When it comes to multi-classification tasks, in particular, the Softmax classification function has shown to be both popular and effective.

1.3 Experimentations

Use cases for industrial ATVs that make use of the edge AI approaches for real-time condition monitoring and defect detection are described in this section. Firstly, the components of the device and the system are described. Afterwards, the detailed experimental setup used in the test situations is described.

1.3.1 System Overview and Devices

The suggested architecture is evaluated in a setting that closely resembles an industrial workplace to the best of our ability. A real-world production deployment of the suggested technology has not yet taken place, nevertheless. The present design presupposes integration without substantial changes to system constituents. In addition, even though the experiments are carried out with a single ATV, the system is expected to scale for a fleet of ATV equipment and other IoT devices monitoring conditions. Any device running ROS is supposed to seamlessly integrate into the system. A static DL model is assumed for inference at the edge, and the system lacks an auto-model updating mechanism. This study does not examine the long-term durability of the edge method for equipment state drifts, assuming manual updates by an operator.

ATVs with middleware ROS are used as IoT devices in investigations. The integrated technology and software safeguard ATVs against accidental accidents during tasks. The used ATV's dimensions are 1027 mm long, 729 mm wide, and 326 mm high, with two 200 mm main wheels and four 81 mm casters. Additionally, each primary wheel has two similar DC motors and two sensor sets to detect humidity, temperature, vibration, sound, and velocity. These smart devices detect the surroundings and collect data at the desired pace. Additionally, FIROS and ROS middleware provide data. Consequently, data collected by the ATV may be accessed by an edge device. Also, you may connect to the internet via Bluetooth, WLAN, and Ethernet on the ATV.

The edge AI device for this study is the NVIDIA Jetson TX2 GPU module. The module, with an inbuilt Ethernet card, 170 mm length × 170 mm width, and minimal energy consumption, is integrated into the bodywork of ATVs. The board has 8 GB 128-bit LPDDR4 Memory and 256 NVIDIA CUDA cores allowing fast and efficient DL algorithms at under 19 V. Thus, real-time conclusions may be drawn from ATV sensor data. Orion CB, MongoDB, data storage, and visualisation tools all make use of virtualised server infrastructure. An open-source microservice called FIROS GE is installed on the computers of the ATVs, and the server establishes a connection to the same network so that it may interact with them. The main workstation is equipped with 64 GB of DDR4 RAM, two Intel Xeon Gold 5218R CPUs, and Ubuntu 20.04.

1.3.2 Experimental Setup

The generic design allows for scaling to multi-robot ATV fleets and expansion into other hardware entities with bespoke data adapters. As shown in Figure 1.2 the experimental scenario mimics a genuine industrial situation for ATVs. The IoT layer consists of a ROS middleware-powered ATV running Ubuntu 16.04. With a Wi-Fi connection, an ATV can go out on its own and do things like explore or carry cargo. While riding an ATV, you may get real-time data about your perceived workspace thanks to the sensors included inside the vehicle. ATV sensor data is published across many topics at a predefined rate via ROS middleware. The topic may be subscribed to and consumed by any device that is connected to the ATV.

The edge layer makes use of an NVIDIA Jetson TX2 card for AI. The board comes preloaded with ROS middleware and an Ubuntu 18.05 operating system.

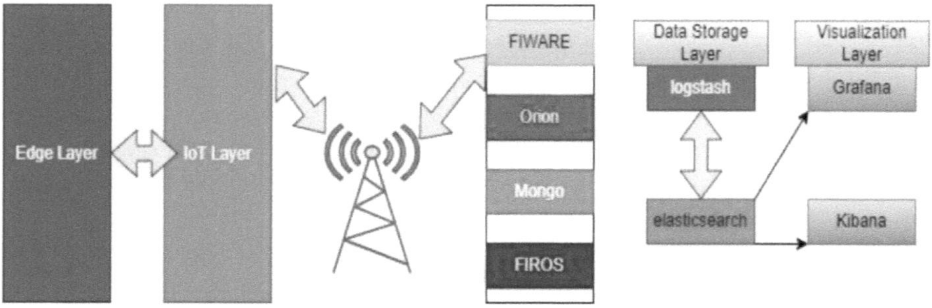

Figure 1.2 Experimental scenario mimics a genuine industrial situation for ATVs.

The ATV's battery supplies the necessary power to the board, which is housed within the vehicle's bodywork; a CAT6 Ethernet cable links the two devices. The powerful GPU unit on the board makes it suitable for deep learning methods and real-time inference. The experiment's inference is based on a DL-based actual time operational defect detection approach that uses sensor fusion. With 25 separate trials, this approach achieves an impressively high level of accuracy of more than 98.62% using the ATV's offline data. The technique takes two DC motors' vibration and sound data as input and uses the STFT to convert irregular signals to the time–frequency domain so that spectrograms may be obtained. For every signal containing 513 data points, spectrograms measuring 64 × 64 were obtained after pre-processing using certain settings. The sensor fusion model, which is based on LeNet-5, is then given four spectrograms that were acquired at the same time and identified for three distinct operational failure scenarios. There are three distinct operating states: normal, low-level anomaly, and high-level abnormality. Sensor data is acquired at 100 Hz for all three operating circumstances, with the ATV being run in identical testing settings several times while being asked to go straight for 3.6 m. The model is sent to the edge AI component for real-time inference-making after training and learning using the 82 observations for every scenario. Microphone and IMU sensor data from each motor are reported in distinct ROS topics, as seen in the picture. Subscribing to the subject, the edge AI board pre-processed data using STFT settings to generate spectrograms.

After that, the inference model receives each synchronous spectrogram. Different master themes publish the inference and use data findings. Therefore, FIROS GE and other IoT devices may use them. Additionally, a sliding window moving 128 data points improves real-time operational diagnosis, even with pre-processed batches of 513 points of data. On average, each input has 385 identical values. Each real-time inference may be accomplished in 1.29 s because of the 100 Hz data gathering frequency. Orion CB, MongoDB, and FIROS are ATV use case GEs in the FIWARE layer. The broker Orion CB supports the FIWARE NGSI v2 API for context information subscriptions. MongoDB stores Orion CB NGSI entities and attributes. FIROS sends ROS communications to the Orion CB as NGSI entities to connect the IoT and FIWARE layers. For such cases, FIROS in the ATV converts printed topics into NGSI objects. FIROS transfers subjects, including real-time analysis findings and inference data, to the Orion CB via Wi-Fi. While MongoDB maintains NGSI entities and attributes, Orion CB publishes them.

Elasticsearch stores data while Logstash processes it. Logstash uses real-time pipelining to receive data, alter it, and deliver it to Elasticsearch. However, Elasticsearch saves data for fast search and analysis. Orion CB data is converted to JSON by Logstash and stored in Elasticsearch for the ATV application case. The visualisation layer includes use case visualisation tools Kibana and Grafana. Kibana integrates well with Elasticsearch and provides sophisticated visualisation. Grafana analyses data, creates real-time dashboards, and alerts on thresholds from several data sources. Both tools have similarities but also substantial variances. Kibana is better at accessing and finding data, while Grafana has a great alerting engine. The engine allows dashboard panel conditional rules to generate alerts and inform users in numerous ways. For querying and searching, Kibana is employed. Data and ATV circumstances may be tracked in real time using Grafana's dashboards, which can alert decision-makers.

1.4 Results and Discussion

The experimental findings acquired when operating the ATV in an area with different permanent barriers that might generate low- or high-level anomalies are described in this section. The testing set and results of the immediate inference are described in detail first. This is followed by the data storage interface. The next step is to display the monitoring system and dashboards that have been created. A comprehensive evaluation of the results is then carried out.

1.4.1 Evaluation and Real-Time Inference

As the ATV follows a predetermined path while carrying out a predefined task, the edge AI unit makes inferences about its status using the pre-trained DL approach shown in Figure 1.3. The ATV edge AI unit runs Python client code that subscribes to ROS topics, runs the DL method to draw inferences based on data released via the ROS master's node, and takes into account the vibration and sound data of both motors. The virtual testing setting is shown in Figure 1.4a. The ATV first comes across three metal items marked from the beginning point that are known to create high-level anomalies in the offline modelling phase; this can be observed in Figure 1.4b. Next, the ATV will be guided to return to its starting place by the path planning assignment, which involves following a rectangle route on the ground. Figure 1.4c shows that on its return journey, the ATV goes over three copper wires that are known to cause minor problems with the ATV's functioning. Recall that the end users of the research may modify the high- and low-level anomalies to fit their

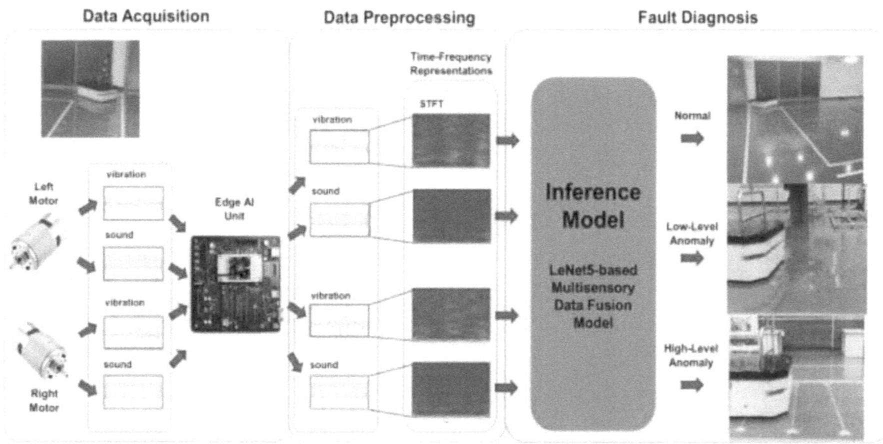

Figure 1.3 Edge AI unit makes inferences about its status using the pre-trained DL approach.

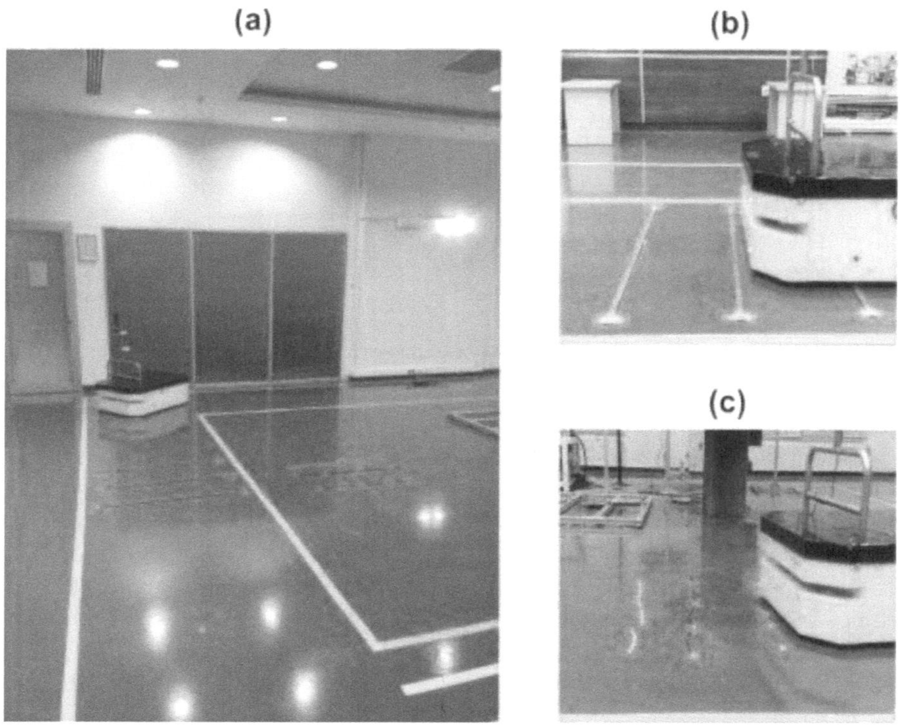

Figure 1.4 (a) Virtual testing setting, (b) create high-level anomalies in the offline modelling phase, (c) return journey.

production environment. Sensors for vibration and sound, which are connected to the DC motors, data collecting while the ATV does the programmed job. The data is then analysed and filtered by the edge AI unit, which determines the ATV's state in real time.

1.4.2 Data Storage Interface

The data is sent, saved, and managed using the Elasticsearch persistent data stack via the middleware platform, which is developed using FIWARE. There is a distinct index pattern at the data storage for every subject that the ATV publishes and that the FIWARE platform subscribes to. Elasticsearch data may be explored and filtered with the help of the Kibana tool. When the ATV completes a specified load-carrying task four times, the testing environment produces results. Assuming the ATV is operational, the quantities of hit (y-axis) and the corresponding timestamps (x-axis) are shown in the initial row of the chart. The middle graph displays the occurrences of low-level anomalies encountered by the ATV. The bottom graph displays the occurrences where the ATV detected high-level irregularities, with timestamps on the x-axis. The field is the primary operating area for the ATV since there are no barriers there. With most data entry results showing healthy hits on the dashboard, it can be concluded that the fused data received from sensors properly represents the field state. Data collected from the ATV's primary wheels' contact with ground-mounted metal bumpers is evaluated and categorised as a high-level abnormality situation. The fact that this physical condition seems to increase the frequency of minimal anomaly hits whenever the ATV passes over copper wires is likely related to the fact that it was first introduced to the system as a low-level anomaly. The time axis emphasises the incidence of anomalies as well as the great capacity of the DL model. The experimental validation of the dashboard figure's timestamp data guarantees that it aligns with the real instances of the anomalous incidents, although somewhat delayed due to network latency. The turning speed, ATV speed, and the distance between the wires determine the time gap between the high and low anomaly strikes.

1.4.3 Condition Monitoring Data Visualisation and Alert Device

Using the inference findings and data, dashboards are built to monitor conditions and make decisions in real time. Grafana can display Elasticsearch data graphically and instantly notify end-user stakeholders by email of any unexpected data changes. Grafana is a versatile tool that can be quickly adjusted to work with many data types. To let end-user operators see the data from the monitored environment and the conclusions from the edge AI inferences, a dashboard was constructed. Figure 1.5 shows the most recent 5 min of an ATV operation, with data updated every 5 s. The two-dimensional graphs provide the raw vibrations and sound information from the

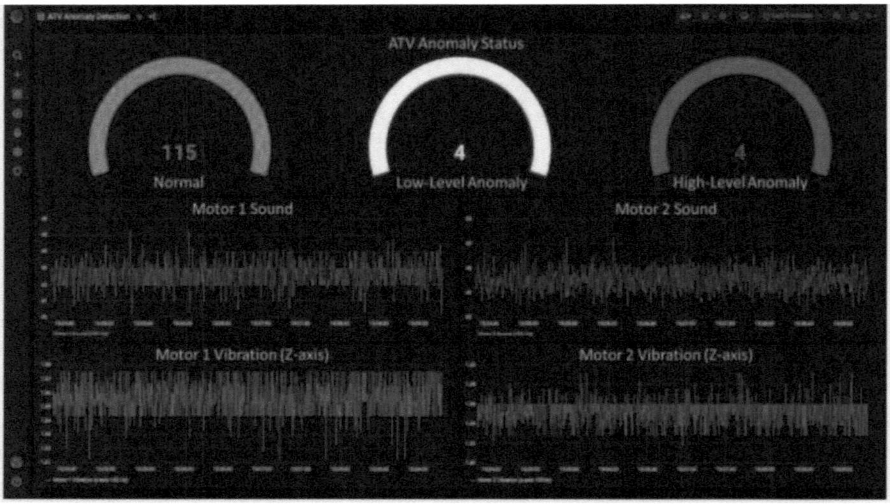

Figure 1.5 Recent 5 min of an ATV operation, with data updated every 5 s.

two motors, while the gauge-style chart shows the present state of the ATV. Each graph displays 30,000 data points since the data collecting frequency is 100 Hz and each graph is refreshed every 5 min. Anomaly hit changes are immediately correlated with the 513 newly added data points as the DL model's conclusions reflect both sets of information simultaneously. Accordingly, the graphs below display the sound and vibration data acquired from each motor, which will help in comparing data gathered from normal and atypical occurrences. Additionally, these graphs might help monitor the health of the system's subscribing hardware.

1.4.4 Discussion

This study develops and assesses an open-source middleware platform based on FIWARE that employs edge AI for condition monitoring and trouble identification in real time. The proposed system architecture is adaptable, so it can be changed and added to meet different needs. This way, several IoT devices may be monitored at the same time. Furthermore, our experimental findings using edge AI technologies have shown that data processing that makes use of edge resources might provide power gains. The edge processing approach significantly improves performance in terms of latency, privacy, communication costs, and storage costs. Several metric values are computed to measure the enhancements offered by edge AI. Two potential outcomes are laid out: (1) sending all the raw data from the vibration and sound detectors to a centralised server for the process, and (2) having the edge AI unit filter and process the data, with only the inference results from the DL approach sent to a centralised server for storing and visualisation. It is evident from the raw form column of the sensor data how much effort was put into uploading and analysing all of the sound and vibration data that the sensors recorded to the main server. The information that was analysed at the edge AI units tab, on the other hand, displays the results of transmitting processed data to the central servers for storage and visualisation. In systems that are time-sensitive and safety-critical, where human operators and autonomous equipment coexist, the computation of such performance measures is of the utmost importance.

There is some partial data flow to the central server from the edge AI units since they pre-process the raw data locally. The primary server just receives the inference results as data. The network's latency and bandwidth usage have been drastically reduced using this strategy. The DL-based defect diagnostic method compares the network bandwidth consumption during the transmission of vibrations and sound information from two similar motors to the server. The data that comes out of the edge AI unit's analysis and filtering also contributes to this usage. The analytics conducted using edge AI have shown a 43-fold reduction in the requirement for network bandwidth, which is based on an average period of 60 s. A different computation compares the data transport delays. Here, the lag is caused by factors such as data density, the state of the network, and the capacity of the central server that reads and writes the data. The overall amount of time it takes to transmit data has been cut in half using the method that was suggested. When looking at the numbers, it's clear that there are major holdups when transferring data to the main server, even when no AI units are located at the periphery. Performing real-time analysis on a central server would be inefficient due to this, which is particularly problematic for applications that need precise timing, such as fault diagnostics. Although there is a performance gain of 68 times when drawing inferences on the centralised server, the delay incurred in data transmission more than cancels out this benefit, as can be seen from the table. In addition, if the system is planned to be used by an ATV fleet, the central server's resources would have to grow linearly with the number of ATVs, as an AI edge unit may be mounted on every ATV according to the design. The dispersed structural design, however, will cause the edge AI device's resource consumption to remain constant. This suggests that edge AI technology, rather than the more conventional method based on a central server, is the way to go in these kinds of situations. While the suggested methodology may have more benefits than previous methods, it does come with certain downsides, such as the need for expensive hardware and specialised knowledge, particularly when scaling the system to accommodate a large number of IoT devices. A large-scale system consumer may be concerned about the financial expense since the outermost layer comprises individual edge AI devices for every piece of equipment. Hence, a more budget-friendly gadget may replace an existing edge unit. Consequently, it is necessary to do an expense analysis on the hardware of the prototype system. To keep low-level technical activities running smoothly in the suggested prototype system, human-level knowledge may still be necessary due to the lack of integration of ML automation tools. Users may tackle this to overcome similar obstacles in the field.

1.5 Conclusion and Work for Future

This study presented a generic architecture that is agnostic of platforms, able to identify and monitor conditions in real time using edge AI and an internal middleware solution based on the FIWARE API. Security systems, logistics, agriculture, and healthcare are just a few of the many possible implementations of the architecture's adaptability. They showed the system in a smart manufacturing solution using an ATV monitoring usage, but it is very adaptable. It is possible to modify the given architecture to work with different IoT devices that use GEs and bespoke data adapters, regardless of whether or not it incorporates ROS. Inference is performed by the edge AI devices using the data obtained from the many sensors mounted on the ATV. Using ATV offline data, the DL technique achieves a very high testing accuracy of 99.95% on average across 25 trials. The next step is to save the data and the findings in a storage location. By adjusting sensor subscriptions in the middleware system, it is simple to disable the network transmission of raw sensor data. Lastly, the data was absorbed by the visualisation tool, which then generated dashboards and, if necessary, alerted the administrators. An ATV was programmed to do a certain task in a controlled environment by introducing several operating anomalies. This was done to show how successful the system was. The findings proved that the system could detect and track ATV problems in real time. In addition, the ATV use case's recommended edge AI solution cuts network bandwidth needs by a factor of 43. The overall amount of time it takes to transport data is reduced by 37 times as compared to the conventional method of using a central server. Though the system's feasibility and experimentation are shown with a single ATV, it's easy to add more IoT devices to the middleware layer by deploying numerous instances of the FIROS platforms to operate on other ATVs. The quantity of accessible ATV units and the amount spent on hardware are both affected by this, however. Smart manufacturing environments may benefit from the system's ability to be incorporated into real-world situations, leading to greater operational efficiency.

Future work will expand this chapter's scope in several ways. When the performance of DL models starts to drop, a new system for training and updating AI models will be put in place using dynamic edges. Consequently, they will use AutoML and a model-based approach to develop our architecture. The suggested architecture allows for the seamless implementation of model auto-update situations by using AutoML approaches. The next stages include adding a variety of equipment monitoring connections to the system and conducting large-scale tests of the proposed architecture.

Bibliography

Ahangar, M. Nadeem, et al. "A survey of autonomous vehicles: Enabling communication technologies and challenges." *Sensors* 21.3 (2021): 706.

Amin, Arslan Ahmed, Muhammad Sajid Iqbal, and Muhammad Hamza Shahbaz. "Development of intelligent Fault-Tolerant control systems with machine learning, deep learning, and transfer learning algorithms: A review." *Expert Systems with Applications* 223 (2023): 121956.

Bathla, Gourav, et al. "Autonomous vehicles and intelligent automation: Applications, challenges, and opportunities." *Mobile Information Systems* 2022.1 (2022): 7632892.

Carballo, Alexander, et al. "End-to-end autonomous mobile robot navigation with model-based system support." *Journal of Robotics and Mechatronics* 30.4 (2018): 563–583.

Cheng, Hong. *Autonomous intelligent vehicles: theory, algorithms, and implementation.* Springer Science & Business Media (2011).

Gómez-Huélamo, Carlos, et al. "Train here, drive there: ROS based end-to-end autonomous-driving pipeline validation in CARLA simulator using the NHTSA typology." *Multimedia Tools and Applications* 81 (2022): 1–28.

Jeong, YiNa, et al. "An integrated self-diagnosis system for an autonomous vehicle based on an IoT gateway and deep learning." *Applied Sciences* 8.7 (2018): 1164.

Kumar, K. Sunil Ratna, et al. "Iot and data mining techniques to detect and regulate solar power systems." *2023 International Conference on Inventive Computation Technologies (ICICT)*. IEEE (2023).

Li, Weihua, et al. "A perspective survey on deep transfer learning for fault diagnosis in industrial scenarios: Theories, applications and challenges." *Mechanical Systems and Signal Processing* 167 (2022): 108487.

Liu, Dongjie, et al. "Data augmentation technology driven by image style transfer in self-driving car based on end-to-end learning." *Computer Modeling in Engineering & Sciences* 122.2 (2020): 593–617.

Ma, Yifang, et al. "Artificial intelligence applications in the development of autonomous vehicles: A survey." *IEEE/CAA Journal of Automatica Sinica* 7.2 (2020): 315–329.

Prathiba, Sahaya Beni, et al. "A hybrid deep sensor anomaly detection for autonomous vehicles in 6G-V2X environment." *IEEE Transactions on Network Science and Engineering* 10.3 (2022): 1246–1255.

Rafique, Fezan, Ling Fu, and Ruikun Mai. "End to end machine learning for fault detection and classification in power transmission lines." *Electric Power Systems Research* 199 (2021): 107430.

Rahat, Irfan Sadiq, et al. "A step towards automated haematology: DL models for blood cell detection and classification." *EAI Endorsed Transactions on Pervasive Health and Technology* 10 (2024): 1–27.

Rahman, Md Arafatur, et al. "A secure and intelligent framework for vehicle health monitoring exploiting big-data analytics." *IEEE Transactions on Intelligent Transportation Systems* 23.10 (2022): 19727–19742.

Shafi, Uferah, et al. "Vehicle remote health monitoring and prognostic maintenance system." *Journal of Advanced Transportation* 2018.1 (2018): 8061514.

Shi, Qian, and Hui Zhang. "Fault diagnosis of an autonomous vehicle with an improved SVM algorithm subject to unbalanced datasets." *IEEE Transactions on Industrial Electronics* 68.7 (2020): 6248–6256.

Tampuu, Ardi, et al. "A survey of end-to-end driving: Architectures and training methods." *IEEE Transactions on Neural Networks and Learning Systems* 33.4 (2020): 1364–1384.

van Wyk, Franco, et al. "Real-time sensor anomaly detection and identification in automated vehicles." *IEEE Transactions on Intelligent Transportation Systems* 21.3 (2019): 1264–1276.

Wang, Zhangu, et al. "A review of vehicle detection techniques for intelligent vehicles." *IEEE Transactions on Neural Networks and Learning Systems* 34.8 (2022): 3811–3831.

Won, Myounggyu. "Intelligent traffic monitoring systems for vehicle classification: A survey." *IEEE Access* 8 (2020): 73340–73358.

Xu, Yan, et al. "A digital-twin-assisted fault diagnosis using deep transfer learning." *IEEE Access* 7 (2019): 19990–19999.

Yaqoob, Ibrar, et al. "Autonomous driving cars in smart cities: Recent advances, requirements, and challenges." *IEEE Network* 34.1 (2019): 174–181.

Yu, Keping, et al. "Deep learning-based traffic safety solution for a mixture of autonomous and manual vehicles in a 5G-enabled intelligent transportation system." *IEEE Transactions on Intelligent Transportation Systems* 22.7 (2020): 4337–4347.

Zong, Wenhao, et al. "Architecture design and implementation of an autonomous vehicle." *IEEE Access* 6 (2018): 21956–21970.

Chapter 2

Data Science and ML Algorithms to Investigate Different Testing Scenarios for Various Anomalies in Driven Electric Motor

Arpit Jain, P. R. Sivaraman, Abhishek Choubey, and Guna Sekhar Sajja

2.1 Introduction

Adapting to the complicated and unpredictable modern market is getting easier and easier with the advent of new technology that makes manufacturing processes more flexible [1]. New production system concepts must be integrated horizontally and vertically, within and across value networks, and inside and outside a manufacturing site [2]. Industry 4.0 developed from the requirement to adapt and exploit new technologies to integrate production systems [3]. Integration of diverse settings may provide a lot of shop floor data about production, methods, and equipment [4]. Processing and analysing this information and data may provide crucial industrial

DOI: 10.1201/9781003479062-2

system insights and assist strategic decision-making [5]. Industry 4.0 and its technology have made predictive maintenance (PdM) a potential new method [5]. Machine learning (ML)-based prediction technologies in PdM management detect defects early [6]. This lets maintenance be done only when needed [7]. One-third of maintenance funds are missed on errors or overtime [8]. Industry 4.0's defining technologies—big data, Internet of Things (IoT), and artificial intelligence (AI), notably ML and data science algorithms—have changed the paradigm [9]. PdM illustrates this development.

This document defines the following goals:

- ML and Data science approaches will be used to concrete industrial equipment PdM to assess deviation detection.
- To determine whether ML and data science approaches may be used to specify the PdM application's instrumentation setup.

The rise of Industry 4.0, according to the researcher, is shifting emphasis towards PdM tactics that increase production equipment availability and decrease downtime costs [10]. To avoid downtime, PdM reports a potential failure using time-based information and expertise [11]. The Industrial Internet of Things (IIoT), big data, and ML are just a few examples of the technologies that have found their way into the industrial setting and made this feasible [12]. Industrial AI has recently demonstrated promise in assisting manufacturers with data-driven predictive analysis and decision-support in extremely complicated processes, both of which are essential as they face the digital change of cyber-physical systems [13]. However, after the experimental pilot's phase, industrial adoption of such solutions is still limited since companies aren't ready to face the unique and challenging problems that exist in the real world [14]. A lack of high-quality monitor data and knowledge about the practicality of analytic techniques have kept PdM from being widely used in the actual world, according to many who have backed this view [15–25]. Based on ML and data science procedures, this chapter's main contribution is a methodology for selecting machine instrumentation configuration variables. The algorithms effectively determine the number of devices needed to sense machine anomalies, which means less time and money spent on defining the instrumentation system. In addition, construct a database that includes entries for the various electric motor application situations, including normal operation and four abnormality scenarios: overloaded, imbalanced, misaligned, and uncoupled.

Below is the outline for this chapter. The approach, tools, and technologies that they suggest, together with the test scenarios that they have developed for PdM applications, are detailed in Section 2.2. The findings and trials of using data science and ML techniques to PdM are presented in Section 2.3. Section 2.4 concludes the study and concludes with some limitations and recommendations for further research.

2.2 Methodology

Figure 2.1 shows the suggested methodological technique that was used to generate the current study. In this procedure,

- Bench experiment: The experimental bench has to be set up such that data may be generated for five different electric motor applications: one for normal operation and four for anomalous cases. All of these possible outcomes line up with the classes in the dataset that stand for potential outliers.
- Pre-Processing/DAQ: The data must be collected from several sources, cleaned up, and then transformed into a format that can be used.
- Engineered features: Generate attributes based on the research, concatenate them using the signals obtained in the previous stage, and pick appropriate characteristics.
- Modelling: Using at least five distinct parameterizations for each approach, optimise classifier models chosen from the available research.
- Selection of Models: Assess the efficiency of the classifiers and choose the optimal model.

The next sections show the findings of the experiments and their analysis, which are based on the steps of this methodical technique.

2.2.1 Tools and Equipment

The experimental benches and ML approaches were put up using resources and technology. The purpose of the bench was to model the potential uses of electric motors for spinning equipment in industrial units. This system may be further divided into an instrumentation system and an electromechanical device.

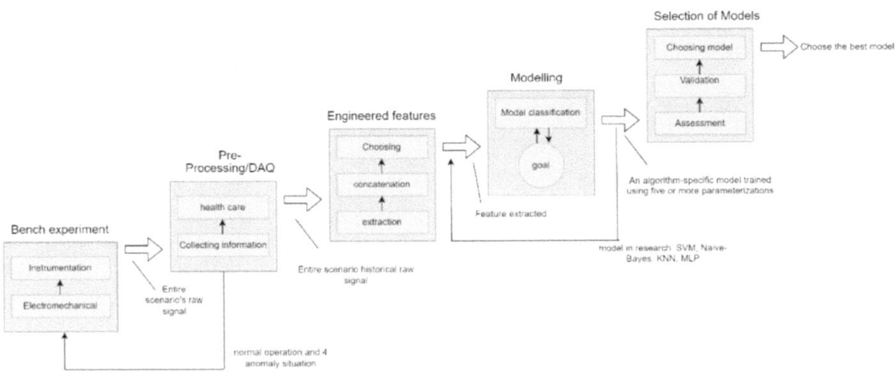

Figure 2.1 Suggested methodological technique.

2.2.1.1 Electromechanical Tool

The researcher's test bench served as the basis for the electromechanical device, with the primary changes being an adjustment to the engine type and an approach to introducing stress to the system. The electric cage motor's shaft is connected to a crown gear in the worm-gear gearbox system. The electric motor was connected to the mechanical gearbox system by a 130 mm steel disc. About 60 millimetres from the shaft's centre, this iron disc was used to place an imbalance mass. The electromechanical device is shown in Figure 2.2.

An electric motor with a nominal speed of 1710 Hz, manufactured by WEG in Brazil and featuring a 0.26 kW, 230 VACS, a three-phase caged design was used. An inverter frequency motor (CFW10 Easy Drive type, CFW100026S2024PSZ, manufactured in Brazil by WEG) served as the motor's gearbox.

2.2.1.2 Instruments

The sensors and data capture devices were used to gather information about the electromechanical device's electric magnitudes and motorized vibrations. The framework for information collecting and storing was built in LabVIEW using acquisition hardware from National Instruments (NI).

Figure 2.3 shows that the "P input [W]" and "I input [A]" signals were captured by a current transducer and a power transducer respectively, earlier to the occurrence inverter. The "I output [A]" signal was obtained from another current transducer following this motor drive. Moreover, the accelerometers were set with one placed vertically with the shaft supporting bearing on top of the electrical motor.

Figure 2.4 displays the experimental bench's system construction in detail. Using the web-based computation atmosphere Jupyter Notebook and the Python language, the procedures of quality abstraction and collection, model training, and presentation assessment and verification were realized in Anaconda's data science

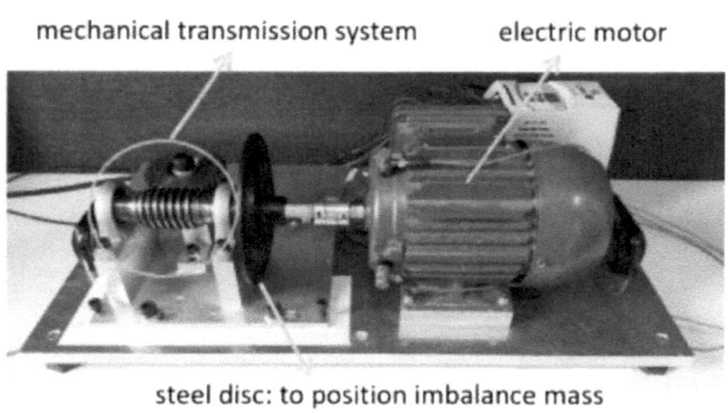

Figure 2.2 Electromechanical device.

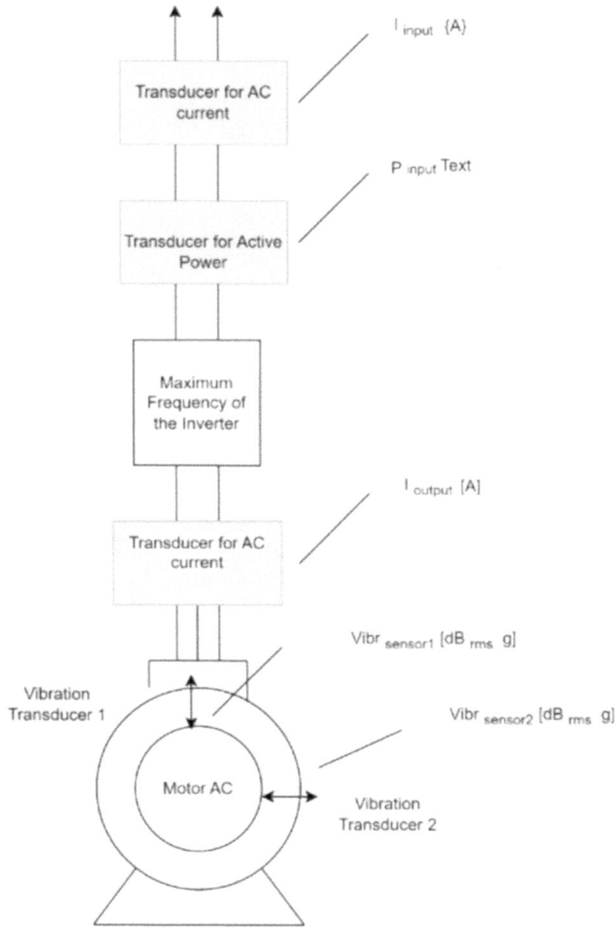

Figure 2.3 Power transducer and current transducer.

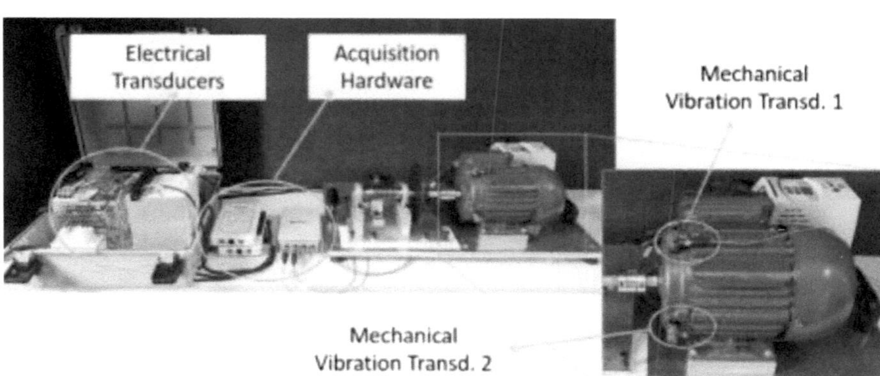

Figure 2.4 Experimental bench's system construction.

stage. TensorFlow, Pandas, Matplotlib, NumPy, Sklearn, and Seaborn were among the libraries that were used.

2.2.2 Data Gathering and Requirements

Normal operating circumstances and fault situations are presented here as scenarios for utilizing the experimental bench. Datasets and fault states are described in-depth, along with the signals used to generate them. In electric motor applications, the electromechanical device may model five distinct states or scenarios: normal operation, abnormal operation, overrun, and underrun.
Various possibilities were examined:

- Standard: The electric motor-driven set is in proper alignment, the crown gear is not overloading the shaft, and there is no imbalance or misalignment (Figure 2.5).
- Decoupled: Figure 2.6 shows the process of removing the top gear from the infinite shaft by guiding it in the path of the red arrow. The crown gear connection generates no torque, therefore the drive shaft rotates "no load."

Figure 2.5 Standard.

Figure 2.6 Decoupled.

It replicates a gearbox interruption from the motor shaft's rotating movement to the driving mechanism.

- Overload: Figure 2.7 shows that the torque acting on the motor's axis is increased when an additional weight is placed on the worm shaft system's crown. This scenario mimics the potential for an increase in torque due to factors such as mechanical gearbox system issues.
- Unbalanced: Figure 2.8 shows the amount of imbalance that is introduced. This scenario mimics potential anomalies in the motor shaft-driven mechanisms.
- Incompatible: Excessive angle misalignment between the motor shaft and the endless shaft, as seen in Figure 2.9. The transmission's mechanical assembly is not properly aligned with the motor shaft, which is mimicked by this condition.

Refraining from adding fluctuations in fault severity was an intentional decision. Even though this choice may influence the trained models' overall performance, it was essential in getting to the heart of the study's objectives. A wider range of possible problem situations was covered by streamlining the fault creation process. Thus, this method allowed for a more thorough evaluation of ML and data science algorithms' efficacy in defining the instrumentation setup in a PdM application and in the context of detecting anomalies inside industrial equipment.

Figure 2.7 Overload.

Figure 2.8 Unbalanced.

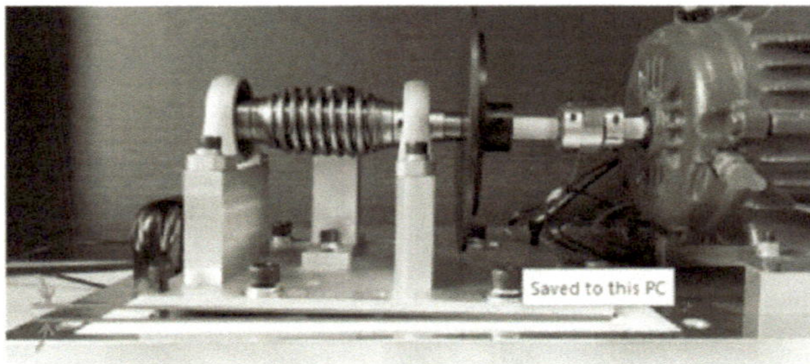

Figure 2.9 Incompatible.

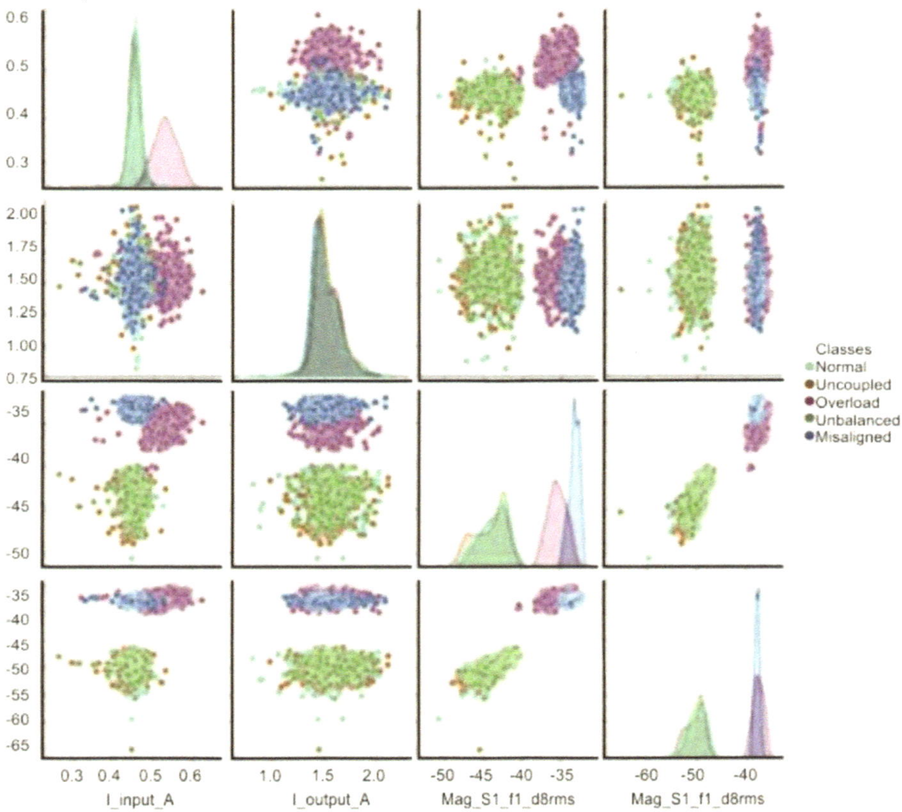

Figure 2.10 The result of creating the scatter plot matrix.

2.3 Results of Experiments

The dataset for simulating various anomaly kinds consisted of the acquired signals and created characteristics for each scenario's sample.

2.3.1 Preparing the Data

To choose the data that is most relevant to identifying the patterns of interest, this part explains the attribute extract and reduction stage.

The first step of the investigation was to visually confirm the basic connections between the classes and characteristics. Figure 2.10 shows the result of creating the scatter plot matrix for this assignment. The four transducers output current, input current, and vibration sensors 1 and 2—operated at the fundamental occurrence f (motor variation) to choose and produce the relevant qualities. The sample clustering for the "misalignment" and "overload" scenarios shows a clear difference. Alternatively, the "uncoupled," "unbalanced," and "normal" patterns have some characteristics.

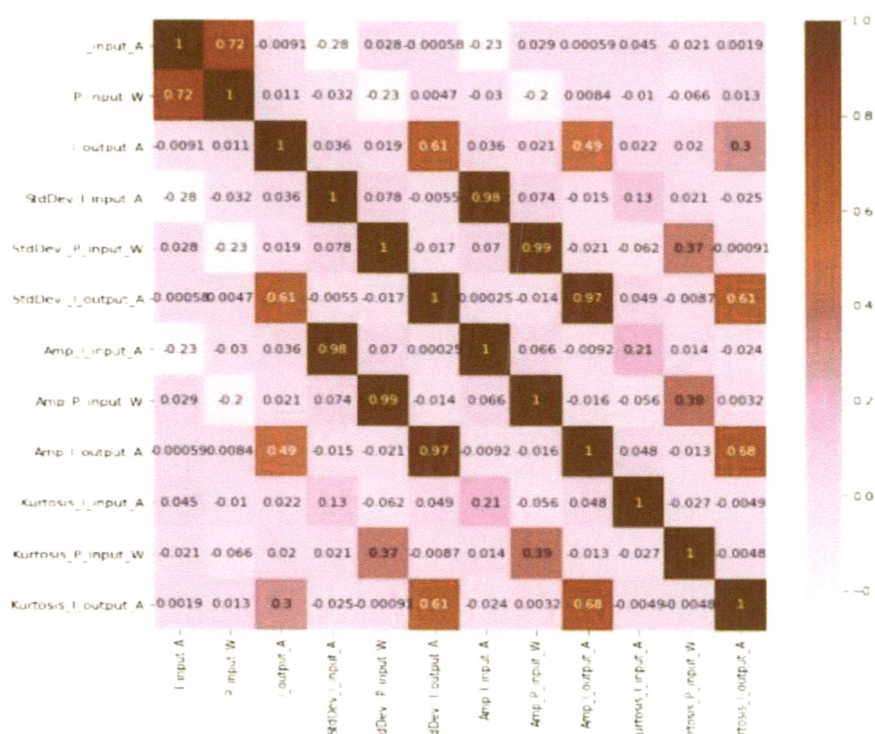

Figure 2.11 Electrical attribute correlation matrix.

After reviewing the five instances, it seems that the qualities are a good bet for finding trends. In light of this, the following paragraphs detail the method that is required to pick qualities.

The correlation matrix was one of the methods used for attribute selection. Box plots and scatter plots were created to certify attribute correlations and detect redundancy. The correlation matrix removed value scaling, revealing the strength of attribute connections. Figure 2.11 illustrates the electrical attribute correlation matrix.

The following qualities derived from electrical values may be redundant, as shown by the significant correlation between them in the correlation matrix:

- P_input_White;
- Kurtosis_I_output_A.
- StandardDev I_output_A;

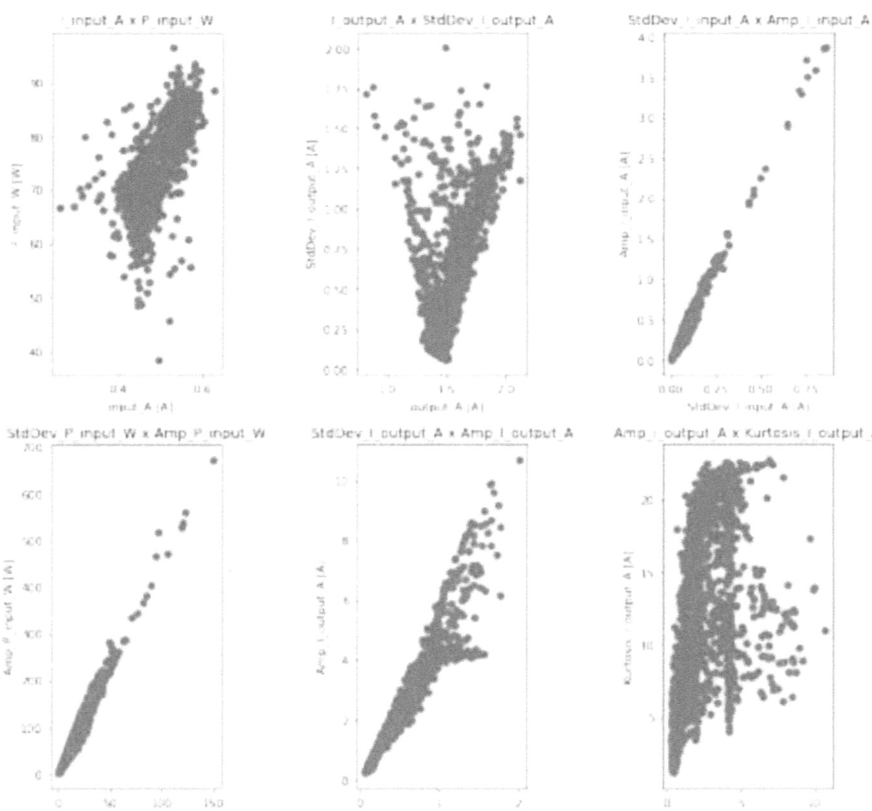

Figure 2.12 Correlation strength.

- Amp P Input W
- Amp I Input A
- Amp I Output A

Scatter plots were created to visually verify the correlation strength, as shown in Figure 2.12, to corroborate this redundancy.

The fact that these characteristics were significantly correlated lends credence to the idea that they are redundant.

The remaining electrical quantities' properties were better understood by creating box plots. The property I_input_A's behaviour in respect to the machine's working circumstances, or the classes, may be seen in Figure 2.13. The median, mean, and ranges between the quartiles vary significantly from the other classes, highlighting the evident usefulness of this property in defining class 2 (overload). Therefore, the characteristic I_input_A was deemed significant.

Figure 2.14 shows the resulting box plot depicting the I_output_A attribute's behaviour, which was constructed using the same procedure. The I_output_A characteristic fails to persuasively prove its importance for class identification, in contrast

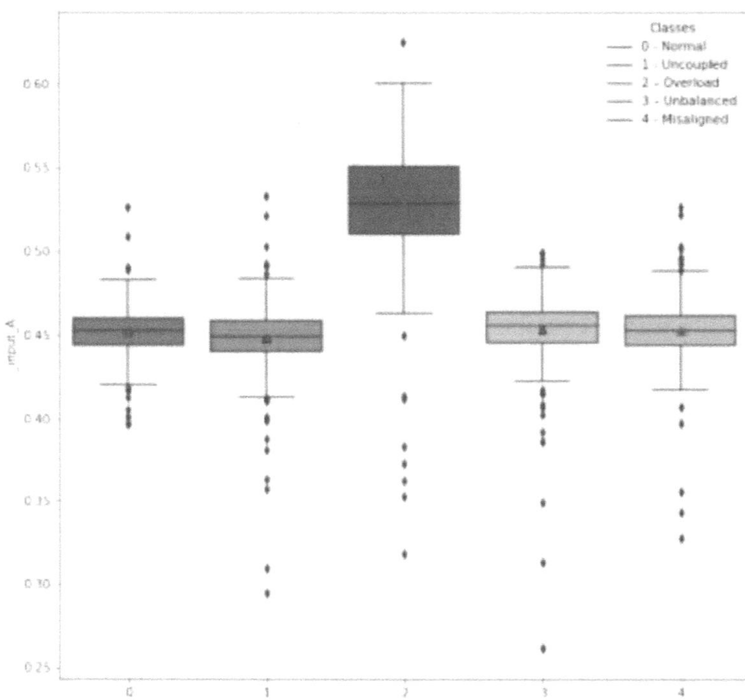

Figure 2.13 Property I_input_A's behaviour.

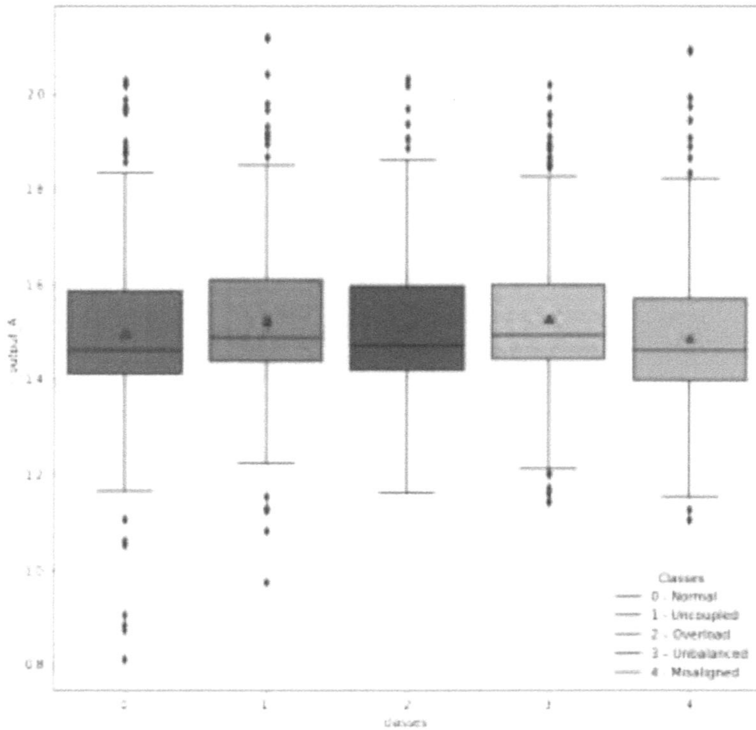

Figure 2.14 I_output_A attribute's behaviour.

to the deduction on the I_input_A attribute. This property and Kurtosis_I_output_A were excluded. Drive output current sensor-derived characteristic. The detection of anomalies system becomes more feasible and cost-effective by eliminating the inverter current output sensor.

Figure 2.15 shows the box plot for the StdDev P input W attribute and Figure 2.16 shows the box plot for the Kurtosis P input W quality; all characteristics were treated in the same way.

The StdDev P Input W quality's box plot displays several extreme values. Furthermore, it seems that the diversity of measurements among the different groups is not given much attention. Therefore, it was determined that it should not be included in the set of characteristics used for training the model.

There is some measure variance among classes, most noticeably in the average, as seen in the box graph of the Kurtosis W input P characteristic. The decision was made to eliminate this feature because it would need an extra power input sensor in a frequency inverter. The system's feasibility and cost-effectiveness are further

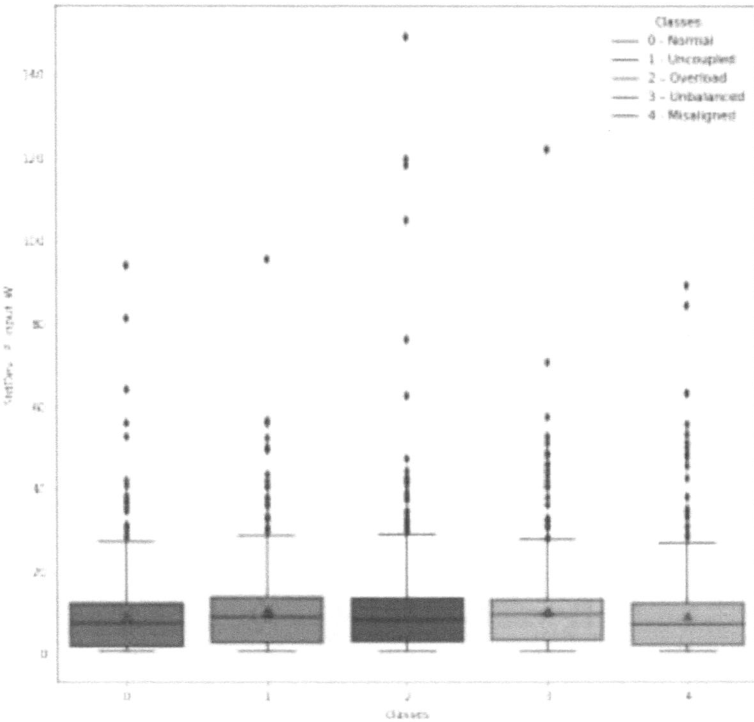

Figure 2.15 StdDev P input W attribute.

enhanced by the fact that the number of electrical amplitude sensors required is reduced as a result.

Thus, electrical quantity attribute selection is complete. The results of the study led us to conclude that the frequency inverter would be best served by using a single input current sensor, producing only the following three attributes: StdDev_I_input_A, I-input_A, and Kurtosis_I_input_A.

2.3.2 Model Training

To make predictions about electric motor-powered application anomaly detection and diagnosis, this sector details the models utilized and the whole algorithm training procedure.

Industrial applications operated by electronic motors for motion were represented in the supervized training, which used a balanced sample of five classes. The class property, which may take on values between zero and four, is responsible for identifying the various scenarios (see Table 2.1).

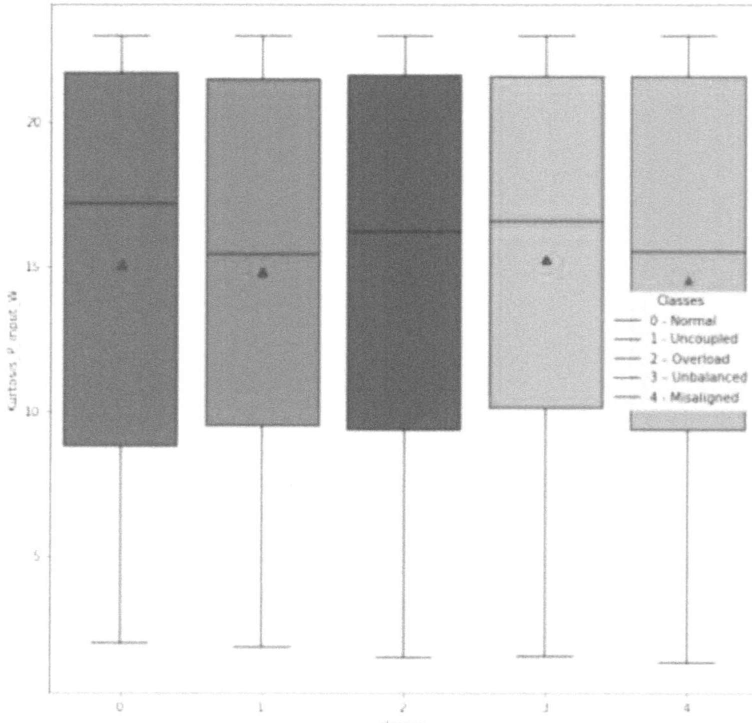

Figure 2.16 Kurtosis P input W quality.

Table 2.1 Classes' Property

Period	*Scenario*
0.0	Standard
1.0	Decoupled
2.0	Overload
3.0	Unbalanced
4.0	Incompatible

2.4 Conclusion

PdM for factories was the focus of this chapter, which evaluated the efficacy of specific machine-learning techniques. Identify and analyse unusual behaviour in an electric motor application, and then determine what causes these abnormalities under different setups. Overload, imbalance, mismatch with the electrical motor shaft, and disconnecting the driving system from the load were among these reasons.

With the results of these tests in hand, it is now possible to detect abnormalities in applications powered by electric motors, diagnose them, and even find out what causes them in the cases described above. In addition, the instrumentation system was defined and its cost was reduced by using data science and ML methods. The accuracy of sensor selection and attribute choice refinement were both greatly aided by these techniques. In this investigation, all data were acquired using a test bench, which may be seen as a restriction. The primary objective of this test bench was to provide an exhaustive simulation of real-world situations involving anomalies in electric motor applications. These anomaly simulations mainly dealt with the gear and worm system, which are mechanical components of the gearbox. Future research may expand the suggested method to include more kinds of electric motor-driven gearbox devices. It would also be possible to look for irregularities that are associated with engine electromechanics.

References

1. Kumar, K. Sunil Ratna, et al. "Iot and data mining techniques to detect and regulating of solar power system." 2023 International Conference on Inventive Computation Technologies (ICICT). IEEE (2023).
2. Rahat, Irfan Sadiq, et al. "A Step towards automated haematology: DL models for blood cell detection and classification." *EAI Endorsed Transactions on Pervasive Health and Technology* 10 (2024): 1–14.
3. Dixit, Palak, et al. "Anomaly detection in autonomous electric vehicles using AI techniques: A comprehensive survey." *Expert Systems* 39.5 (2022): e12754.
4. Murphey, Yi Lu, et al. "Model-based fault diagnosis in electric drives using machine learning." *IEEE/ASME Transactions on Mechatronics* 11.3 (2006): 290–303.
5. Lang, Wangjie, et al. "Artificial intelligence-based technique for fault detection and diagnosis of EV motors: A review." *IEEE Transactions on Transportation Electrification* 8.1 (2021): 384–406.
6. Drakaki, Maria, et al. "Machine learning and deep learning based methods toward industry 4.0 predictive maintenance in induction motors: State of the art survey." *Journal of Industrial Engineering and Management (JIEM)* 15.1 (2022): 31–57.
7. Gonzalez-Jimenez, David, et al. "Machine learning-based fault detection and diagnosis of faulty power connections of induction machines." *Energies* 14.16 (2021): 4886.
8. Martin-Diaz, Ignacio, et al. "An experimental comparative evaluation of machine learning techniques for motor fault diagnosis under various operating conditions." *IEEE Transactions on Industry Applications* 54.3 (2018): 2215–2224.
9. Principi, Emanuele, et al. "Unsupervised electric motor fault detection by using deep autoencoders." *IEEE/CAA Journal of Automatica Sinica* 6.2 (2019): 441–451.
10. Al-Saud, Mamdooh, et al. "An intelligent data-driven model to secure intravehicle communications based on machine learning." *IEEE Transactions on Industrial Electronics* 67.6 (2019): 5112–5119.

11. Nikfar, Mohsen, Julia Bitencourt, and Konstantinos Mykoniatis. "A two-phase machine learning approach for predictive maintenance of low voltage industrial motors." *Procedia Computer Science* 200 (2022): 111–120.
12. Al-amri, Redhwan, et al. "A review of machine learning and deep learning techniques for anomaly detection in IoT data." *Applied Sciences* 11.12 (2021): 5320.
13. de Paula Monteiro, Rodrigo, et al. "A hybrid prototype selection-based deep learning approach for anomaly detection in industrial machines." *Expert Systems with Applications* 204 (2022): 117528.
14. Sobie, Cameron, Carina Freitas, and Mike Nicolai. "Simulation-driven machine learning: Bearing fault classification." *Mechanical Systems and Signal Processing* 99 (2018): 403–419.
15. Wang, Zhenpo, et al. "A data-driven method for battery charging capacity abnormality diagnosis in electric vehicle applications." *IEEE Transactions on Transportation Electrification* 8.1 (2021): 990–999.
16. Nikfar, Mohsen, Julia Bitencourt, and Konstantinos Mykoniatis. "A two-phase machine learning approach for predictive maintenance of low voltage industrial motors." *Procedia Computer Science* 200 (2022): 111–120.
17. Sobie, Cameron, Carina Freitas, and Mike Nicolai. "Simulation-driven machine learning: Bearing fault classification." *Mechanical Systems and Signal Processing* 99 (2018): 403–419.
18. Purarjomandlangrudi, Afrooz, Amir Hossein Ghapanchi, and Mohammad Esmalifalak. "A data mining approach for fault diagnosis: An application of anomaly detection algorithm." *Measurement* 55 (2014): 343–352.
19. Olivotti, Daniel. "Digital transformation in the manufacturing industry: technologies and architectures" (2020).
20. Madubuike, Obinna C., Chimay J. Anumba, and Rana Khallaf. "A review of digital twin applications in construction." *Journal of Information Technology in Construction* 27 (2022): 1–16.
21. Song, Dengqin, et al. "Leveraging ISMOTE-KPCA-STACKING algorithm for enhanced vascular vertigo/dizziness diagnosis and clinical decision support." *IEEE Access* 11 (2023): 99735.
22. An, Zenghui, et al. "Adaptive cross-domain feature extraction method and its application on machinery intelligent fault diagnosis under different working conditions." *IEEE Access* 8 (2019): 535–546.
23. Sinha, Akruti, et al. "AI-Assisted Big Data Analytics for Smart Healthcare Systems." *Intelligent Internet of Things for Smart Healthcare Systems*. CRC Press (2023). 81–100.
24. Sharif Bidabadi, Shiva. *Human Gait Model Development for Objective Analysis of Pre/Post Gait Characteristics Following Lumbar Spine Surgery*. Diss. Curtin University (2019).
25. Zarkogianni, Konstantia, et al. "A review of emerging technologies for the management of diabetes mellitus." *IEEE Transactions on Biomedical Engineering* 62.12 (2015): 2735–2749.

Chapter 3

A Data Fusion Technique to Detect and Assess Electromechanical Damage

Arpit Jain, Bala Ganesh N., Abhishek Choubey, Pramoda Patro, and Lakshman Kumar Kanulla

3.1 Introduction

The process of corrosion, exhaustion, wear, delamination, and other structural flaws shorten the lifetime of any man-made civil, aeronautical, or mechanical structure [1–3]. Appropriate SHM methods are used for monitoring these structural problems. When evaluating a structure's local health in the high-frequency region, one SHM methodology used is the electromechanical impedance (EMI) method. Developing a data-driven technique to properly detect the precise position and degree of structural flaws is a major research topic in EMI-based detection. To identify deterioration in reinforced concrete (RC), the investigator used effective structural mechanical impedance [4–7]. To determine which is more effective for damage detection, compare electromechanical admittance, efficient structure mechanical impedance signatures, and structure mechanical impedance. The researcher used a battery of sensors to look for cracks in the pipeline system. They used the raw admittance's active portion to generate the damage-sensitive feature. A recently suggested active

DOI: 10.1201/9781003479062-3

monitoring approach makes use of a revised electromechanical impedance analytical model to extract signatures based on mechanical impedance via a direct connection [8–10].

The RMSD index was used to identify and measure the disband's extent inside the structures. In addition to advancements in damage detection based on electromagnetic interference (EMI), data fusion-based approaches have also seen a resurgence of attention in recent years [11–13]. Using the DS theory and the Rotation Forest (RF) approach, the researcher proposed a hierarchical ensemble system for data fusion. After RF was used to build a precise and diverse foundation, DS aggregated the output from the multiple RF facts bases [14–16].

The use of deep knowledge, artificial neural networks (ANNs), and support vector machines (SVMs) as fusion-level machine learning methods is on the rise in data fusion. Researchers recently put out a data fusion idea that relies on deep learning to identify and pinpoint fissures in the reactors' metal surfaces, which are used in nuclear power plants [17]. Also getting some buzz is the idea of using PCA, or principal component analysis, to categorise and identify damages. To identify and fix issues with the production of polyester film, the researchers employed T2 and PCA-based Q statistics [18]. Using these statistical methods, the researcher investigated the possibility of damage detection and differentiation in turbine blade and steel plate constructions. Researchers have proposed a statistical approach that utilises PCA to classify damage in region piezoelectric active sensor networks while analysing time-domain vibrational structural responses. The researcher used the PCA model to locate loose bolts in the bolted aluminium plate construction of wireless SHM by analysing impedance data [19–21]. They used a macro-fibre composite sensor system and impedance measurement chips as part of an active sensor system that ran on board. Even though evidence merging for EMI is a relatively new method, the concept is slightly well-established in other academic domains. One definition of data fusion is the process of combining datasets into a single, more complete dataset to improve representational efficiency and lower uncertainty [22]. There are three tiers of data fusion: quality of data, degree of decision-making, and level of features. By preventing the post-processing noise and measurement mistakes from propagating and amplifying, data source data fusion enables pure data merging [23]. Additionally, it improves the signal-to-noise ratio, which is explored in this chapter. A need to provide a unified method of structural damage identification motivates the usage of data fusion [24]. To enhance damage detection via vigorous decision-making, data fusion enables the procedure of extraction of knowledge from frequency domain data [25].

This work grants a method for detecting damage using electromagnetic interference (EMI) by fusing low-level sensor data. The chapter's originality is in the implementation of variable data fusion approaches for EMI-based SHM methodology, an area that has seen less research despite the major work done on data fusion. This study implements variable data fusion in the electromagnetic interference (EMI) approach to identify damage in both isotropic and anisotropic materials, namely aluminium and composite plates. To achieve data fusion, the resistance and conductivity of the restrained impedance spectra are combined to form a fused parameter (F). By using the RMSDk and MAPD losses indices, quantify the discrepancy in the fusing parameter F. Here, RMSD stands for the mean of the root difference with the k-th condition as a point of reference. The results show that the fusion variable outperforms both conventional and conductance-based approaches in damage identification, proving the principles behind it.

3.2 Concepts and Approaches

The electrical resistance of a connected piezoelectric transducer (PZT) is the voltage (V) split by the current (I) going through it. To determine the EMI (Z(w)), the electric current (I) that has been measured is used.

$$Z(w) = \frac{V}{I} = R + jX = \frac{1}{Y} = \frac{1}{(G + jB)}$$

$$= \frac{1}{jw \in_{33}^{-T}} \frac{h_p}{\pi a^2} \left[1 - \frac{2d_{31}^2}{S_{11}^{-E} \epsilon_{33}^{-T} (1 - v)} \left[1 - \frac{1}{\frac{Z_a(w)}{Z_s(w)} + 1} \left(\frac{2}{\varnothing} \right) \left(\frac{J_1(\varnothing)}{J_0(\varnothing)} \right) \right] \right]^{-1} \quad (3.1)$$

$\varnothing = \frac{w}{c_p^-} a, \bar{c}_p = \frac{1}{\sqrt{\rho_{S_{11}}^{-E}(1 - v^2)}}, \bar{S}_{11}^{E} = S_{11}^{E}\,(1 - j\eta), \; \bar{\epsilon}_{33}^{T} = \in_{33}^{T}(1 - j\delta), J_0$. The following variables are defined in the PZT equation: Z_a the disrupted mechanical impedance of the actuator, Z_s the structure's mechanical impedance, h_p the transducer's thickness, a the transducer's radius, ρ the density, S_{11}^{E} *is coefficient of compliance*, d_{31} the piezoelectric coefficient for course 3-1, j the complex symbol, η is the mechanical loss factor and δ is the dielectric loss factor, and they both contribute to the $\in_{33}$ complicated permittivity of the piezoelectric components transducer. The EMI parameters R and G have shown to be very useful in several applications when it comes to structural damage identification. By looking at what electrical resistance (Z) and admittance (Y) are, deduce the relationship between R and G:

$$Z = R + jX = \frac{1}{Y} = \frac{1}{(G + jB)} = \frac{1}{(G + jB)} \times \frac{(G - jB)}{(G^2 + B^2} = \frac{(G - jB)}{[|Y|]^2}, \quad (3.2)$$

X represents the system's reactance, while B stands for its susceptance. The relationship between R, G, B, and X may be determined by combining the actual and fantasy sections of Equation (3.2)

$$R = \frac{G}{(G^2 + B^2)}; X = \frac{-B}{G^2 + B^2}; G = \frac{R}{R^2 + X^2}; B = \frac{-X}{R^2 + X^2} \quad (3.3)$$

Take G and multiply it by G, or R with R, or B and X—Equation (3.4)—combine non-dimensional variable that is dependent on two variables. For damage detection, this multiplicative dimensionless parameter could be useful.

$$F = G \times R = \frac{G}{G^2 + B^2} = \frac{R^2}{R^2 + X^2} \quad (3.4)$$

Damage detection skills are anticipated to be enhanced by the F, which enhances the typical peak of G and R. For the EMI-based damage evaluation within a rather small frequency range, researchers have so far only used a single quantity. The resistance's improved damage sensitivity across a small occurrence group, namely 10–80 kHz for the bolted-joint aluminium sample and 16–40 kHz for the beam with added mass injury model, were highlighted. When testing carbon composites, the researcher also looked at resistance in the low-band frequency of 10–80 kHz. Therefore, the suggested approach enhances the responsiveness of the variables to a wider range of frequencies by merging the two variables. Consequently, indices that use the value of F are more suited to evaluating damage and may be used over a wider frequency range. They created a variable of intermediate sensitivity throughout the full domain by combining two variables with greater sensitivity in a limited region in their education on vibration-based recognition of injury in constructions.

Damage detection utilising electromagnetic interference (EMI) relies on measuring the discrepancy between the structure's measurements at a certain moment and a known prior state. The two most common damage detection indices used for this measurement are RMSD and MAPD. To measure damage to a glowing condition of the construction, the EMI approaches use Equations (3.5) and (3.6):

$$RMSD = \sqrt{\sum_{i=1}^{n} (D_i - D_i^0)^2 \_\_\_\_\_\_\_\_\_\_ \sum_{i=1}^{n} (D_i^0)^2} \quad (3.5)$$

$$MAPD = 100 / n \times \sum_{i=1}^{n} \left| (D_i - D_i^o) / (D_i^o) \right|. \quad (3.6)$$

Damage detection performance has been greatly improved with the recent use of RMSDk. Using these characteristics, this research additionally explores the damage index RMSDk, which is used to distinguish and assess the absolute changes in damaged intensity at each level. The formula for computing the RMSDk is given by Equation (3.7). In contrast to the RMSD, the RMSDk compares the measured information to the prior state instead of the healthy state, which is the main distinction between the two.

$$RMSDk = \sqrt{\sum_{i=1}^{n}(D_i - D_i^k)^2 \underline{\qquad\qquad\qquad\qquad} \sum_{i=1}^{n}(D_{i-1})^2}, \tag{3.7}$$

this is where the i-th damaged state is represented by D_i, n is the amount of spectrum of frequencies models, and k is the k-th state. The D_{i-1} squared value represents the frequency divided single-sample spectrum for the (i–1)-th damaged state. To remove the frequency trend from the data, multiply the values of each R measurement by the normalised frequency and divide by each G measurement, following Equation (3.8), as it is apparent that G and R exhibit a frequency movement. The G, R, and F units are preserved by the aforementioned process.

$$G_{rescaled} = \frac{G \times f_{max}}{f}; \; R_{rescaled} = \frac{R \times f}{f_{max}}; \; F_{rescaled} = \frac{F \times f_{max}}{f}. \tag{3.8}$$

It should be mentioned that the rescaling process was previously used for EMI measurements of R and G. Figure 3.1 shows the observed connection to the frequency, which was used to choose to rescale variable F. F, G, and R will stand in for $F_{rescaled}$, $G_{rescaled}$, and $R_{rescaled}$, respectively, for the sake of brevity. The rescaling has a noticeable effect of flattening the F plot. As a result of rescaling, the G plot also shows more points in the low occurrence region, but the R plot shows less low-frequency contribution.

Figure 3.1 shows the rescaled ranges of G, R, and F for both the strong and damaged scenarios. The F spectra stand out with their smoothness compared to the G and R spectra. One common source of noise from measurements and other errors is non-dominant peaks; when computing F, this influence is minimised. F produces more consistent findings since these uncertainties' impact on the difference-quantification index is less. The F also increases the contribution of the shared peak in the G and R by amplifying them.

3.3 Methodology and Purpose of the Experiment

Metals and composite with attached piezo actuators were used in the trials to validate the suggested approach. PZTs manufactured by CeramTec using SONOX P502

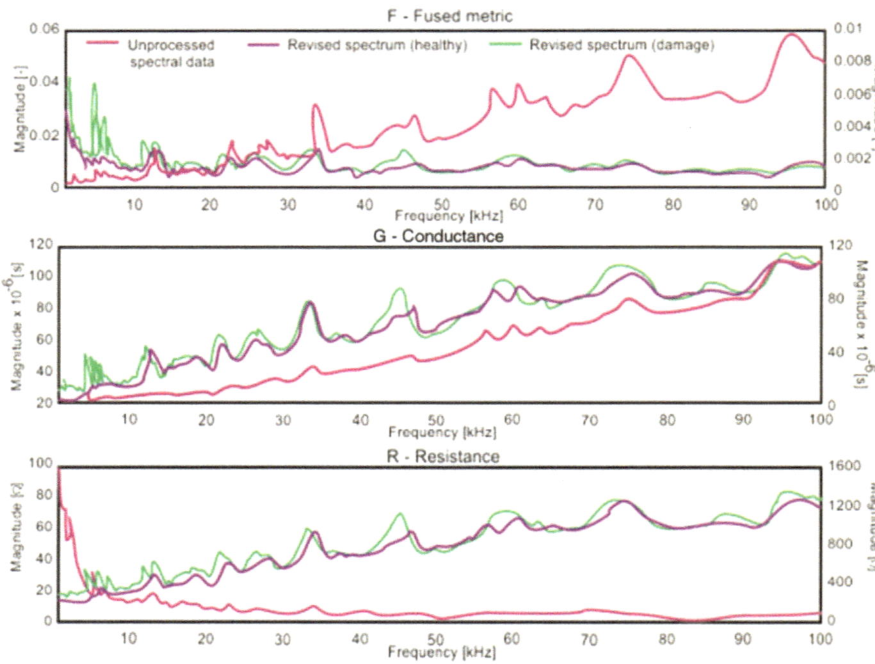

Figure 3.1 Scan results for F, G, and R, both in their raw form and after rescaling (healthy and damage scenarios).

material, with a disc form and a deepness of 0.05 cm then a diameter of 1 cm, were used to measure the electromagnetic interference. The measurements were taken using a Hioki IM3570 analyser in the investigation. Cyanoacrylate glue secures the sensors to the metal and composites, while soldered wires connect the sensors to an impedance analyser via the sample surfaces' tops. The research used chirp excitation having a peak value of 1 V. Based on engineering judgement, we selected damage simulation areas that fall within the sensitisation range of the relevant sensor. The three buildings that were examined were

1. A little aluminium plate with holes already punched into it.
2. A delaminated GFRP beam made of glass fibres.
3. The surface of the GFRP plate has been hit many times.

An aluminium square plate of 10 cm × 10 cm × 0.1 cm was used in the first investigation. The holes were systematically introduced to the core of the PZT plate from both the highest point and bottom in a consecutive manner. The holes are 0.3 mm in diameter. There is about 1 cm of space between each hole. About 1.5 cm away from the piezo-actuator's centre are the closest holes. There were two sets of measurements taken: one before each hole was punched in the plate, and one after the reference (healthy condition). The temperature of the room (23 ± 1°) was used for the measurements. Figure 3.2 shows the drilled holes and their succession. The tests were taken with a 200 Hz step size and throughout the frequency variety from 1 kHz to 5 MHz.

Figure 3.3 shows the consequences of the second succession of tests conducted on a thick composite beam that was rectangular and had one PZT connected to it. The dimensions of the beam were 50 cm × 9.5 cm × 0.3 cm. Three damage states (through progressive delamination) and a healthy condition were used to gather EMI responses from the composite plate. The simulated delamination was introduced using the scalpel. Specifically, (S1) measured 1 cm × 0.5 cm, (S2) 2 cm × 1 cm, and (S3) 3 cm × 1 cm in terms of delamination dimensions. Two sets of measurements were obtained at every delamination stage. The measurements were taken with a 10-Hz step size and over a frequency range of 1–100 kHz.

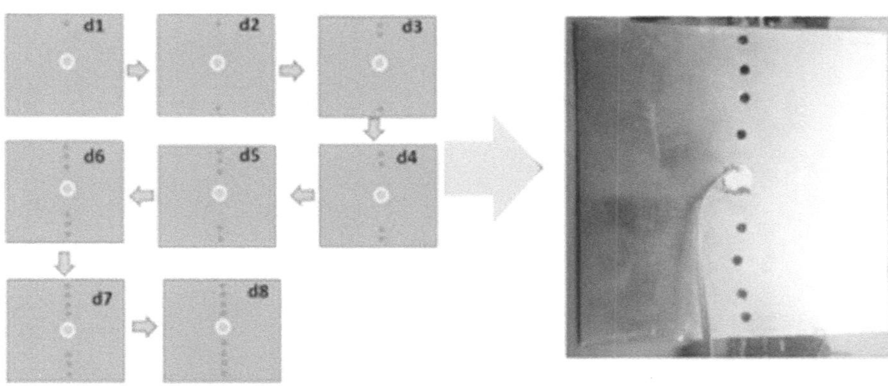

Figure 3.2 The processes involved in making holes on an Al square plate and the finished product.

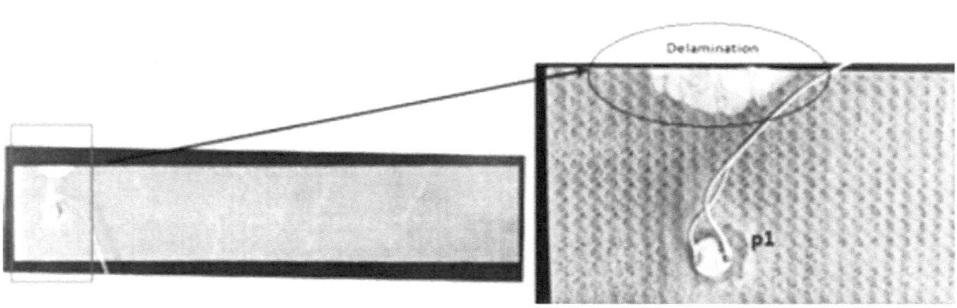

Figure 3.3 An enlarged view of the composite beam reveals the delaminated section (S3).

A woven GFRP plate of 50 cm × 50 cm × 0.3 cm was the third sample examined (Figure 3.4). All measurements were taken in a controlled environment at room temperature, with an audio frequency span from 1 kHz to 5 MHz and a 200 Hz step. A projectile with a sphere-shaped finish was dropped on the external of the sample to create the impact. A projectile with a total weight of 1.3 kg, consisting of a steel ball merged with a cuboid framework, was released from a height of 2.39 m. A single collision was calculated to have 30 J of energy. Table 3.1 details all four impacts that were applied to the GFRP sample, as well as the corresponding surface characteristics of the samples. Despite the presence of four PZT sensors in the picture, the data from sensor P2 is the only one shown here. Despite their apparent insensitivity to damage, the other detectors were located distant from the impact sites.

The three constructions analysed were included in separate investigations, resulting in variations in occurrence ranges and step sizes used for the capacities.

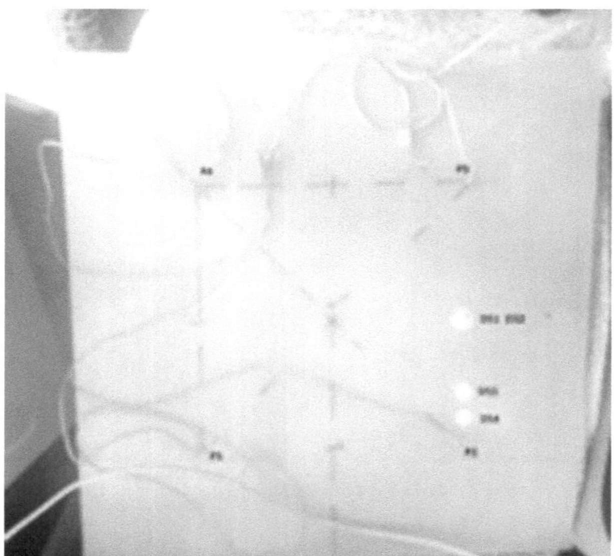

Figure 3.4 The GFRP plate has been exposed to impact spots.

Table 3.1 Effective Damage on the Surface of GFRP Plates

Potential Situation	*Full Description*
DS1	30J effect of P2 and P3 at a distance of 130 mm from P2
DS2	DS1 + 30J effect of P2 and P3 at a distance of 130 mm from P2
DS3	DS2 + 30J effect of P2 and P3 at 55 mm from P2
DS4	DS3 + 30J effect near P2 at a distance of 25 mm from P2

In command to provide consistent measurements and an extensive analysis of all three platforms, this research only reports results from an occurrence range of 1–100 kHz. Be advised that every reading is the mean of 50 readings collected at each occurrence. Hence, these numbers are measured to be statistically constant. Differences in the surrounding environment could explain why two observations of the same scenario might provide different results. The threshold value is determined by comparing the data of two known healthy conditions, and it accounts for the variations in various readings.

3.4 Findings and Analysis

3.4.1 Aluminium Plate Damage Detection

Measurements were taken twice for each of the eight damage scenarios and the healthy scenario for the aluminium plate. As previously indicated, the difference between the healthy and damaged instances was measured using three separate indices: RMSD, MAPD, and RMSDk. They are shown in Figures 3.5–3.7 in that

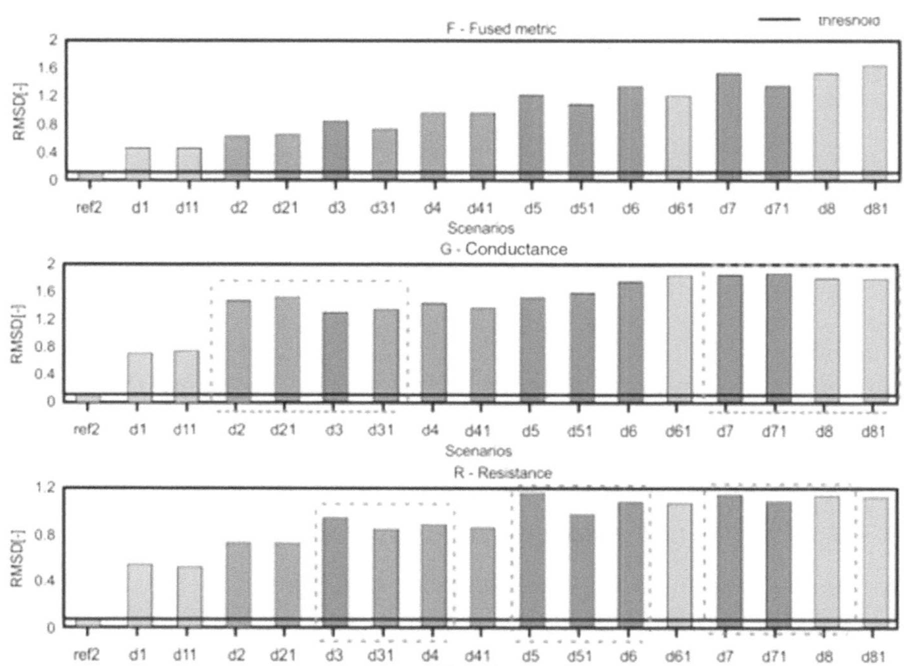

Figure 3.5 Regarding the aluminium plate, F, G, and R were determined using the root mean square deviation (RMSD) index.

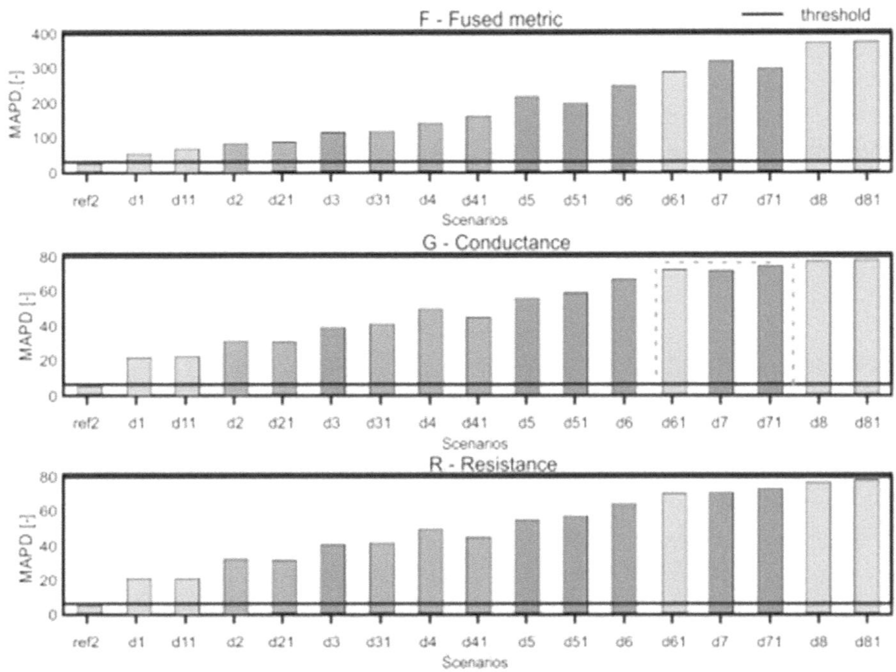

Figure 3.6 The aluminium plate MAPD guide for the F, G, and R environmental variables.

order. Because it reflects the inherent uncertainty in the observations, every reference measurement was used to estimate the damage detection threshold. The harm is considered to have been discovered if the directory rises beyond the threshold value.

There is a clear upward trend in measure F for the MAPD and RMSD indices with rising hole counts, indicating that the damage is becoming more severe. Notably different for the G and R spectra, as seen by the red outlined boxes, is this pattern. No rising trend is discernible since the RMSDk compared the current dimension to the prior state; at this point, the threshold worth established by the ref2 situation is exceeded in all damage scenarios.

3.4.2 Examining Composite Beam Delamination

For the S3 health and damage situation, the rescaled spectra are shown in Figure 3.1. It is evident from the figure that the F spectra amplify the standard peak in the R and G spectra, which in turn contribute more to the harm quantification index. Measurements taken from the composite beam yielded the same three catalogues for the F, G, and R spectra as those from the aluminium plate; these are shown in

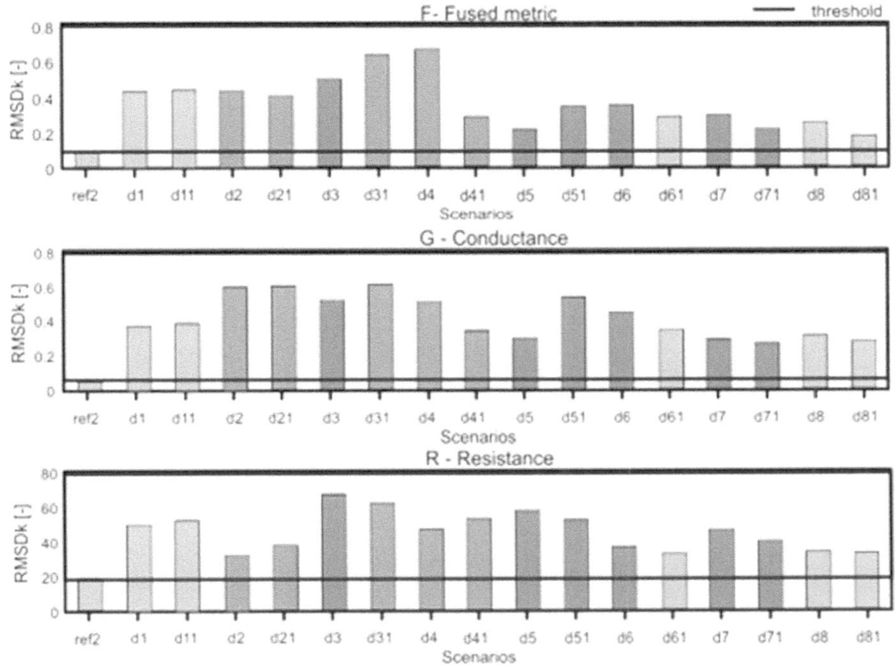

Figure 3.7 The RMSDk guide for aluminium plate variable quantity F, G, and R.

Figures 3.8–3.9. Damage is detected by all three indices for all three measurements. In addition, the tendency for the severity of damage is growing according to F-based RMSDk. This does not show in the RMSDk values that are based on G and R.

3.4.3 Evaluation of Compound Plate Impact Damage

The next structure examined was the GFRP plate that had incurred damage due to impact.

The three indexes for F, G, and R are shown in Figures 3.10 and 3.11, respectively. None of the three indices F, G, and R recognise the scenario of damage DS1. The first collision caused little damage to the building and occurred at a considerable distance from the sensors, which explains why it went undetected. The three variables successfully identify all the following damage possibilities. It is worth noting that in instances DS2 and DS3, the variable F exhibits a larger disparity compared to the G and R variable quantity. This suggests a better compassion in distinguishing among damages of varying magnitudes. Because of its heightened sensitivity and capacity to distinguish among diverse harm situations, the F measure is more favourable than the G and R metrics.

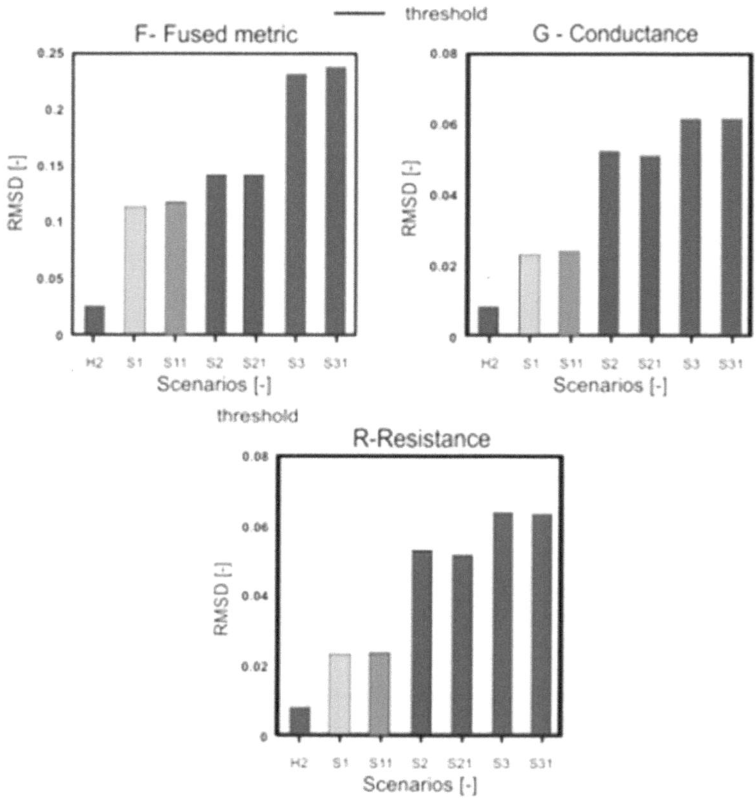

Figure 3.8 Factors F, G, and R contribute to the MAPD of the GFRP beams index.

3.4.4 Overview

Table 3.2 summarises the findings of the proportional analysis of the damage index's presentation that were reported in Sections 3.4.1–3.4.3. It should be noted that the damage assessment index determines how well each variable performs. For this reason, it is necessary to compare the variables' performances in the quantification measure.

In instances when the damage increases, the F measure indicates a trend, but neither the G nor the R metrics for the aluminium plate. Parameters R and F demonstrate an upward trend in the MAPD index, however the measure based on G does not. All three parameters work well when the RMSDk index is utilised for damage detection exclusively. As the damage to the GFRP beam increases, all three approaches reveal that the indices are trending upwards. Therefore, it follows that they perform adequately. On the GFRP plate, not a single one of the three characteristics or indices picked up on the tiniest damage, DS1. In contrast, the F

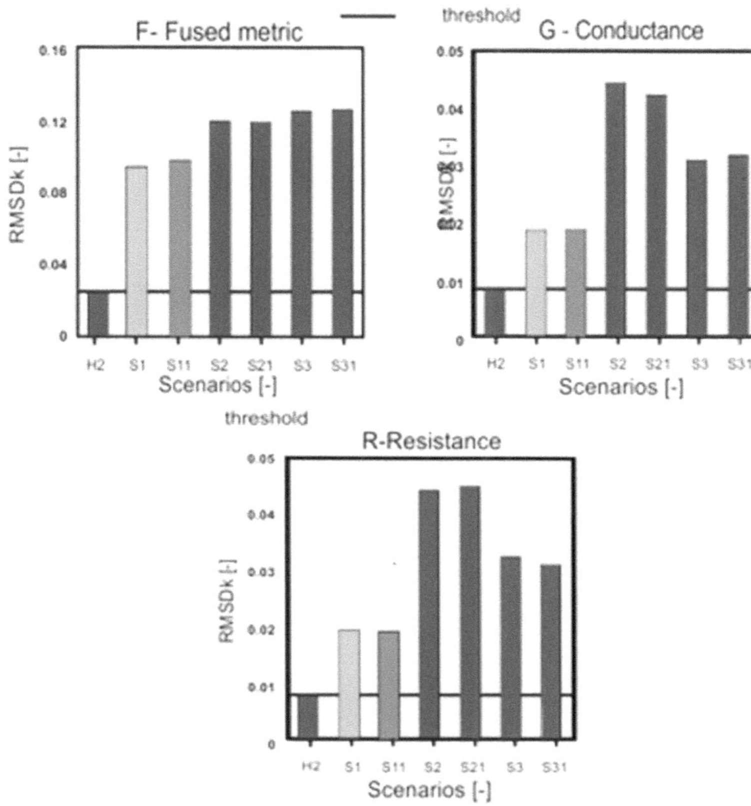

Figure 3.9 The GFRP beam's RMSDk index, where F, G, and R variables.

parameter performs better than the G and R variables in the other harm scenarios, such as DS2 and DS3, since it can differentiate between the two scenarios' increasing harm. Using case DS2 as a baseline, just one of the settings related to the RMSDk metric can identify the damage DS3. So, it seems that all of the factors apply to this situation. It is more likely that the RMSDk metric's shortcomings than the values of G, R, or F are to blame for this parameter shortfall.

3.4.5 Analyses of Sensitivity

To facilitate comparison, the occurrence band from 1 to 100 kHz was used, even though the three samples' measurements were collected across various ranges of frequencies. They used two distinct frequency bands 100–300 kHz and 1 kHz–1.3 MHz to test the resilience of the fusion parameter F for the damage detection. Figures 3.12 and 3.13 show the outcomes for the three variables using the RMSD index on the aluminium plate.

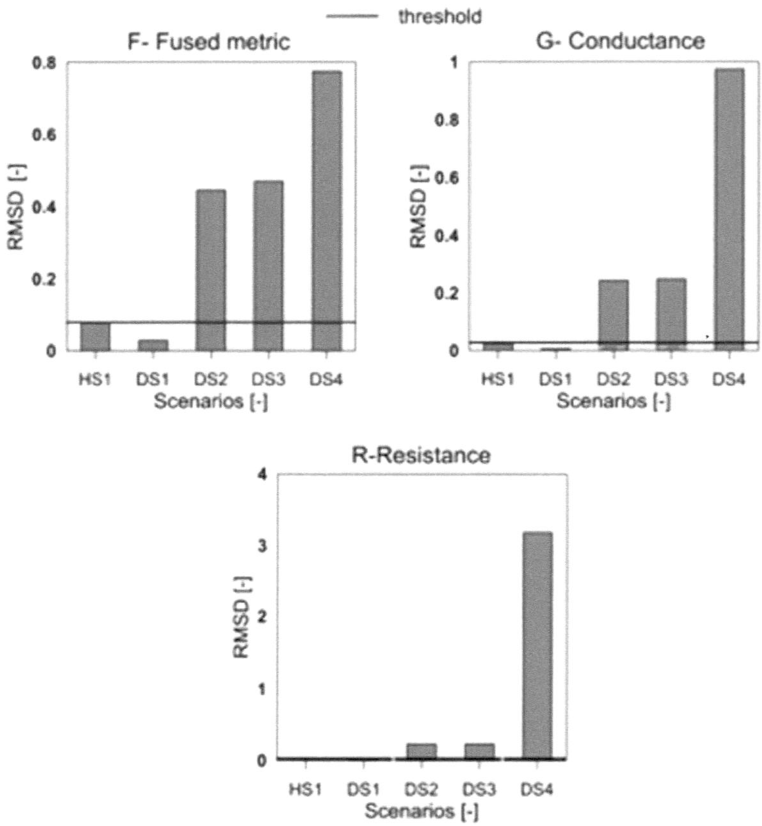

Figure 3.10 F, G, and R's RMSD index as it pertains to the GFRP plate.

A predicted pattern of higher values with a hike in damage severity is seen by the F in these frequency ranges. However, the G and R variables are not trending upwards. Therefore, across a wider frequency range, the bonded limit F performs better than the R and G limits.

3.5 Conclusions

In detecting electromagnetic interference damage, the RMSD, RMSDk, and MAPD indices are often used. This research aims to suggest ways to make these indexes better at detecting damage. The chapter's main emphasis was on the F parameter, which proved to be more accurate for damage detection when the G and R parameters, derived from EMI dimensions, were combined.

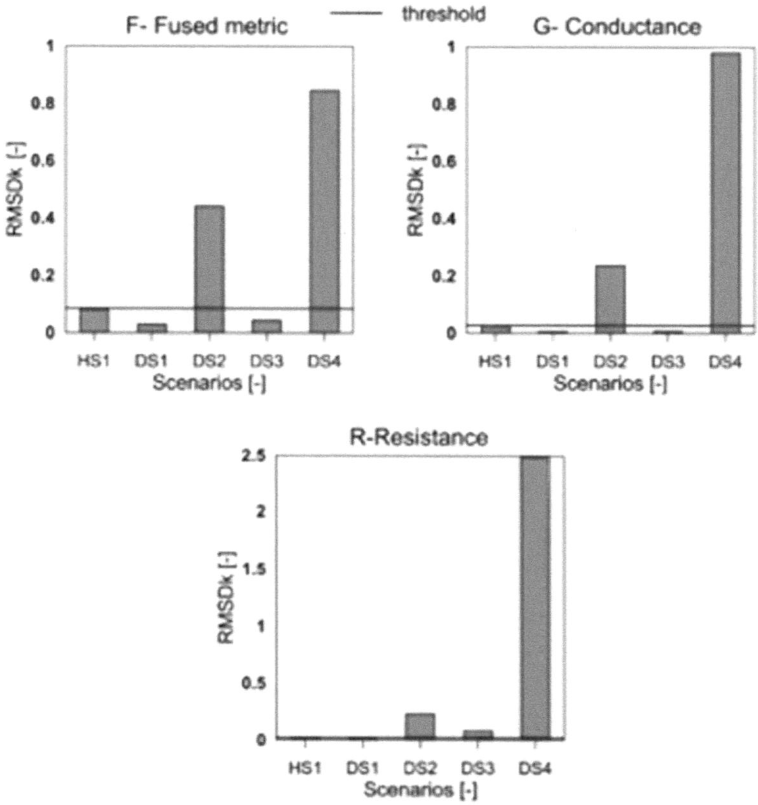

Figure 3.11 The GFRP plate's RMSDk index, where F, G, and R are variables.

Table 3.2 Research Comparing the G, R, and F Variables' Performance with that of Al, GFRP Beams, and GFRP Plates

Case Studies	*RMSD*	*MAPD*	*RMSDk*
Hole-drilled Al plate	F	F, R	F, G, R
Cracks appearing in the GFRP beam	F, G, R	F, G, R	F, G, R
A damaged GFRP plate from an impact	F	F, G	none

This study assesses the efficacy of the parameters F, R, and G for three distinct constructions: an aluminium plate through progressively punched holes, a GFRP column with delamination inserted, and a GFRP plate with impression injury presented. The analysis was conducted within the occurrence band of

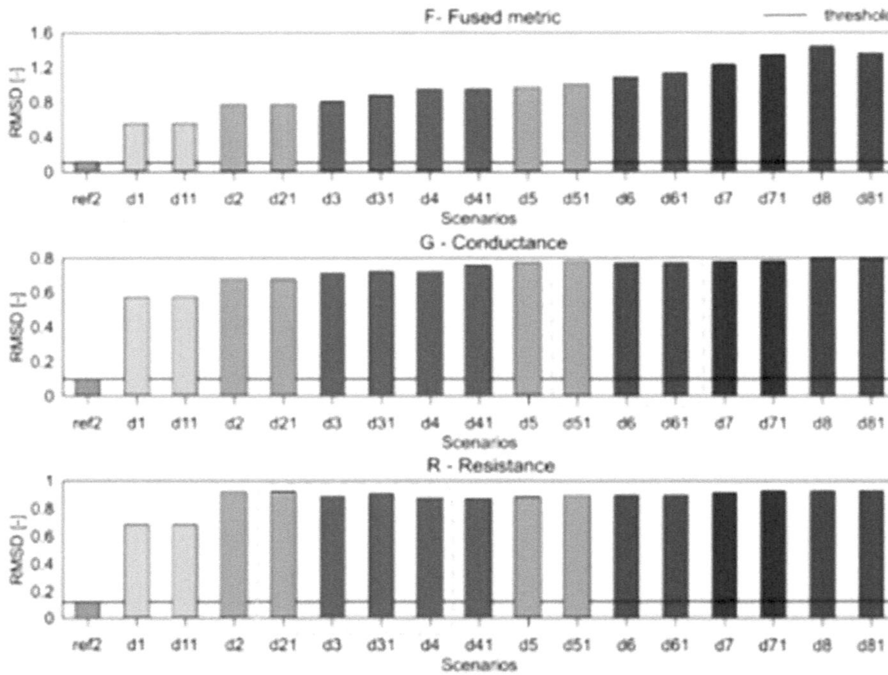

Figure 3.12 Over the occurrence variety of 100–300 kHz, the RMSD index for the aluminium plate's F, G, and R parameters is shown.

1–100 kHz to make it easier to compare various samples. On top of that, one of the results demonstrated that the parameter F is resilient across a wider frequency range.

When it comes to detecting damage and its increasing severity, the findings show that the F factor is more resilient. Integrating the G and R variables increases the sensitivity, as can be shown. One possible explanation is that by multiplying the G and R spectra, the F spectrum becomes smoother. On top of that, unlike existing approaches, F spectra may identify damage throughout a wide frequency range, not only the narrow variety where the selected limit (G or R) is most delicate. It is difficult to apply specific settings since this sensitivity zone is not known a priori. A wider range of frequencies for sensitivity makes the F parameter more practical, which simplifies its use.

This study does provide some evidence of the concept and presents the idea of the F parameter, which the authors themselves admit. It has been determined that the next stage is to expand the F variable for a sensor network and study its potential use in damage localisation. Furthermore, it is necessary to examine how the parameter is affected by environmental factors.

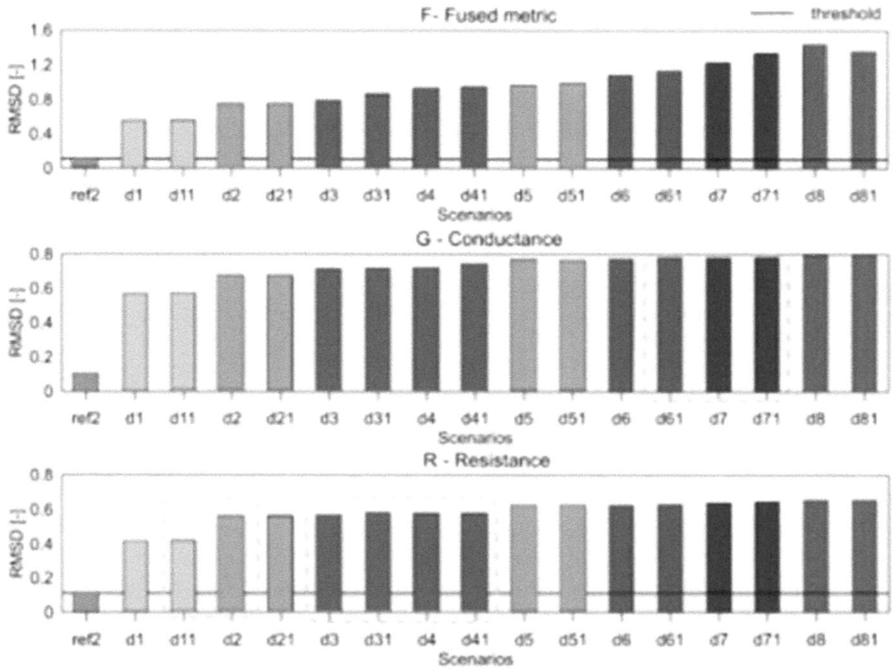

Figure 3.13 A measure of the root mean square deviation (RMSD) for the aluminium plate's F, G, and R parameters from 1 kHz to 1.3 MHz.

References

1. Singh, Shishir K., and Pawel H. Malinowski. "An innovative data-driven probabilistic approach for damage detection in electromechanical impedance technique." *Composite Structures* 295 (2022): 115808.
2. Kumar, K. Sunil Ratna, et al. "Iot and data mining techniques to detect and regulating of solar power system." *2023 International Conference on Inventive Computation Technologies (ICICT)*. IEEE (2023).
3. Hassani, Sahar, et al. "A systematic review of data fusion techniques for optimized structural health monitoring." *Information Fusion* 103 (2024): 102136.
4. Janapati, Vishnuvardhan, et al. "Damage detection sensitivity characterization of acoustic-ultrasound-based structural health monitoring techniques." *Structural Health Monitoring* 15.2 (2016): 143–161.
5. Sha, Ganggang, et al. "A novel method for single and multiple damage detection in beams using relative natural frequency changes." *Mechanical Systems and Signal Processing* 132 (2019): 335–352.
6. Fan, Xingyu, and Jun Li. "Damage identification in plate structures using sparse regularization based electromechanical impedance technique." *Sensors* 20.24 (2020): 7069.

7. Wang, Hao, Giorgio Barone, and Alister Smith. "A novel multi-level data fusion and anomaly detection approach for infrastructure damage identification and localisation." *Engineering Structures* 292 (2023): 116473.
8. Zhao, Xiukuan, et al. "Hierarchical ensemble-based data fusion for structural health monitoring." *Smart Materials and Structures* 19.4 (2010): 045009.
9. Byington, Carl S., and Amulya K. Garga. "Data fusion for developing predictive diagnostics for electromechanical systems." *Handbook of Multisensor Data Fusion.* CRC Press (2017). 721–758.
10. Elsamanty, Mahmoud, Abdelkader Ibrahim, and Wael Saady Salman. "Principal component analysis approach for detecting faults in rotary machines based on vibrational and electrical fused data." *Mechanical Systems and Signal Processing* 200 (2023): 110559.
11. Patro, Pramoda, et al. "A hybrid approach estimates the real-time health state of a bearing by accelerated degradation tests, machine learning." *2021 Second International Conference on Smart Technologies in Computing, Electrical and Electronics (ICSTCEE).* IEEE (2021).
12. Zheng, Yuebin, et al. "Lamb waves and electro-mechanical impedance-based damage detection using a mobile PZT transducer set." *Ultrasonics* 92 (2019): 13–20.
13. Bandara, Senaka, Madhubhashitha Herath, and Jayantha Epaarachchi. "Sensory methods and machine learning based damage identification of fibre-reinforced composite structures: An introductory review." *Journal of Reinforced Plastics and Composites* 42.21–22 (2023): 1119–1146.
14. Kullu, Omer, and Eyup Cinar. "A deep-learning-based multi-modal sensor fusion approach for detection of equipment faults." *Machines* 10.11 (2022): 1105.
15. Si, Liang, and Zongfeng Li. "Online structural state assessment for aerospace composite structures using an acousto-ultrasonics-based multi-damage index identification approach." *Structural Health Monitoring* 19.6 (2020): 1790–1807.
16. Lu, Yinghui, and Jennifer E. Michaels. "Feature extraction and sensor fusion for ultrasonic structural health monitoring under changing environmental conditions." *IEEE Sensors Journal* 9.11 (2009): 1462–1471.
17. Kessler, Seth Stovack. *Piezoelectric-Based in-Situ Damage Detection of Composite Materials for Structural Health Monitoring Systems.* Diss. Massachusetts Institute of Technology (2002).
18. Mujica Delgado, Luis Eduardo, José Rodellar Benedé, and José Vehí Casellas. "A review of impact damage detection in structures using strain data." *International Journal of COMADEM* 13.1 (2010): 3–18.
19. Chelidze, David. "Multimode damage tracking and failure prognosis in electromechanical systems." *Component and Systems Diagnostics, Prognostics, and Health Management II.* Vol. 4733. SPIE (2002).
20. Zhang, Chengdong, and Yongli Xu. "Structural damage identification via multi-type sensors and response reconstruction." *Structural Health Monitoring* 15.6 (2016): 715–729.
21. Singh, Shishir Kumar, Rohan N. Soman, and Paweł H. Malinowski. "A novel multi-damage localization method for polymers and composites based on electromechanical impedance." *Mechanical Systems and Signal Processing* 216 (2024): 111508.

22. Yang, Hongjuan, et al. "Ultrasonic detection methods for mechanical characterization and damage diagnosis of advanced composite materials: A review." *Composite Structures* 324 (2023): 117554.
23. Loyola, Bryan R., et al. "Detection of spatially distributed damage in fiber-reinforced polymer composites." *Structural Health Monitoring* 12.3 (2013): 225–239.
24. Jacob, Geo, and Florian Raddatz. "Data fusion for the efficient NDT of challenging aerospace structures: A review." *NDE 4.0, Predictive Maintenance, and Communication and Energy Systems in a Globally Networked World.* Vol. 12049. SPIE (2022).
25. Ai, Demi, Duluan Zhang, and Hongping Zhu. "Damage localization on reinforced concrete slab structure using electromechanical impedance technique and probability-weighted imaging algorithm." *Construction and Building Materials* 424 (2024): 135824.

Chapter 4

AI: Classifications and Protection of the Smart Grid Systems

Akshaya Kubba, D. Prabakar, N. M. G. Kumar, and Shruti Bhargava Choubey

4.1 Introduction

The smart city (SC) enhances the infrastructure and longevity of existing cities by collecting, analysing, and combining crucial data about their core systems via the use of information technology (IT), artificial intelligence (AI), and the Internet of Things (IoT) [1]. The idea of a smart city is to manage resources, provide services, and solve issues that affect everyone's daily lives by analysing massive amounts of information collected by sensors, and power systems [2]. The demands of municipal services, public safety, everyday life, and industrial and commercial operations are met by SC in a timely, efficient, and intelligent manner [3–5]. Power or energy is the foundation of all systems. This means it's a good starting point for implementing the SC idea.

Conventional electricity networks are getting smarter [6]. Traditional grids mislay electricity production, transmission and spreading, and due to unidirectional communication. The introduction of smart grid (SG) offers free bi-directional communiqué and overcomes typical grid concerns [7]. Information systems, control systems, and bi-directional communication devices were incorporated into safety

DOI: 10.1201/9781003479062-4

data sheets. Safety data in sheets is crucial for industry and research [8]. However, erroneous SDS information might lead to serious accidents. The complex equipment includes modern phasor networks [9]. Managers might check grid stability using these sensors. SG has added feeder switches to reroute energy during grid failures [10], a computerised system for greater information and feedback, issue recognition, and additional batteries for future needs [11]. Rapid electric grid changes both opportunities and challenges for power distribution system improvement. Predicting electricity was essential for an effective SG and reduced operating costs [12]. Future trends are predicted using load and pricing data.

Advancements in data technology and access to diverse data are making intelligent algorithms increasingly practical. Machine Learning (ML) methods excel in generalisation compared to older computing methods [13]. Controlling energy and connectivity net with actual energy, cost, quality, and consumption data is a major challenge for SG implementation [14]. Additionally, the goal is to develop intelligent systems using ML approaches to make judgements in uncertain scenarios. The stability between power output and consumption determines grid resilience. Increased power production and decentralisation have led to greater electricity flow in both directions [15]. These advances require A sophisticated control centre to manage output and use of energy. ML approaches assess the link between network stability metrics and components of input for diverse users [16]. However, convolutional neural networks (CNNs) can correctly depict the structural localisation using the feature space. In addition to handling interruptions and small changes in the feature space, they reduce translational variation by using local frequency area pooling [17]. The long-term links between speech frames may be exploited by making use of prior knowledge about speech signals. Unfortunately for ASR systems, CNNs aren't up to the challenge of working with dirty data. In contrast, recurrent neural networks (RNNs) may enhance identification accuracy, especially in noise-resistant applications, by gathering extensive contexts. Nevertheless, RNNs were limited in their ability to understand temporal connections due to the vanishing and growing gradient problem.

Long short-term memory (LSTM) was created to solve these problems; it controls the flow of information via a special unit called a memory block. Because LSTM-RNNs are so sensitive to static input, feature-related target delays do occur. The use of deep neural networks (DNNs) has been very beneficial in acoustic modelling. An enhanced version of DNNs, also known as DNNs, is called a convolutional neural network, or CNN with a relative word error rate (WER) increase of 5–13% over DNNs. When the voice stream contains spectral variations and local correlations, CNNs are better at recognising speech. The texts need to be machine-readable for digital technologies to be used for text research. Digital corpora of texts, sometimes built from digital texts and made publicly available, exist for major contemporary languages like English. The Revised Encoding Scheme (RES) system allows for the stacking of signs. Despite the need for higher-level processing, RES includes new sign operators. Additionally, RES has font-independent semantics and an encoding lifetime. According to the latest research, RES is also much better. An existing parsing mechanism may be used to immediately transform the relative arrangement of scanned signs to the relevant encoding primitives. Text mining uses AI to extract useful information from text. Natural language processing (NLP) is used to convert unstructured input into systematic data. This is needed for ML algorithms and analysis. All data is unorganised and raw. Without robotic text mining, it is impossible to search, find trends, or get keywords. Starting text mining and organising data may improve your company. Review customer evaluations and common words to adjust communication or pricing.

4.1.1 Problem Statement

A large electrical power network is encircled by the underlying communication system to securely connect increasingly dispersed SG components into the power system, which is a result of the fast expansion of the contemporary power system. These parts automate and enhance the SG's efficiency by allowing a broad variety of applications: distributed energy management, cyber-attack security, system status forecasting, and so on. The data generated by these applications is immense. A lot of people are interested in AI-based alternatives since conventional computer technology can't manage the vast amounts of data produced by SG systems. To distinguish between LSTM changes and real-time threats, this study will develop recurrent neural networks and LSTM to handle this problem. The recurrent neural networks will incorporate the evolving characteristics of the energy system to improve the model of the dynamic features of modern SGs. These grids are affected by the RES or method reconfiguration. Specifically, they offer an effective, holistic educational approach for users to share capacity data related to the power grid (PG) with the help of edge clouds.

4.1.2 Contributions

This study aims to evolve repeating Neural Networks (RNNs) and LSTM to address this issue. By including the attributes of an adaptively opportunity-cultivating strength structure, we can correct our model of the vital features of new SGs, that are concerned by changes in the RES or structure reconfiguration. This will allow for a change between LSTM changes and original-period warnings. In particular, we have developed a teaching method that focuses on generating and sharing capacity data for PG supported by edge clouds. This method is designed to address the challenges of accurately identifying Fast Dynamic Impact Analysis (FDIA) in real time. It integrates energy system characteristics that are unique to time into a unique category learning method. Using deep learning, or DL, techniques, this work is the first to develop a novel real-time detection method that adequately and correctly captures the dynamic features of modern power systems. Deep autoencoders, recurrent neural networks, and convolutional neural networks are a few examples of DL techniques that have significantly improved methods over time in a variety of areas.

4.1.3 Organising Manuscripts

The chapter's remaining parts are arranged as follows. The materials and procedures are described in Section 4.2. Section 4.3 delves into the outcomes of conducting experiments to test the predictions made by the CNN approach using the MNIST datasets. Section 4.4 contains the debate. As a conclusion, they outline plans for the future in Section 4.5.

4.2 Methods and Materials

The CNN method's predictive power is compared to that of Federated Learning (FL), the MNIST databases, and the centralised Artificial Intelligence IoT (AIoT) models. To accomplish the goal, the produced data was randomly split into training and test sets. To protect privacy and improve communication, SG offers an FL framework for energy data analysis. This framework supports cloud-edge communication. ESP and EDO optimisation methods as well as a local storing valuation strategy for cost models were developed using the theoretical framework. These techniques looked at the non-IID distribution of heterogeneous customers. To swiftly transfer global/local model updates between clouds and edge device operators, traditional cloud-based federated training neglected integrators. This work proposes a novel method for separating LSTM alterations from impending threats using neural networks and lengthy short-term recollections. To improve the model of a modern SG's dynamic characteristics affected by an RES or systems reconfiguration, an adaptively time-developing energy system will be included. They provide a federated education system for PG data exchange that is cloud-based, communication-efficient, and consumer-privacy-protected. Next, they developed two optimisation issues for energy-related operations and Energy Data Owners (EDO), Results showed that Model 3's power systems were sufficient, and it replicated the average training loss's detailed behaviour superbly. LSTM has four secret levels, a longer training session, and greater training loss. In many cases, the suggested strategy successfully detects FDIA. Lastly, the comprehensive simulation demonstrates that the proposed method has the potential to encourage EDOs to use the local method to enhance ESP payout and reduce task latencies.

4.2.1 Cyber-Attack

The block diagram of the attack is shown in Figure 4.1. The increasing use of technology and advancements leads to an exponential growth in the energy needed. Emerging cyber security threats to physical power systems are a direct outcome of SGs' fast development of communication infrastructure. Part of the old school of thought when it comes to protecting PGs from cyber-attacks is to keep the two domains apart. Connectivity between the cyber and physiological PGs is, hence, important. A solution developed to deal with these issues is the cyber-physical power system. The cyber technique of the Cyber-Physical Production Systems (CPPS) is tightly linked to the core physical power system. SG monitoring and regulation are the primary goals of CPPS. The CPPS consists of many steps, including transmission of power, generation, distribution, consumption, Supervisory Control and Data Acquisition (SCADA), and energy usage. Building, analysing, and maintaining track of the many cyber-attack methods on CPPS is essential as these SG stages are susceptible to cyber-attacks. The SC would provide the wrong state of the system bxa

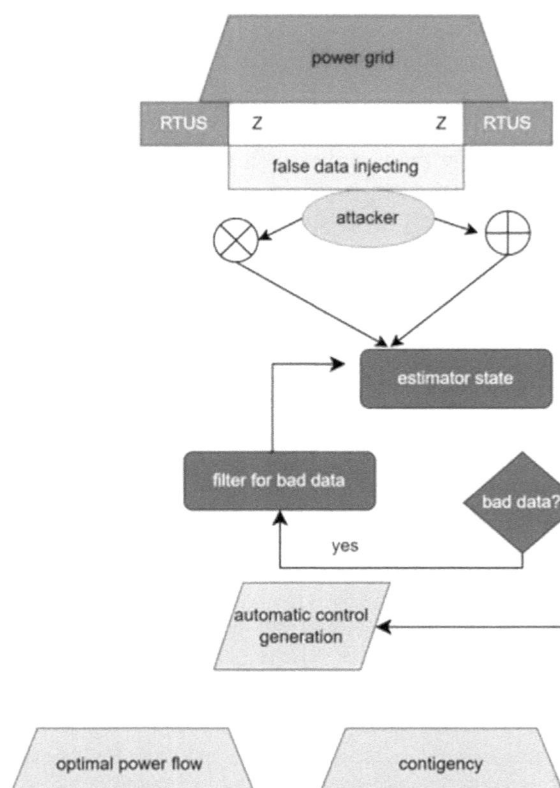

Figure 4.1 Block diagram of attack.

= bx + c when executed on the changed measurement data z_a. The attacker may add random error into the measurement residual to FDIA.

$$z_a = Hx + a + e = z_a \tag{4.1}$$

4.2.2 Modern Power Systems Using LSTM-RNN

An internal memory feed-forward neural network may enlarge a recurrent neural network. Recurrent neural networks (RNNs) are defined by their reliance on the computation results of previous inputs and the use of an identical function for every data input. After the output is generated and returned to the recurrent network, a duplicate of it is made. Not only does it take the present input into account, but it also remembers the outcomes of past inputs. By relying on their internal state, or memory, RNNs can process input sequences internally. This means they are useful for tasks like linked, unsegmented handwriting recognition, and voice recognition. There is no correlation between the inputs to different neural networks.

On the other hand, RNNs rely on interconnected inputs. A recurrent neural network may be fine-tuned to produce an LSTM network, which improves the recall of previously learned material. This fixes the RNN's disappearing gradient issue. Time series with unclear delays may be processed, categorised, and predicted using LSTM. To train the model, backpropagation is used. Which input value modifies the memory is decided by the input gate. Learn which blocks' worth of data has to be erased by applying the forget gate. That decision is made by the sigmoid function. By using the input and memory of the block, the output is produced.

The ability of RNNs to form temporary linkages between the most recent and earlier states is a result of feedback connections at the buried level. Power system measurements, on the other hand, are time-stamped continuous quantities that may be thought of as sequence data organised in the period domain. Therefore, RNNs are a great choice for portraying SG's dynamic nature.

4.2.3 Modelling Power System Dynamics

The input layer's upper hidden units receive the data that is supplied from the database. An LSTM concealed layer is employed to replicate the nonlinear connections seen in the data. As the linkages became more intricate, the need for remote teams to handle classification increased. The fully connected (FC) layer uses data from the preceding hidden layer to determine group probabilities for a sample. The final completely linked layer sends probabilities to the resulting layer, which finds cases. The size of the energy system, proposed network design, and renewable energy source deployment affect the number of concealed equipment at the FC and long-term memory levels. Studies also establish the number of distant units needed for LSTM and FC levels, while taking into consideration LSTM RNN efficiency and computational needs.

4.2.4 A Proposed Framework

The experimental results will describe this framework in detail, and it can detect all three types of assaults. That is to say, with proper architecture, an LSTM may mimic the dynamism of contemporary PGs. The technique of work that was suggested for the study is shown in Figure 4.2.

4.2.4.1 Federated Learning

Since the linked energy information in the Extended Data Output is limited, FL and short-term load forecasting need to collect high-quality power data from households to get an accurate STLF forecast and optimise power delivery efficiency. Figure 4.3 shows the SG edge-cloud integrated FL paradigm. It has a cloud, a plethora of EDOs, and ESPs spread out around the globe.

In edge-cloud integrated SG, the FL formula must be solved up to the nth GTE (where n = 0, 1, 2…) by completing the following steps Task publication, model

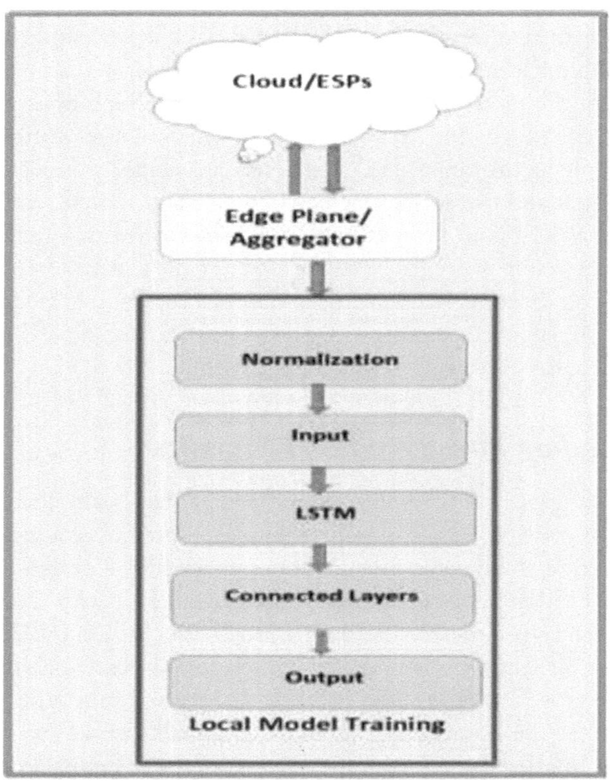

Figure 4.2 Suggested technique.

setup, local updating, global collection, and edge aggregation are all prerequisites for the completion of training.

Some emerging needs may be too complex for edge computing to handle in complex and scattered environments like SGs. A practical way to maximise the benefits of edge computing has therefore garnered a lot of attention: putting AI to the network's edge. The result of fusing these two ideas is edge intelligence (EI), which stands for edge computing with AI. One of the benefits that EI enables is a dramatic boost in the processing speed of IoT devices. One other perk of EI is that it reduces reliance on the Internet cloud for computing by making use of distributed network resources to boost the energy efficiency of different AI applications.

A collection of diverse electrical consumption data produced by smart meters, home appliances, electric cars, solar panels, etc. would be owned by each Energy Data Owner (EDO) in a smart home or SC. Si should represent the owner's power database. To further lessen the impact of integrators' and the cloud's implicit privacy loss, cryptography-based safe-gathering techniques are being used. Additionally, this system might take advantage of it. In particular, the suggested solution may safeguard the structure against subsequent attacks: free-riding and low-quality local method updates.

4.3 Results

4.3.1 Numerical Research

Figure 4.3 compares the CNN method's prediction performance as it pertains to different numbers of interaction and communication rounds on the MNIST databases, the FL, and centralised AIoT models. After around 31 communication cycles, the FL framework-developed global model could achieve coverage. With complete engagement in the 100th connection cycle, the FL and AIoT solutions may achieve 96.9% accuracy, which is an improvement above the centralised AIoT strategy's 99.79% predictive performance.

In Figure 4.4, the graph depicts the total payment of the ESP and EDO in the Deep Q-Network (DQN) as the time slot changes, with the ESP type set to 0.9. If the utility grid's pricing includes a feed-in tariff for dispersed PV electricity, subsidies may vary between half a kilowatt to 1 megawatt-hour. For transmission distribution pricing, ESP indicates the internal prices of many ESRs. Figure 4.6 shows ESP's decision to cut payment to boost the future high return after seeing EDOs lead to IHPs performing well. In general, EDO reduces payouts to pursue larger benefits later.

Figure 4.5 shows the combined communication latency of the two approaches for jobs ranging from one to ten. In the standard FL cloud-based approach, integrators were disregarded and changes to the global or local models were promptly transmitted between the cloud and EDOs. Figure 4.6 displays the potential impact of

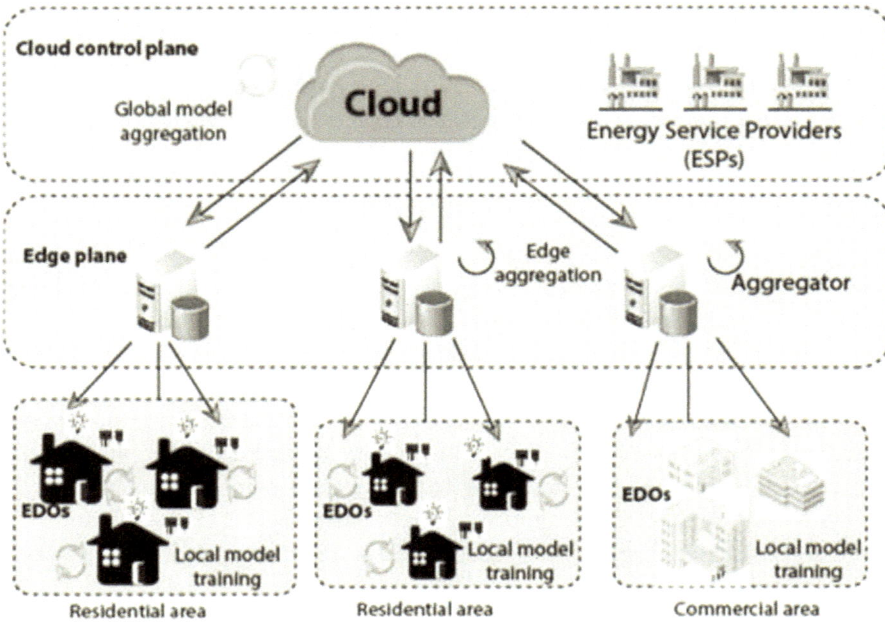

Figure 4.3 CNN method's prediction performance comparison.

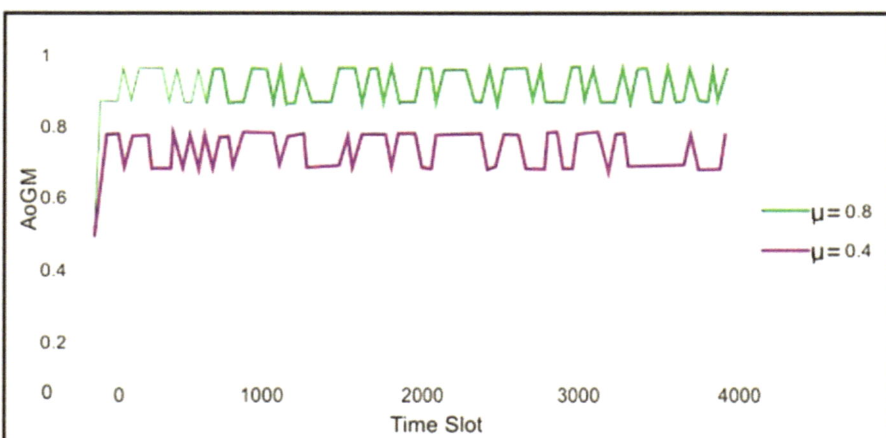

Figure 4.4 Total ESP/EDO payment.

the proposed edge-cloud cooperation on reducing communication time for different work counts. Edge layer intermediaries improve network capacity and connectivity, reducing the impact of intermittent and unpredictable wireless connections on users as well as the internet. Through edge aggregation actions, the aggregators might

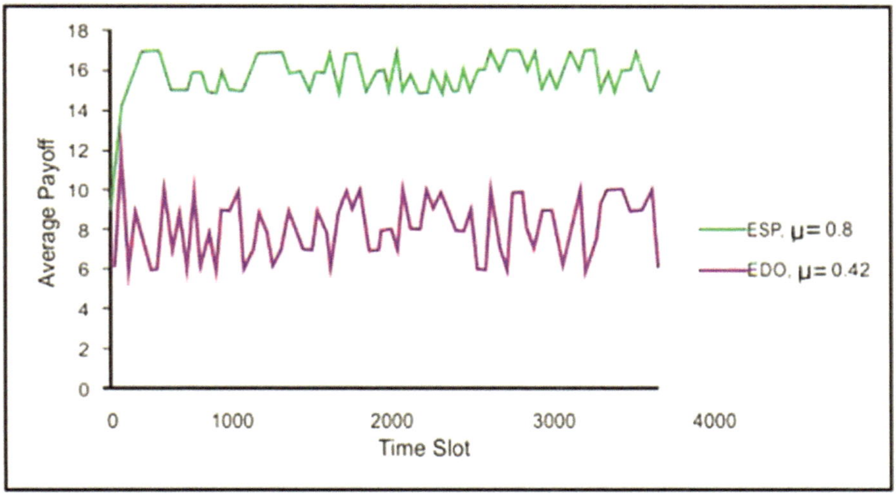

Figure 4.5 Combined communication latency of the two approaches.

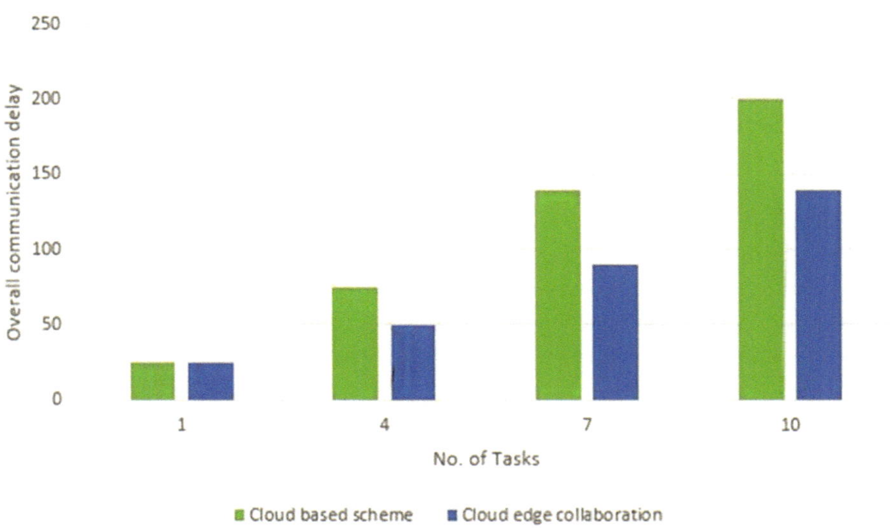

Figure 4.6 Training loss for all four models.

collaborate on network FL. This means our edge-cloud collaboration architecture has a better chance of improving the efficacy of communication.

4.3.2 Results of Simulation

Finding the F1 score in the confusion matrix allowed us to confirm that the suggested LSTM RNN-based FDI detectors worked as intended. Here is the formula for F1.

$$F1 = \left(\frac{2 \times P_r \times R_e}{P_r + R_e} \right) \tag{4.2}$$

Here, R_e stands for recall and P_r for precision; the following defines the way these two metrics are calculated:

$$P_r = \left(\frac{True\ positive}{True\ positive + False\ positive} \right) R_e = \left(\frac{True\ positive}{True\ positive + False\ positive} \right) \tag{4.3}$$

A true negative represents the correctly detected assault measures, while a true positive indicates the accurately discovered regular observations. False positive refers to the mistakenly observed assaults, and false negative denotes the missed assaults. The F1 value 1 indicates that any measurement labelled as typical is average, but the sample labelled as assaulted was considered to be under control. These structures are evaluated using a portion of the data from the case study simulation. Results showed that Model 3's power systems were manageable and that it accurately simulated the complicated behaviour of averaged training loss. This research scenario uses Model 3 to construct a system for real-time attack detection. Both models failed to achieve LSTM convergence, and training loss was statistically significant. LSTM models with four hidden layers exhibit more training loss and need more time to train. Overfitting is more likely in a more complex LSTM framework as well. The training loss for all four models is shown in Figure 4.7.

The information on the database now constitutes the highest concealed units in the data input layer. The LSTM concealed layer is used to replicate the nonlinear connections seen in the cases. The need for remote teams was determined based on the level of intricacy of the linkages that needed to be categorised. The FC layer assesses the probability of each group in the collection based on the data received from the preceding hidden layer. The output layer identifies instances using top FC layer data. The amount of FC and LSTM hidden components depends on power system size, network architecture, and RES integration. Studies of LSTM RNN efficiency and computational burden are typical to identify the number of remote units for LSTM and FC levels.

Figure 4.7 Simplified version of the simulation findings.

To find the optimal settings, the selected Framework has been utilised to assess several factors. To do this, two sets of data were randomly generated: one for testing purposes and another for training. Batch size, learning rates, optimisation technique, and phases are some of the other parameters to keep in mind. But their main effect is to change the time required for the procedure. The results were obtained by using the pre-existing detection techniques together with the Smart-designed LSTM RNN-based erection. The results were obtained by using the pre-existing detection techniques together with the Smart-designed LSTM RNN-based model. The majority of approaches effectively detected every assault sample.

Figure 4.8 shows a simplified version of the simulation findings. Current sensors see a considerable decline in accuracy when compared to the suggested technique, which excels in identifying FDIA in a variety of contexts. It can withstand the stresses of today's electricity grids with ease. Figure 4.4 shows a comparison of the proposed system with the present one.

4.4 Discussion

Smart city operations are largely reliant on the contemporary electrical grid. Nevertheless, catastrophic occurrences with large effects but low probability pose a significant threat to the safety of urban electricity systems. The increasing focus on these risks has made ensuring the robustness of urban power systems an essential

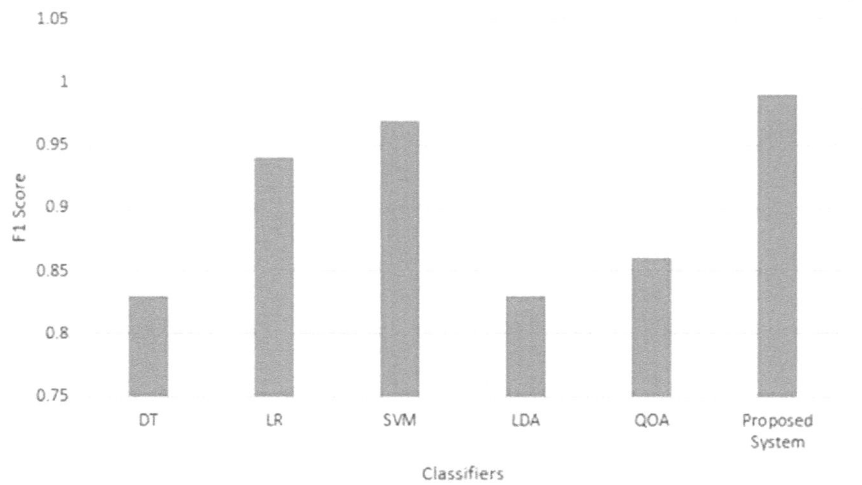

Figure 4.8 Comparison of the suggested system.

component of any modern smart city. The ability to endure, adjust to, and swiftly recover from disruptions is a hallmark of a resilient electrical grid. Adaptability, anticipating, absorbing, and recovering are its four distinguishing features. The rapid advancement and extensive usage of communication and information technology have led to a dramatic rise in urban intelligence levels. Academics and corporations alike are interested in the concept of a "smart city" that is generated by smart urban infrastructures. Promoting urban economic development and improving the quality of life for its residents are the main objectives of a smart city. If a city is to be considered smart, it must possess human-level intellect and be able to ensure its citizens' safety while also being environmentally friendly, economically viable, ecologically sound, and resilient. Stronger, safer, less carbon-intensive, and environmentally friendly urban electricity infrastructure is essential.

A novel category learning algorithm was developed to overcome these challenges and detect FDIA in real time by combining the characteristics of the energy system that define periods. One example is the lack of a revolutionary real-time detection method that accurately and efficiently uses DL methods to represent the dynamic aspects of contemporary energy methods Conventional wisdom is the bedrock of DL. After using DL methodologies, the majority of research domains that DL has influenced have seen substantial enhancements in their methodology. Deep auto-encoders, RNNs, and CNNs are just a few of the newer types of neural networks that have appeared in the last few years. Here, they provide an effective and safe AIoT method of FL that may be used in cooperative SGs that exchange private energy data via edge clouds. More specifically, they demonstrate an FL architecture that allows SG users to exchange energy data while protecting their privacy and saving communication. Afterwards, they consider non-IID repercussions while designing two optimisation tasks for ESPs and EDOs, they build a local information assessment technique in FL. An incentive system based on two layers of deeper reinforcement learning was developed to inspire EDO engagement and high-quality method input, as well as to improve understanding of multidimensional personal information provided by users in real-world settings. As the simulation results demonstrate, the suggested method increases communication efficiency and encourages EDOs to provide high-quality localised updates to the model.

The complex interplay of various variables in urban power networks makes aggregating dynamically distributed resources reliably and securely essential to PG resilience. Modern urban distribution grids require future aggregation technology to monitor and use resource flexibility. Cyber–physical–social systems (CPSS) combine cyberspace, physical, and social systems. CPSS's three-dimensional marketing of energy and information resources will increase urban infrastructure operators in smart cities. Modern technologies are strengthening the urban electrical grid. To make distribution networks more resilient to cyber-attacks, soft open points are recommended. A self-organising map-based technique for resource allocation and dispatch may overcome the limitations of subjective weighting systems. Distribution networks employ P2P energy trading to regulate renewable energy production volatility and reduce uncertainty from extreme events.

By enhancing safety technology, decreasing costs, and boosting battery capacity, wind-hydrogen storage and battery storage facilities may make distribution systems more resilient. In addition, important loads may get localised help from mobile energy storage during an intense event. Its portability allows for more operational flexibility to accept loads that are geographically spread, in comparison to permanent storage. Urban power networks in smart cities may be better prepared for severe events by integrating a wide range of interdisciplinary and state-of-the-art technologies that enhance resilience.

4.5 Conclusion and Future Work

Consequently, the disparities between the current electrical system and a potential data security breach are now being investigated. The chapter's substantial volatility and unpredictability were effectively addressed by using an architecture that relies on an LSTM recurrent neural net. The constructed detector model accurately depicted the future dynamics of the energy system in both the training and assessment datasets. A framework for FL was suggested to analyse energy data in Singapore, to optimise communication efficiency and ensure privacy. This framework integrates cloud and edge computing with ease. Two ESP and EDO optimisation techniques, a local storage assessment strategy for cost modelling, and an approach to account for the non-IID allocation of heterogeneous consumers were all developed using the theoretical framework. Lastly, thorough simulation of the proposed method may reduce task latencies, improve the ESP's payout, and route EDOs to excellent local models. In the early stages of the contemporary grid, there is still a great deal of opportunity for the creation and implementation of this concept. How much more SG research is needed to make this a possibility is still unclear. Some recent innovations that show promise in this field are smart metres, self-healing, big data, and demand-side management systems. They are excited about the CPPS because it presents a chance to think outside the box when it comes to grid design and development.

References

1. Rahat, Irfan Sadiq, et al. "A step towards automated haematology: DL models for blood cell detection and classification." *EAI Endorsed Transactions on Pervasive Health and Technology* 10 (2024): 1–24.
2. Kumar, K. Sunil Ratna, et al. "Iot and data mining techniques to detect and regulate solar power systems." *2023 International Conference on Inventive Computation Technologies (ICICT)*. IEEE (2023).
3. Nath, Dhruv Chakravarty, et al. "Internet of Things integrated with solar energy applications: A state-of-the-art review." *Environment, Development and Sustainability* 26.10 (2024): 1–56.
4. Kabir, M. Humayun, et al. "Explainable artificial intelligence for smart city application: A secure and trusted platform." *Explainable Artificial Intelligence for Cyber Security: Next Generation Artificial Intelligence*. Cham: Springer International Publishing (2022). 241–263.
5. Al Hammadi, Ahmed Y., et al. "Explainable artificial intelligence to evaluate industrial internal security using EEG signals in IoT framework." *Ad Hoc Networks* 123 (2021): 102641.
6. Wang, Shen, et al. "Applications of explainable AI for 6G: Technical aspects, use cases, and research challenges." *arXiv preprint arXiv: 2112.04698* (2021).
7. Javed, Abdul Rehman, et al. "A survey of explainable artificial intelligence for smart cities." *Electronics* 12.4 (2023): 1020.
8. Islam, Mir Riyanul, et al. "A systematic review of explainable artificial intelligence in terms of different application domains and tasks." *Applied Sciences* 12.3 (2022): 1353.
9. Sohail, Shaleeza, et al. "Explainable and optimally configured artificial neural networks for attack detection in smart homes." *arXiv preprint arXiv: 2205.08043* (2022).
10. Jagatheesaperumal, Senthil Kumar, et al. "Explainable AI over the Internet of Things (IoT): Overview, state-of-the-art and future directions." *IEEE Open Journal of the Communications Society* 3 (2022): 2106–2136.
11. Abou El Houda, Zakaria, Bouziane Brik, and Sidi-Mohammed Senouci. "A novel IoT-based explainable deep learning framework for intrusion detection systems." *IEEE Internet of Things Magazine* 5.2 (2022): 20–23.

12. Bose, Bimal K. "Artificial intelligence techniques in smart grid and renewable energy systems—Some example applications." *Proceedings of the IEEE* 105.11 (2017): 2262–2273.
13. Kotsiopoulos, Thanasis, et al. "Machine learning and deep learning in smart manufacturing: The smart grid paradigm." *Computer Science Review* 40 (2021): 100341.
14. Ali, Syed Saqib, and Bong Jun Choi. "State-of-the-art artificial intelligence techniques for distributed smart grids: A review." *Electronics* 9.6 (2020): 1030.
15. Ghiasi, Mohammad, et al. "A comprehensive review of cyber-attacks and defence mechanisms for improving security in smart grid energy systems: Past, present and future." *Electric Power Systems Research* 215 (2023): 108975.
16. Mohseni, Sina, Niloofar Zarei, and Eric D. Ragan. "A multidisciplinary survey and framework for design and evaluation of explainable AI systems." *ACM Transactions on Interactive Intelligent Systems (TiiS)* 11.3–4 (2021): 1–45.
17. Radoglou-Grammatikis, Panagiotis I., and Panagiotis G. Sarigiannidis. "Securing the smart grid: A comprehensive compilation of intrusion detection and prevention systems." *IEEE Access* 7 (2019): 46595–46620.
18. Mirzaee, Parya Haji, et al. "Smart grid security and privacy: From conventional to machine learning issues (threats and countermeasures)." *IEEE Access* 10 (2022): 52922–52954.
19. V.S., Arjun. "Multi Label Deep Learning classification approach for False Data Injection Attacks in smart grid." *KSII Transactions on Internet & Information Systems* 15.6 (2021): 2168–2187.
20. Diaba, Sayawu Yakubu. "The role of machine learning algorithms in smart grid cybersecurity." *2024 International Workshop on Artificial Intelligence and Machine Learning for Energy Transformation (AIE)*. IEEE (2024).
21. Amin, Irtiqa, and Mithilesh Kumar Dubey. "Artificial intelligence techniques in smart grid." *Introduction to AI Techniques for Renewable Energy System*. Boca Raton, FL: CRC Press (2021). 347–366.
22. Xu, Chongchong, et al. "Review on interpretable machine learning in smart grid." *Energies* 15.12 (2022): 4427.
23. Szczepaniuk, Hubert, and Edyta Karolina Szczepaniuk. "Applications of artificial intelligence algorithms in the energy sector." *Energies* 16.1 (2022): 347.
24. Algarni, Abdulmohsen, Zulfiqar Ahmad, and Mohammed Alaa Ala'anzy. "An edge computing-based and threat behavior-aware smart prioritization framework for cybersecurity intrusion detection and prevention of IEDs in smart grids with integration of modified LGBM and One Class-SVM models." *IEEE Access* 12(2024): 104948–104963.

Chapter 5

An Artificial Intelligence-Based Solar Radiation Prophesy Model for Green Energy Utilisation in the Energy Management System

Pradeep Kumar Mishra, Amit Kumar Pathak, Abhishek Choubey, and Sesikala Bapatla

5.1 Introduction

The demands for resources have increased due to evolution's progress and growth [1]. Fossil fuels like natural gas, oil, and coal are the conventional energy sources that cause many problems for the environment, including acid rain, greenhouse gases, ozone depletion, and more [2]. Renewable, clean energy sources that are pure and unpolluting are essential for environmental preservation [3]. This aim may be achieved via hydroelectricity, wind, geothermal energy, and biomass [4]. Considering their pros, downsides, economics, and feasibility, these assets may meet demand as alternative energy producers [5]. Renewable energy's inconsistent production is its

 DOI: 10.1201/9781003479062-5

biggest drawback [6]. Aside from several unidentified characteristics of renewable energy sources, the need for loads is also ambiguous [7]. Renewable energy generation has temporal and geographical variability, unlike traditional power sources [8]. Basic renewable energy sources like the sun and wind affect power production [9]. Insufficient sources of clean energy may cause blackouts or other challenges by causing a power generation-demand gap [10]. Conversely, the excessive generation of renewable energy sources results in energy wastage.

Recent decades have seen remarkable advancements in wind turbines, photovoltaic (PV) solar technologies, and the likelihood of effectively capturing the power of tides and waves [11]. Solar thermal solutions, especially those that include heat storage, hold great promise in sunny regions [12]. As a result of government subsidies, the per-kilowatt-hour expenses for wind and solar power are now comparable to those of fossil fuel power plants, particularly in areas where these sources are subject to carbon emissions taxes [13]. However, the intermittent nature of wind and solar power does not align with the majority of energy demand [14]. Additionally, due to government incentives, several countries have built substantial capacity, resulting in occasional instances of high-power production [15]. These processes provide superior modelling, design, monitoring, issue detection, and estimation of renewable energy in smart grids. Artificial intelligence (AI) encompasses several methodologies that enable machines to acquire knowledge from existing data and develop a system capable of autonomously adjusting its intelligence to a specific subject [16–20].

AI technology has the potential to predict solar energy production, minimising the possibility of financial loss and maintaining the reliability of the system. AI can efficiently address a wide range of issues in energy systems. Various methods may be used depending on the specific requirements and architecture of the system [21–25]. The use of these technologies is likewise challenging; however, most issues may be resolved by technical developments and algorithms. To use solar and wind power in an independent system, it is necessary to have a battery-operated or another kind of storage. The potential for widespread use of hydrogen as a fuel for transportation in the future enhances the opportunities for both renewable power sources and the provision of energy during periods of low demand.

This study explores the use of AI in the field of prediction, after providing a concise introduction. This section covers numerous approaches for forecasting solar system performance. Various AI methodologies were also investigated in this context.

5.2 Various Methods for Predicting Solar Radiation

There are two main methods for making predictions: indirect and direct methods. Indirect forecasting makes use of a PV manufacturing process model to assess solar radiation before moving on to the production of PV power. The exact amount of power that the plant will produce is then determined by the direct prediction. There are four main kinds of PV output prediction, distinguished by the periods used.

- Brief duration
- Limited duration
- Intermediate duration

Solar forecasting addresses the issue of variability and uncertainty in resources by considering various periods for predicting future conditions. This study examines several periods for forecasting and the corresponding areas of use in solar energy structures, with a specific emphasis on their practical utility. Various forecasting strategies are available to predict sun irradiation, which will be addressed below.

- Brief duration (ranging from a few seconds to minutes):
  This form of forecasting may be used for PV storage management and power sales. In the current environment of smart grids and microgrids, accurate and immediate prediction of PV electricity has become crucial, especially in the solar business. Purpose: Real-time surveillance.
- Limited duration (within the next two and three days):
  This forecasting approach is becoming more important for various decision-making tasks in the energy industry and power grid management, such as economical cargo shipments and load prediction. Subject: Unit commitment application.
- Intermediate duration (within the next seven days):
  Repairs of solar power plants, conventional generators, transformers, and transmission lines may be better planned with the use of medium-term forecasts. Scheduled maintenance is one such use.

5.3 Models for Predicting Future PV Generation

One method of preparing for the future is via the process of forecasting, which entails speculating about potential conditions, techniques, or incidents. Forecasts of future development or events, based on a variety of design methodologies, fall under this umbrella.

When developing forecasting strategies for PV plants, the wide range of factors that impact their efficiency involves the use of several distinct methodologies. Researchers in this discipline have the challenging challenge of selecting a paradigm that offers the highest level of accuracy while minimising computational overhead.

The enhanced PV energy forecast method has two main categories: direct and indirect. One model focuses on combining previous performance and meteorological information to estimate energy production, while the other model specifically looks at solar radiation as the main factor for prediction, such as using numerical weather forecasts. Comparative analyses of both indirect and direct solar energy forecasting techniques demonstrate the superior accuracy and reliability of the direct strategy. Nevertheless, the availability of potential incoming data and the extent of predicting constraints restrict the applicability of a specific prediction.

Solar energy is regarded as the most influential factor among the climatic components affecting the performance of PV panels since it has a direct correlation with the generation of PV energy. The impact of the PV performance benefit is uncertain, since it may vary depending on general conditions, geographic location, and design aspects of technology advancements. Additionally, particular parameters may also influence this impact. The significance of a solitary factor is emphasised.

5.3.1 PV Projection

Power generation using PV systems is mostly dependent on solar energy. Forecasting the generation of PV systems relies on accurate predictions of solar radiation; hence, it is crucial to examine it throughout the process to prevent a disparity among demand, storage, and renewable electricity. There are two main components to AI forecasting:

a. The modelling uses past data to create future predictions, while real-time forecasts use present data to estimate the necessary variables.
b. Several methods exist for developing PV system predictions, including RF, vector support, and artificial neural networks (ANNs).

The elimination of system parameters, reduced processing requirements, and succinct resolution of multivariable problems are some of the advantages of ANNs over conventional models. An appropriate model for practical applications is a neural network, which can be taught in complex nonlinear interactions.

5.3.2 Wind Prediction

The unexpected nature of wind energy production sets it apart from regular wind energy output. Managing the balance between generation and power consumption is a challenge that the wind-generating system forecasts aim to solve. In several studies, the neural network a standard multi-layer arrangement of nodes is the most effective method for wind prediction. For short-term wind speed predictions, multi-coater perceptron are the neural networks of choice.

5.4 Machine Learning for Power Control Systems

AI methods were used to achieve efficient energy management in both the demand and production sectors. AI approaches may be used for renewable resources, whether they are linked to the grid or operate independently, depending on the specific circumstances and restrictions involved. The main areas where AI technology may be used include demand forecasting, power projections, control of renewable energy systems (RES), and improvement of system performance. The following section provides a concise explanation of the primary applications of AI methods for RES.

There could be three primary parts to it. The left part of Figure 5.1 displays energy generation, such as a PV facility. The Emergency Medical Services (EMS) system, which is the second part of the power storage system, includes the power dispatching predictor model's output signal, the power source's recharging state (SOC), and the block. EMS and related signal control make up the bulk of the second segment. When the energy produced goes over the limit, the excess may be transmitted back to the grid.

An energy system receives input from renewable energy sources, which then create electricity via the corresponding power method. Contrarily, the accelerometer and pyranometer devices pick up on and record information from the time series renewable sources. Also, using the history of renewable energy sources, the forecasting model projects how much solar radiation will be available in the future. The energy storage system (ESS) takes these expected results into account thereafter. The system for energy management is a crucial component of renewable power systems since it controls all aspects of renewable energy generation for a vital power distribution process. Consequently, it is critical to accurately forecast the state of

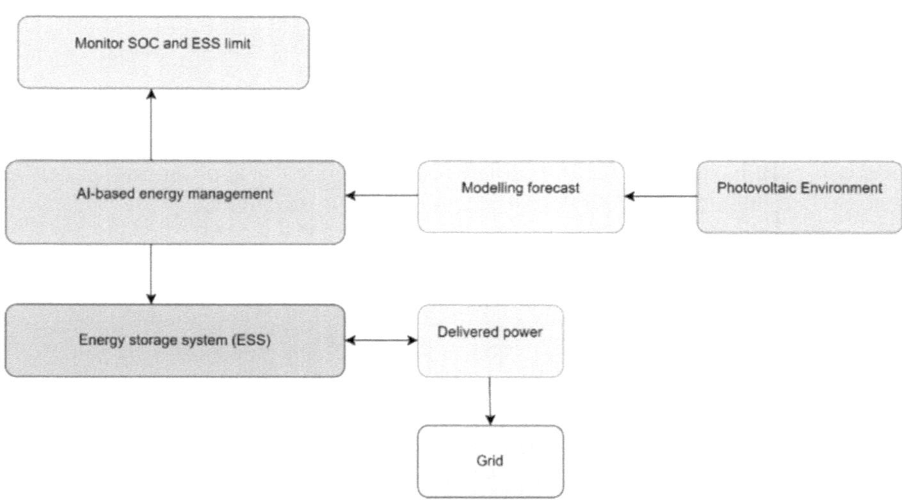

Figure 5.1 Energy generation of PV facility.

potential renewable sources to forecast power generation and energy dispatch. The reliability of the power system is significantly enhanced by it.

5.4.1 In What Ways May Artificial Intelligence (AI) Help the Renewable Energy Industry?

The electric grid is one of the most intricate systems seen in nature. Nevertheless, the incorporation of many renewable energy sources is rapidly progressing. The current system has several challenges in accommodating the diverse range of renewable energy sources, primarily owing to the inherent unpredictability of solar and wind power. The utility industry requires sophisticated technologies capable of enhancing the integration of renewable power into the existing organisation and ensuring that renewable energy functions seamlessly. The reliability of renewable energy may be enhanced by the use of AI technology and comprehensive grid renovation.

- **Intelligent, Consolidated Command Centres:**
Connecting the grid to sensing applications allows for the collection of massive amounts of data. Grid operators get more insight into improved control processes when AI is used with this data. It allows power companies to be more adaptable in their demand-supply strategies. Complex load control methods may be implemented using equipment like blast furnaces and huge AC units, which can be easily shut down when intensity is lowered. Modifying smart storage systems is another possible application of the supply flow. One example of how smart machines and advanced equipment may improve the incorporation and effectiveness of green sources is by producing weather and load predictions.

- **Greater Efficiency in Microgrid Integration:**
AI can facilitate microgrid connection and the management of dispersed electricity. Energy flow management is difficult when integrating distributed renewable power sources into the centralised power grid. Reliability and traffic congestion may need the AI management system.

- **Enhancements to Security and Dependability:**
AI focuses on managing renewable power fluctuation and improving security, productivity, and dependability. It can explain energy usage, power inefficiencies, and device status. To detect turbine damage, the AI-driven prediction assessment may gather windmill sensor data. The system monitors everything and alerts the control unit when repairs are needed.

- **Increase the Market Size:**
AI may help green energy companies grow by creating and marketing new services. AI-powered systems analyse energy-gathering data and provide energy utilisation

insights. This data helps organisations improve and provide new services. It may help retail providers find and target underserved client segments.

- **Intelligent Storage-Enabled Smart Grid:**

AI and intelligent energy storage may provide the solar sector with a reliable answer. Fast and accurate energy distribution decisions are made possible by this smart grid's ability to analyse vast amounts of data from many sensors. In addition, it facilitates the continuation of the power dynamic with the power network, which allows microgrids to efficiently manage local energy demands.

5.5 Solar Forecasting Model Utilising an AI Algorithm Proposed System

The solar radiation prediction system, as seen in Figure 5.2, is constructed using a three-stage technique. For this, the solar radiation intake is checked, cleaned, and filtered. Afterwards, the data undergoes pre-processing and is split into separate sets for training and testing. A solar model is trained using training data and various weights that are adjusted using AI algorithms. Training the model entails feeding it test data so that it may make predictions and evaluate how accurate those predictions are. To determine how well the solar forecasting technique worked, the researcher matched the results to the test data. A simple method for training approach is an additional way to forecast sun irradiation; it's an algorithm where the weights are adjusted according to the unwelcome reaction it generates.

5.5.1 Gathering Facts

To build a suitable solar-producing model, the gathered information on the relevant outputs of PV solar energy from PV solar method suppliers and included in the model. Data collected by solar share solar equipment that is integrated into the smart power grid is maintained in a record electronically. Researcher analysed a database of weather updates maintained by the Taiwan Central Climate Bureau to determine the objective PV solar system's proximity to local meteorological conditions. This information is then used to develop a similarity evaluation approach that places the PV solar system inside a regional weather area. After that, a naturally accessible dataset is created by using web crawlers to collect prior data about the same time range as the PVs dataset.

5.5.2 Processing Steps

In the past researchers could build AI models with better accuracy, but they desired to conduct an information screening step. This is because time series data

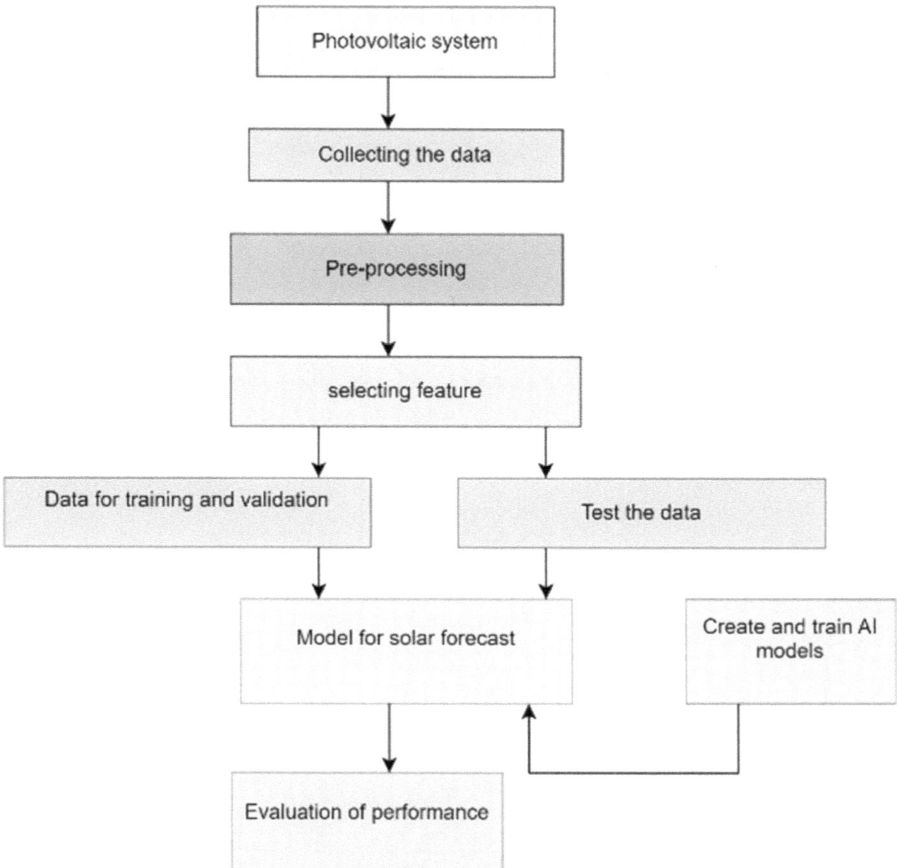

Figure 5.2 Solar radiation prediction system.

acquired through the various channels mentioned earlier often contains a few missing indicators and particular interpretations that indicate abnormality readings. Following this procedure, the researcher managed to detect and eliminate out-of-the-ordinary case measurements and make educated guesses about data points that were missing. To find out-of-the-ordinary amounts of weather data, they look at relevant regional weather data in conjunction with a weather database that is connected to professional meteorological evaluation based on forecasts. Additionally, the quantity of practical solar energy generated is limited by the readings that the PV generator can receive as input. The unpredictable nature of network failure and the degradation of renewable energy due to technological age are two additional factors that might lead to measurements being unreliable. The inclusion of errors in the observational data is still acknowledged, and they expect that in the future, more advanced detection methods will be used to evaluate the network's stability and the gathered data's reliability.

It is possible to impute missing time series data points using a variety of methods: the assertion of an unchanged value, its replacement with statistical standards and principles, as well as methods for approximation, restoration, and complex estimation processes. To restore a complete set of information that may be used in AI modules, this research employs many interpolation algorithms. While most weather history segments often have hundreds of missing data points every day, many of the segments that have been lost just have one missing data point. To fill in the gaps while using weather data that was incomplete from locations close to the target observer stations, the reverse distance weighting method was used. The researcher also used the inverse of the distances measured from the target position to fill in the gaps.

Equation (5.1) shows the solution to this notion, which is called Inverse Distance Weighting.

$$R'_j = \frac{\sum_{i=1}^{S} R'_i \left(\frac{1}{y_{ij}}\right)}{\sum_{i=1}^{S} \left(\frac{1}{y_{ij}^{n}}\right)} \tag{5.1}$$

where R'_i is the interpolated value of the meteorological condition at the target station 'j'. R_i are the weather values from the 'S' meteorological stations that are close to the j station, with the distance between them being as y_{ij}.

Parameter 'n', which is the inverse coefficient, is often set to 2.

5.5.3 Techniques for Selecting Features

There are two main types of feature selection processes: filtering type and wrapper type. Both involve using basic filters to eliminate certain characteristics. The filter method relies on selecting just those attributes that match predefined criteria, with the variables that boost the algorithm's efficiency being randomly selected when the wrapper approaches them. Meanwhile, there's no need for an analytical procedure while using the filter approach. A learning strategy for improvement determines the sort of wrapper, nevertheless.

5.5.4 Initial, Preliminary, and Final Test Results

There is no separation of testing, training, and validation of data for PV power and weather. The breakdown was 60%–20%–20%, as mentioned in the study. Table 5.1

Table 5.1 Three Subsets

Data Set	*Ratio Split*	*Quantity of Remarks*
Data for training	60% of data	6,400
Data for validation	20% of data	4,056
Data for testing	20% of data	4,056
Total	100%	14,512

provides more details about the three subsets. They trained the prediction models and determined the optimal number of nodes with the help of the training package. The prediction models used the validation set to choose the best parameters. To evaluate the efficacy of the prediction models, test data has been used.

It is crucial to highlight that the weather data used in the test set consists of actual data with added noise, rather than predicted weather data. The lack of weather predictions for 2019 and 2020 is due to the meteorological stations near the PV site not providing us with this information. To provide a fair evaluation, they introduced a 20% level of noise to the authentic meteorological data in the set that was tested.

5.5.5 Models for Artificial Intelligence

- **Artificial Neural Networks:**

To better handle nonlinearity data and account for weather fluctuations, ANNs were developed for use in solar system forecasting. Machine learning (ML) refers to a collection of robust algorithms that can evaluate data, acquire new knowledge from it, and then use that knowledge to uncover intriguing patterns. In contrast, a neural network is a set of procedures for model data using cell graphs in learning algorithms. An ANN relies on three main layers: the input, hidden, and output layers. The input layer is where the data is obtained. Several layers of input data are analysed by the hidden layer. The output layer gives the results once the hidden layer has analysed them. Figure 5.3 shows the ANN model's scheme.

ANNs are based on theoretical models of biological neurons. One way in which artificial neurons work is by taking in data in the form of inputs and then sending out emissions to other neurons. Photos, articles, or the results of several neurons may all serve as inputs or values from a selection of databases. Such as when an image of an item is identified, the job is finished by the results of the last neurons in the neural network. An individual neuron's output may be derived by first determining the total input and then dividing it by the strength of the connections between the neuron and the inputs. Here, the researcher includes a correction factor in the total.

Predictions of wind and solar power production in the medium to long term are crucial for power grid planning, optimisation of dispatching, management growth, and consumption enhancement. To achieve power mutual help and supplemental

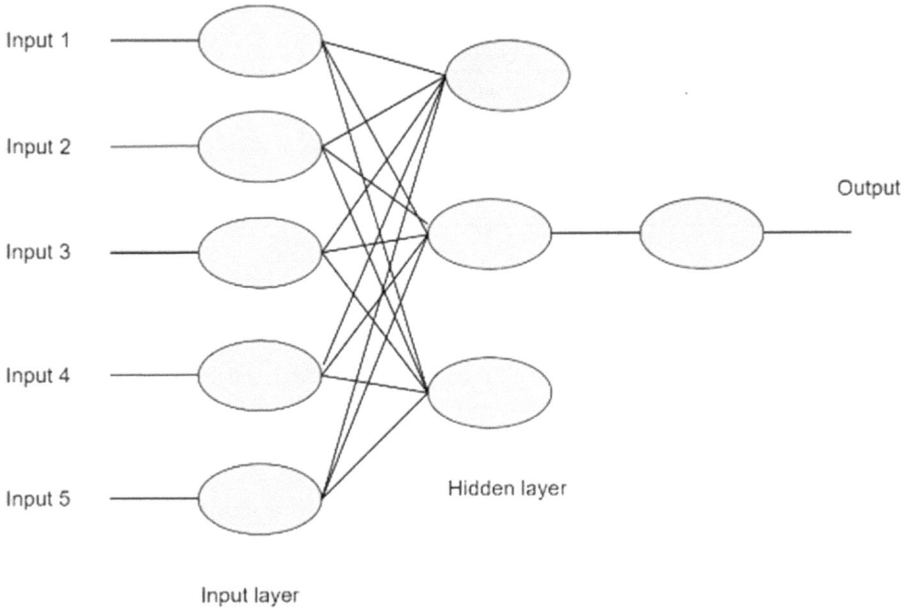

Figure 5.3 ANN model's scheme.

dispatching of energy output, they are key variables in the wide domain of renewable energy. However, due to factors such as the vast difference between accurate electrical prediction and energy production prediction, the small data samples used in chronological power generation, the poor accuracy of weather forecasting, and the lengthy time frames involved in mid- to long-term prognostication, short-term power forecasting technology cannot be physically replicated. Consequently, there is currently no reliable way for businesses to foretell how much wind and solar electricity will be generated in the medium to long term.

Common forecasting strategies for calculating PV power generation include the sigmoid function, hyperbolic tangent sigmoid, and radial basis function (RBF). A common method to define the ANN model is as

$$v_m = c + \sum_{k=1}^{m} \left(P_k \times H_k\right) \tag{5.2}$$

where v_m, P_k, H_k, c, m are the concluding network output, bias weight, input number, connection weight, and several inputs, respectively.

- **Vector Machine Assistance**

SVMs are a kind of supervised learning used in AI and the reduction of structural hazards. Its overfitting issue is reduced, which increases its capability for wider adoption. For the use of linear projection and nonlinear data translation, the SVM idea has a basic premise in some domains. For the use of linear projection and non-linear data conversion, the SVM idea has a basic premise in some domains. Support vector regression is an additional method that has been developed to tackle different types of regression issues. The principles of structural risk minimisation and learning statistics form their foundation.

Essential components of the SVM for data mapping into higher-dimensional space are the kernel purposes. The nonlinear kernel operation is defined as

$$g = exp\left(-\frac{1}{\sigma^2}\left\|d - d_i^2\right\|\right) \tag{5.3}$$

the vector of issues derived from the training models is denoted as d_i, whereas the vector in input space is d. The objectives in greater dimensional space of input are similar to those in lower or similar level input space.

- **Random Forest Methods**

Regression and classification problems are both amenable to RF, an ensemble learning method. It provides predictions by combining the outputs of several fundamental models, such as decision-making bodies or SVMs. Once each base model is trained on a new data sample, it undergoes replacement using an aggregation bootstrap technique.

- **Measures for Evaluating Forecasts**

This study uses four evaluation metrics to quantify the performance of the prediction. There are benefits and drawbacks to each. Based on these metrics, they can look at the results. The following describes the four measurements if the time series data is defined as t = 1 ~ n, with the actual observed values of the sequence z_t and the anticipated values of the sequence z_t'.

It tracks the difference between two successive time series without taking their instructions into account. Disparities between the actual data and predictions, considering both the average and absolute values, with equal weighting. Equation (5.4) shows the Mean Absolute Error (MAE) calculation equation.

$$MAE = \frac{1}{n}\sum_{t=1}^{n}\left|z_t - z_t'\right| \tag{5.4}$$

- Root mean square error (RMSE): This is a common way to evaluate the differences between the actual and expected values. It shows the lowest possible sample standard deviation. The Mean Square Error (MSE) may be traced back to it. Data that has not been properly pre-processed is susceptible to anomalies, as large errors have a bigger effect than little ones. Equation (5.5) displays the RMSE calculation equation.

$$RMSE = \sqrt{\frac{1}{n}\sum_{t=1}^{n} | z_t - z_t' |^2} \tag{5.5}$$

- Mean forecast error (MFE): An average positive as well as negative biological assessment may be used to determine whether a forecast under- or overestimates the actual values seen; it is also a method for evaluating how well a prediction matches the data pattern. Equation (5.6) shows the formula for MFE.

$$MAE = \frac{1}{n}\sum_{t=1}^{n}\left(z_t - z_t'\right) \tag{5.6}$$

- Mean absolute scaled error (MASE): The conventional unit of measurement addresses difficulties in data gathered through observation, such as scale dependence and imbalance in absolute sizing. Equation (5.7) shows MASE's formula.

$$MASE = \frac{\sum_{t-1}^{n}\left(z_t - z_t'\right)}{\frac{n}{n+1}\sum_{t-2}^{n}\left|z_t - z_{t-1}'\right|} \tag{5.7}$$

5.6 Results

Table 5.2 displays the evaluation metrics for the solar PV training and testing phases, comparing the predicted and actual values. When compared to RF and SVM, ANN performs better during training and testing. There is an improvement in ANN's performance when compared to other measurements for RMSE, MAE, MASE, and MFE.

The examples of forecasting output are shown in Figures 5.4a and b, 5.5a and b, and 5.6a and b which represent a comparison between the values that were observed and those that were anticipated using the SVM, ANN, and RF approaches, respectively. The previous discussion showed that ANN model projections are more in line with the actual data trend than SVM and RF models.

Table 5.2 Solar Photovoltaic Training and Testing Phases

Assessment Metrics	Train Phase			Test Stage		
	ANN	SVM	RF	ANN	SVM	RF
MAE	0.9659	1.6346	1.8732	1.31656	2.6722	3.5611
MASE	0.8976	1.4569	1.5603	1.5208	2.0178	2.6718
MFE	0.4457	0.9654	0.8238	0.5622	1.0127	0.9177
RMSE	1.4568	2.2348	1.3166	1.9874	3.0157	3.4518

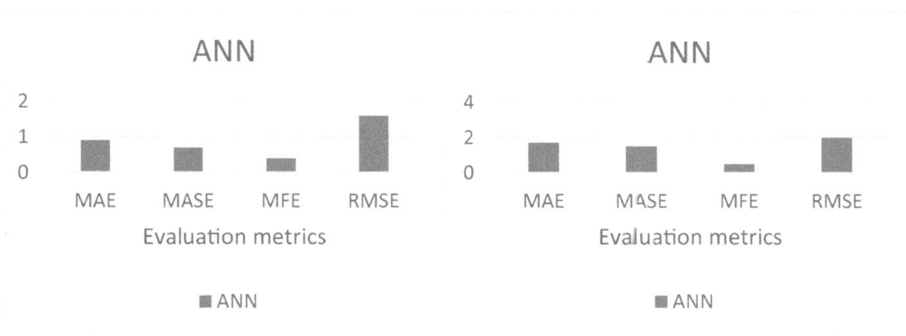

(a) (b)

Figure 5.4 (a) Training data and (b) testing data.

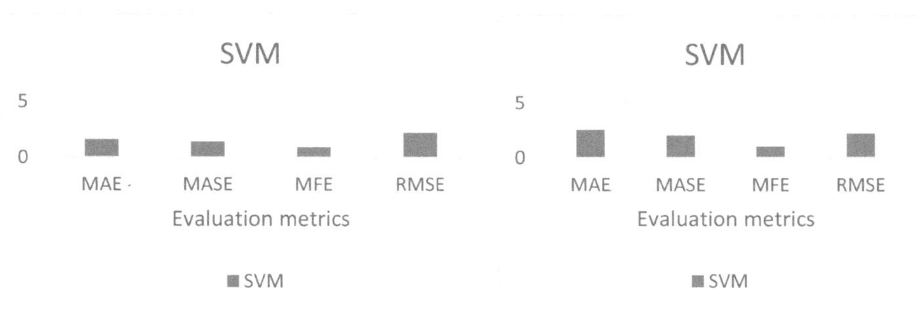

(a) (b)

Figure 5.5 (a) Training data and (b) testing data.

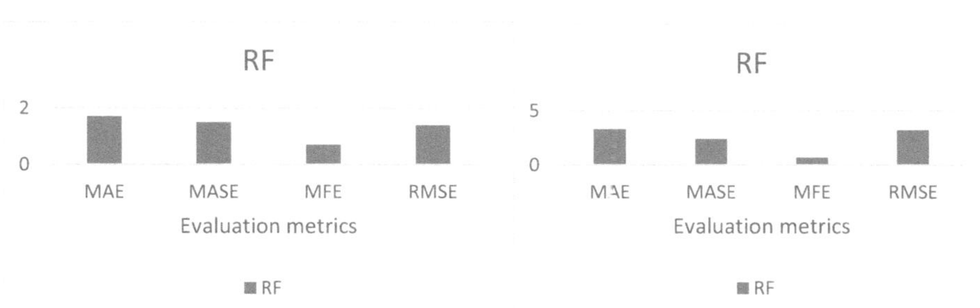

(a) (b)

Figure 5.6 (a) Training data and (b) testing data.

5.7 Conclusion

The most significant challenge in solar energy production is the random and unpredictable fluctuations in PV power output, which are mostly caused by weather conditions. To mitigate the effects of energy price volatility and uncertainty and to incorporate PV systems properly into an intelligent grid, precise forecasts of solar output are thus crucial. Accurate predictions of solar energy were made in this study using an AI model. Considering the potential and forecast issues, it is essential to accurately estimate solar light to understand the renewable power perspective of a location. Renewable energy is critical for product development and provides a viable economic option in light of resource depletion and the need to safeguard the environment. The real estate industry may now make use of AI's growing economic clout, leading to more efficient and accessible outcomes. The evaluation set's RMSE, MAE, MASE, and MFE measure the efficacy of each approach. Compared to SVM and RF, ANN performs better during testing and training. Patterns for hydropower, wind, and thermal power production may also be generated using the suggested stochastic approach. It is also possible to use the methods in this research to model the possible disastrous effects of transmission or distribution network outages, given the necessary reference data.

References

1. Rahat, Irfan Sadiq, et al. "A step towards automated haematology: DL models for blood cell detection and classification." *EAI Endorsed Transactions on Pervasive Health and Technology* 10 (2024): 1–21.
2. Kumar, K. Sunil Ratna, et al. "Iot and data mining techniques to detect and regulating of solar power systems." *2023 International Conference on Inventive Computation Technologies (ICICT)*. IEEE (2023).
3. Kumari, Pratima, and Durga Toshniwal. "Deep learning models for solar irradiance forecasting: A comprehensive review." *Journal of Cleaner Production* 318 (2021): 128566.
4. Nematirad, Reza, and Anil Pahwa. "Solar radiation forecasting using artificial neural networks considering feature selection." *2022 IEEE Kansas Power and Energy Conference (KPEC)*. IEEE (2022).
5. Zhu, Tingting, et al. "Solar radiation prediction based on convolution neural network and long short-term memory." *Energies* 14.24 (2021): 8498.
6. Qazi, Atika, et al. "The artificial neural network for solar radiation prediction and designing solar systems: A systematic literature review." *Journal of Cleaner Production* 104 (2015): 1–12.
7. Wang, Yuhan, et al. "Hybrid solar radiation forecasting model with the temporal convolutional network using data decomposition and improved artificial ecosystem-based optimization algorithm." *Energy* 280 (2023): 128171.

8. Eşlik, Ardan Hüseyin, Emre Akarslan, and Fatih Onur Hocaoğlu. "Short-term solar radiation forecasting with a novel image processing-based deep learning approach." *Renewable Energy* 200 (2022): 1490–1505.
9. Zhou, Yong, et al. "A review on global solar radiation prediction with machine learning models in a comprehensive perspective." *Energy Conversion and Management* 235 (2021): 113960.
10. Wang, Jianzhou, et al. "Forecasting solar radiation using an optimized hybrid model by Cuckoo Search algorithm." *Energy* 81 (2015) 627–644.
11. Olatomiwa, Lanre, et al. "A support vector machine – Firefly algorithm-based model for global solar radiation prediction." *Solar Energy* 115 (2015): 632–644.
12. Halabi, Laith M., Saad Mekhilef, and Monowar Hossain. "Performance evaluation of hybrid adaptive neuro-fuzzy inference system models for predicting monthly global solar radiation." *Applied Energy* 213 (2018): 247–261.
13. Ghimire, Sujan, et al. "Hybrid convolutional neural network-multilayer perceptron model for solar radiation prediction." *Cognitive Computation* 15.2 (2023): 645–671.
14. Dong, Na, et al. "A novel convolutional neural network framework based solar irradiance prediction method." *International Journal of Electrical Power & Energy Systems* 114 (2020): 105411.
15. Alrashidi, Massoud, Musaed Alrashidi, and Saifur Rahman. "Global solar radiation prediction: Application of novel hybrid data-driven model." *Applied Soft Computing* 112 (2021): 107768.
16. Ghanavati, Farideh, João CO Matias, and Gerardo J. Osório. "Towards sustainable smart cities: Integration of home energy management system for efficient energy utilization." *Sustainable Cities and Society* 111 (2024): 105579.
17. Sassi, Abderrahim, et al. "Energy management of PV/wind/battery hybrid energy system based on batteries utilization optimization." *2017 International Conference on Green Energy Conversion Systems (GECS)*. IEEE (2017).
18. Williams, Daniel O., Tamer F. Megahed, and Sobhy M. Abdelkader. "Increasing green energy penetration and efficient utilization through a finite control set model predictive-based virtual inductance droop control." *Electric Power Systems Research* 210 (2022): 108176.
19. Ahmed, Rehan, et al. "Optimal power management with guaranteed minimum energy utilization for solar energy harvesting systems." *ACM Transactions on Embedded Computing Systems (TECS)* 18.4 (2019): 1–26.
20. Tala't, Mohammad, et al. "Optimal transmission policy for maximizing green energy utilization in small cell networks." *2019 IEEE 89th Vehicular Technology Conference (VTC2019-Spring)*. IEEE (2019).
21. Tala't, Mohammad, et al. "Optimal transmission policy for maximizing green energy utilization in small cell networks." *2019 IEEE 89th Vehicular Technology Conference (VTC2019-Spring)*. IEEE (2019).
22. Lee, Chien-Chiang, Jafar Hussain, and Yongxiu Chen. "The optimal behavior of renewable energy resources and government's energy consumption subsidy design from the perspective of green technology implementation." *Renewable Energy* 195 (2022): 670–680.

23. Midilli, Adnan, Ibrahim Dincer, and Murat Ay. "Green energy strategies for sustainable development." *Energy policy* 34.18 (2006): 3623–3633.
24. Hussain, Irshad, et al. "Optimizing energy consumption in the home energy management system via a bio-inspired dragonfly algorithm and the genetic algorithm." *Electronics* 9.3 (2020): 406.
25. Kim, Jimin, et al. "An optimization model for selecting the optimal green systems by considering the thermal comfort and energy consumption." *Applied Energy* 169 (2016): 682–695.

Chapter 6

Two-Channel Convolutional Neural Networks for Rolling Bearing Fault Diagnosis in Unbalanced Datasets

Ramesh Babu Pittala, Bhuvaneswari M. S., and Pramoda Patro

6.1 Introduction

Nowadays, industrial machinery is more sophisticated and self-aware than ever before, with a plethora of sensors built into the system to monitor its vital signs [1]. These sensors gather a mountain of data about how the system is running. Researchers have made significant strides in developing data-driven fault diagnostic approaches, and intelligent fault diagnosis algorithms may delve into the detailed aspects and apply them to problem detection [2]. When systems involving rotating equipment fail, rolling bearings are usually to blame. System functioning might be disrupted due to a failure of rolling bearings, resulting in time and cost losses. As a result, the whole apparatus depends on the rolling bearings' dependability and safety

DOI: 10.1201/9781003479062-6

[3]. Additional study is necessary to address the issue of defect diagnostics in rolling bearings.

There has been an explosion in the development of intelligent defect detection algorithms, such as Deep Belief Networks (DBNs), auto-encoders (AEs), convolutional neural networks (CNNs), and many more. Researchers were able to identify problems with gearbox chains by using a DBN that had been optimized using a genetic approach [4]. Built a deep neural network (DNN) using many AEs; fed the network the time domain vibration signal directly; used it to learn features and troubleshoot problems. Since this technique ignores features in other domains in favour of time domain features, it fails to make full use of the signals' characteristics [5]. A model for diagnosing rolling bearing faults was developed by the researchers. It features multiple stacked auto-encoders (SAE). The frequency domain and the time domain are all simultaneously utilized in several SAEs to extract the signal's properties. Afterwards, the features from the various domains are merged [6]. This approach outperforms both LDA + NN and PCA + neural network (NN). When it comes to handling pictures and time series data, CNNs are superior to AE and DBN. Researchers have been directly extracting discriminative features using CNN in the last several years [7]. To decrease the computational complexity, Levent integrated CNN's feature extraction and classification phases. To make defect detection more resilient, mixed CNNs with second-generation wavelet transform [8]. Using an adjustable weight vector, the researchers made CNN more sensitive to defect characteristics. The rolling bearing vibration signal was transformed into a greyscale picture. Then, fault characteristics were extracted from the greyscale image, leading to an improved fault diagnostic outcome [9].

The diagnostic success of the aforementioned CNN models is contingent upon their use of balanced datasets [10]. The reliability and accuracy of fault diagnosis are compromised in real-world operating conditions due to the low failure frequency of rolling bearings, which leads to an imbalance between normal and failure data. Due to the data imbalance, CNN features learned are more biased towards features found in normal state samples and under-fit to features found in fault samples. The standard approach is to increase the number of flawed samples to create a balanced dataset, which will eliminate the negative impacts of imbalance learning. To create balanced virtual samples for use in training, presented the synthetic minority oversampling technique (SMOTE). Using a superposition denoising AE, Generative Adversarial Network (GAN) may produce fresh samples and identify errors [11]. To prepare the collected data for GAN-generated synthetic minority class samples, the fast Fourier transform (FFT) is used. The automated expansion of the picture collection is suggested by Multiview GAN (MV-GAN). Results for defect identification using GAN-generated empirical mode decomposed (EMD) the energy spectrum data outperformed those using conventional oversampling methods, used a combination of adversarial learning and CNN to make it more resilient [12]. The first one is several techniques for extracting time-frequency features from bearing failure signals; then, utilizes a single-image GAN to create new datasets. The ACGAN-SN, an auxiliary classifier GAN using spectrum normalization, was used by the researchers to identify bearing faults. The stability of ACGAN-SN is superior to that of GAN, according to the experimental data. The GAN generator was tweaked by the researchers using the CNN fault diagnostic findings.

Further, to ensure that fault samples are comprehensive, the envelope of spectrum error is included as an additional corrective term [13]. While the aforementioned methods can address the issue of fault diagnosis in imbalanced datasets to a certain degree, they do so with the following drawbacks: training time is increased and valuable sample information is lost when new samples are generated, extremely unbalanced data can generate useless noise, and outcomes based on noise are not reliable for fault diagnosis. Whether or whether the created data is available also depends on how good the source data is. There is a lack of defect data since problems do not happen very often. It could be difficult to learn the features and patterns of fault data if there is a lack of data to work with.

To fix these issues, the authors of this study suggested a Two-channel convolutional neural network (TC-CNN) model. Because it can extract fault characteristics from the signal vibration in both the frequency and time domains at the same time, the TC-CNN model finds more information. Fault diagnosis is based on feature-finding engineering. Good results can be easily obtained regardless of the strength of the data imbalance if the separation of data characteristics is sufficient. Another good technique to tackle issues is to add additional features or make them simpler to master. By boosting the feature of dimensions utilized for extraction of feature, EMD, and by expressing time-frequency structures as a collection of Intrinsic Pattern Functions (IMFs), the suggested method can concurrently gather error features in two feature domains, making fault diagnosis easier than with the Denoising Autoencoder (DAE) method projected in. The suggested method, on the other hand, employs generalized linear transformation (GST) to extract time-frequency data and graphically represent it, leveraging the convolutional structure of CNN for image feature withdrawal. Final thoughts about the main takeaways from this research: (1), present a TC-CNN method that integrates 1D and 2D convolutional neural networks (2D-CNNs). This model extracts features in the time-frequency domain in addition to the frequency domain, thereby increasing the dimensionality of the features. (2) The frequency spectrum is inputted into the 1D-CNN using FFT, and the time-frequency image is inputted into the 2D-CNN using GST. (3) Two CNN channels are used for feature extraction, independently. The fusion layer is responsible for merging the two feature vectors. In TC-CNN, the characteristics learned by frequency spectra and time-frequency pictures are combined and utilized during preparation and knowledge [14].

What follows is a synopsis of the remaining content of this article: the feature extraction of two different approaches used by TC-CNN is presented in Section 6.2. In Section 6.3, the theory surrounding CNN is introduced, and the TC-CNN model's framework is provided. The effectiveness of TC-CNN is confirmed in Section 6.4 using many experiments. In addition, the benefits of the proposed TC-CNN method are shown by comparing it to both current and approximation models [15]. Section 6.5 provides a concise overview of the method's shortcomings and identifies areas for further investigation.

6.2 Extracting Features Approach

6.2.1 FFT

Many signal processing tasks, such as feature extraction and radar signal processing, make use of FFTs.

6.2.2 GST

Using IMFs as inputs for deep learning in EMD is a challenge since the frequency independence of IMFs makes it difficult to establish the relationship between fault characteristics and IMFs. Short-Time Fourier Transform (STFT) acquires the temporal-frequency spectrum by using a sliding time frame. The width of the window affects both the high temporal accuracy and low-frequency resolution. The pertinent time-frequency resolution is also determined after the identification of the window function by the STFT. Vibration Dynamics (WVD), like STFT, is used for analysing signals that exhibit non-stationary behaviour. WVD is the study of the frequency-time characteristics of signal vibrations. However, WVD faces cross-interference application complications in engineering. In different human analyses, the automated differentiation of cross-term interruption using deep learning algorithms is difficult. When compared to the wavelet transform, WPT can deconstruct the high-frequency components, leading to better resolution in the high-frequency band. Bearing failures in rotating equipment often occur less frequently than ten times the rotation frequency, which means that WPT might lead to an increase in computational cost. Finally, the frequency domain-based feature extraction approach that was selected is GST.

6.3 Framework for the TC-CNN Model

CNN is used as a technique for theoretical research because of its impressive track record of success in the extraction of features and pattern detection across numerous domains. Before outlining the steps to diagnose a problem with a rolling bearing using the TC-CNN model, this part provides a quick overview of the CNN concept.

6.3.1 CNN

A network of CNNs consists of four layers: the input layer, hidden layers, fully connected layer, and output layer. The hidden layer consists of alternating convolutions and pooling layers. A fully connected layer and an output layer make up a CNN classifier. The input layer's feature map and the convolution layer's output layer undergo a convolution process with the help of convolution kernels. The stimulation function is then used to produce the output feature. Within the pooled layer, the attribute map undergoes filtration. Figure 6.1 displays a standard design of a CNN.

6.3.1.1 Layer for Convolution

Multiple convolutional kernels make up the convolution layer. The activation function is used to translate the new feature data to the feature map inputs or vectors of features from the previous layer, which are then convolutionally processed using convolutional kernels in the convolutional layer. When doing a convolution operation, CNN follows this formula:

$$M_n^l = f\left(\sum X_m^{l-1}.W_{mn}^l + B_n^l\right) \tag{6.1}$$

the f is the activation function, B_n^l is the bias vector, M_n^l is the feature matrix of the lth layer output, X_m^{l-1} is the feature matrix of the mth or previous layer input, W_{mn}^l is the weight matrix of the mnth convolution kernel, and l marks the lth layer.

6.3.1.2 The Role of Activation

An essential and crucial component of CNN is the nonlinear activation function. This work takes advantage of the widely used Rectified Linear Unit (ReLU) function to avoid gradient explosion and gradient dispersion:

$$f(x) = RELu(x) = (0, X) \tag{6.2}$$

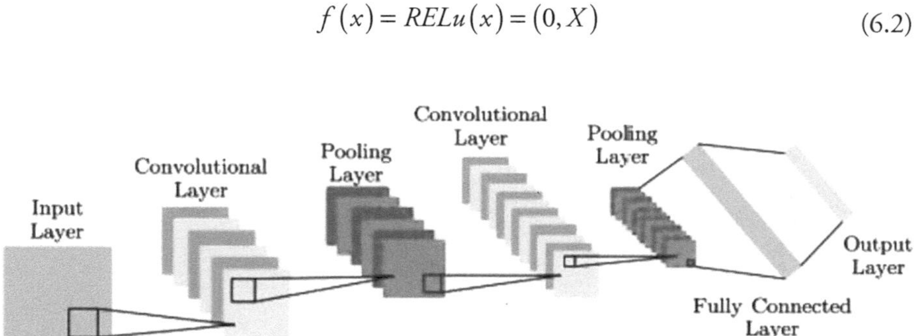

Figure 6.1 A conventional CNN architecture.

6.3.1.3 Layer Combining

To decrease the dimensionality and capture more feature information, CNN set up a layer pooling to execute down-sampling processes. This simplifies and refines the output feature info. The following formula is used by CNN in its pooling operation:

$$P_m = M_{n\in S} max M_n^l \tag{6.3}$$

The size of the pooling layer is denoted by *S*, and the output matrix is *Pm*.

6.3.1.4 Layer with Full Connection

To classify the features that the network has learned, the layer that is fully connected must first map them to the sample labelling space. It is the result layer of the network. As a classifier, Support Vector Machine (SVM) is used in this study. Classification is accomplished by processing the retrieved features with SVM.

6.3.2 The TC-CNN Method under Consideration

Figure 6.2 shows the suggested TC-CNN outline for diagnosing rolling bearing faults; the model integrates FFT, GST, and CNN.

To extract features, 1D-CNN and 2D-CNN take advantage of their parallel convolution structure. The model integrates a frequency spectrum-based 1D convolutional structure with a GST time-frequency image-based 2D convolutional structure. Using the sample fault data to its fullest potential and ensuring that the fault information complements each other are both capabilities of the suggested technique. A fully connected layer, two pooling layers, and two convolutional layers make up each CNN model. Layer $C^*$of the feature fusion system stitches together the features obtained by FC_1 and FC_2. Between C^* and FC_1^* is where the dropout operation is implemented to enhance the generalizability.

There are four steps to the fault diagnostic procedure: (1) the original sampled data is processed using FFT and GST to create frequency spectra and GST time-frequency pictures. (2) To extract fault features, a feature extractor is fed a frequency spectrum and the GST time-frequency picture. A feature fusion layer then combines the fault features. (3) Supervised formation of combined defect features is carried out using the convolutional network, with the model's parameters and weights being trained and updated. (4) Using the fault dataset, the confirmation happens that TC-CNN performs well in fault classification.

6.3.3 Criteria for Assessment

NNs often use accuracy as an assessment metric. Nevertheless, if the dataset is imbalanced, accuracy alone cannot be used for the assessment. For the model to achieve a high level of accuracy, it will be trained to favour the majority class in

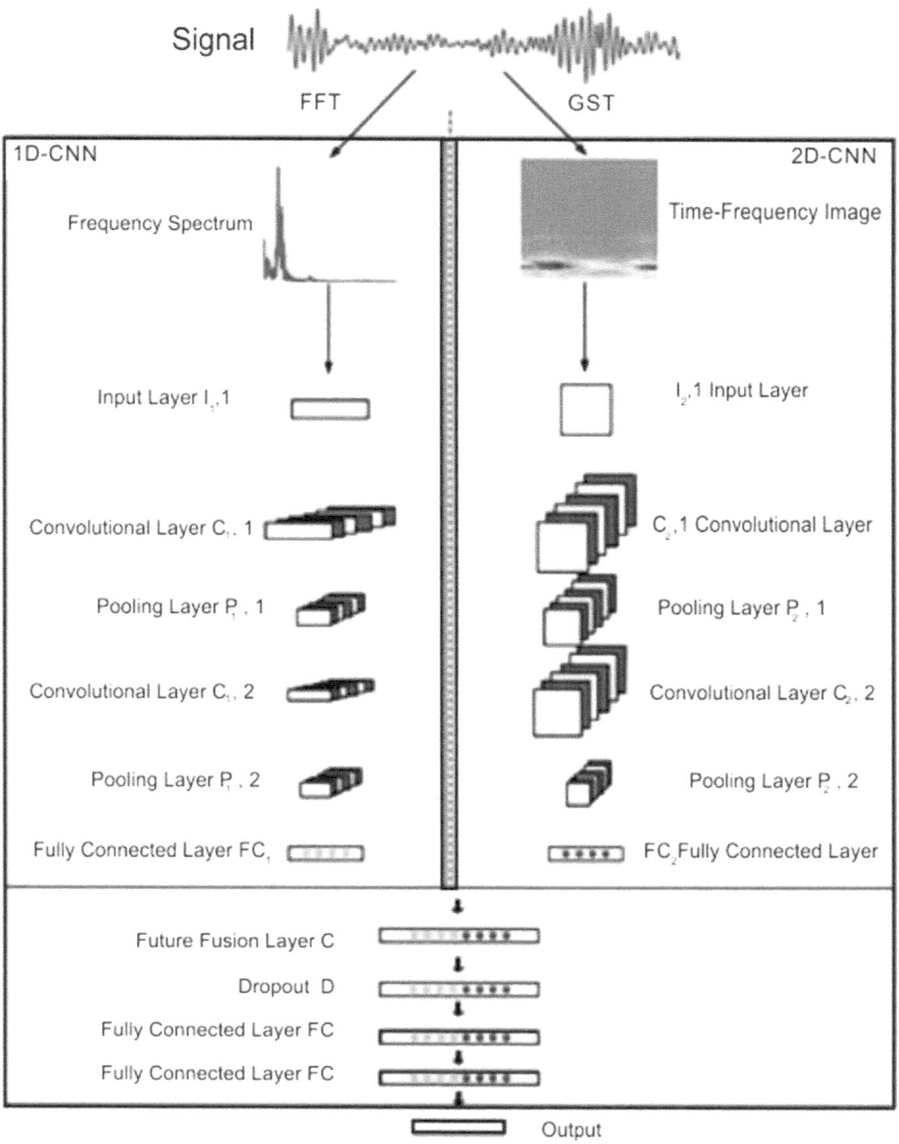

Figure 6.2 Diagram depicting the planned TC-CNN architecture for identifying problems with rolling bearings.

its classification results. Having said that the results of minority class samples' classifications are more significant. Having a high level of accuracy does not guarantee a better categorization outcome. For this reason, the F1 value also serves to provide a more thorough and accurate evaluation of the model in the study. Here is the most fundamental equation for the confusion matrix:

6.3.3.1 Accuracy

The ratio of accuracy is properly anticipated outcomes to the whole sample, expressed as a percentage.

$$Accuracy = \frac{tp + tn}{tp + fn + tn + fp} \tag{6.4}$$

6.3.3.2 Precision

One measure of accuracy is the proportion of true samples relative to all projected true samples; this is known as precision.

$$Precision = \frac{TP}{TP + EP} \tag{6.5}$$

6.3.3.3 Recollect

Recall measures the likelihood of accurately predicting a positive sample from an actual positive one.

$$Recall = \frac{TP}{TP + FN} \tag{6.6}$$

6.3.3.4 F1 Rating

Greater levels of accuracy and recall indicate superior performance. However, accuracy and recall can have different values, with one being high and the other being poor. To address this, the F1 rating is introduced as a way to syndicate precision and recall. The F1 rating quantifies the capability to detect positive models.

$$f1 = \frac{2 \times Precision \times Recall}{Precision + Recall} \tag{6.7}$$

This study employs the macro F1 rating computation for the multiclassification issue in the following manner:

$$F1 = \frac{\sum_{i=1}^{n} F1_i}{n} \tag{6.8}$$

The variable n represents the entire amount of fault classes. $F1_i$ refers to the F1 score of class i, where i ranges from 1 to n.

6.4 Analysing Experiments

6.4.1 Data of Experiments

To test how well the TC-CNN fault diagnostic model works, this research uses the Case Western Reserve University (CWRU) bearing failure dataset. In Figure 6.3, see the experimental facilities.

A dynamometer, a torque transducer, and a 2-hp motor make up the bulk of this experimental setup. The motor shaft is supported by the test bearing. With no horsepower applied to the motor and an average speed of 1724 revolutions per minute, the data is sampled at 48,000 hertz from the drive's end bearing. Most of the faults were located in the outer ring (OR), inner ring (IR), and ball (BD) flaws. Electrical machining produces faults in this bearing database with 0.008-inch, 0.015-inch, and 0.022-inch fault sizes. Each point of failure has one of these three damage states, which indicate a varying degree of severity. As a result, Table 6.1 contains the information for the ten data classes.

In Table 6.1, the direction of the fault is indicated by the notation @6:00.

6.4.2 Parameters of the Model

In the context of GST, the study used a dataset containing 1025 data points, a probability value (p) of 0.6, and a lambda value (λ) of 0.6. (1) A frequency spectrum measuring 1 × 514 is produced using the FFT of the original test signal, which has a size of 1 × 1025. (2) A time-frequency image of 65 × 65 × 4 is produced by compressing the original test indication, which has a size of 1 × 1025, using GST. The primary goal of compression is to highlight the most important feature info while reducing the influence of irrelevant information, reducing background data disturbance, and emphasizing the main characteristics. For the 2D-CNN, the input

Figure 6.3 Experimental facility for testing rolling bearings.

Table 6.1 Information on Faults in Rolling Bearings

Location	Fault Diameter (Inch)	Fault Orientation	Label
f1	Ball	0.008	–
f2	Ball	0.015	–
f3	Ball	0.022	–
f4	Inner race	0.008	–
f5	Inner race	0.015	–
f6	Inner race	0.022	–
f7	Outer race	0.008	Centre @6.00
f8	Outer race	0.015	Centre @6.00
f9	Outer race	0.022	Centre @6.00
f0	Normal	–	–

is the time-frequency picture; for the 1D-CNN, the input is the frequency spectrum. (3) To get the final fault characteristics, the recovered characteristics from the two models based on CNN are combined. The model's network parameters are obtained from actual test data. The parameters allotted to both of the networks are shown in Table 6.2.

An SVM classifier with a Gaussian kernel function as the last fully linked layer FC_3^*. C is the penalty factor and γ is the kernel parameter, both of which are set to 2. The rest of the parameters are these: the loss function is cross-entropy, the total number of epochs is 500, the learning rate is 0.006, the dropout ratio is 0.6, and the mini-batch size is 8.

6.4.3 Findings from a Balanced Dataset Investigation into Fault Diagnosis

For each class, an extra hundred samples are created at random, for an overall number of 2000 samples. Apply t-distributed Stochastic Neighbour Embedding (t-SNE) technique to demonstrate TC-CNN's hierarchical feature knowledge process. The taught features from the first input indication, completely linked layers FC1, FC2, and FC_1^*, are transferred to two-dimensional features in the test dataset in the following sequence using t-SNE. Observe the plotted properties of the various layers in Figure 6.4. As the model learns, the feature maps for each class alter, as shown in Figure 6.5. Figure 6.4 shows that the first data characteristics for each class are dispersed and hard to differentiate. The learned characteristics of the fully linked layers FC1 and FC2 are shown in Figures 6.5 and 6.6, respectively. The samples get more clumped following the pooling and convolution processes when compared to

Table 6.2 Specifications of the TC-CNN

Name of Layer	*Parameter*	*Size of Layer*
$I_{1,1}$	–	1×514×1
$I_{2,1}$	–	65×65×3
$C_{1,1}$	Conv (1×7), kernel size=7	1×509×1
$C_{2,1}$	Conv (4×7), kernel size=6	61×61×7
$P_{1,1}$	kernel size=5	1×128×7
$P_{2,1}$	kernel size=3	31×31×7
$C_{1,2}$	Conv (7×17), kernel size=6	1×124×17
$C_{2,2}$	Conv (7×17), kernel size=6	27×27×7
$P_{1,2}$	kernel size=4	1×42×17
$P_{2,2}$	kernel size=3	14×14×17
FC_1	–	657
FC_2	–	2705
C	–	3361
D	–	3361
FC_1	3361×85	85
FC_2	SVM	11
O	–	1×11

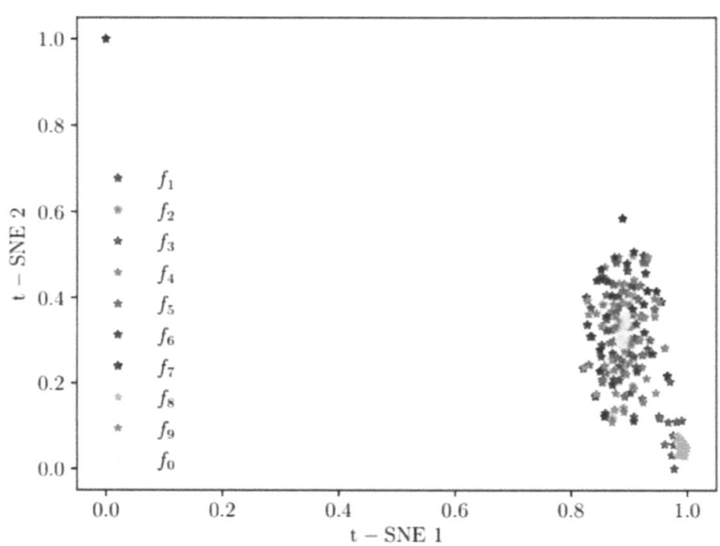

Figure 6.4 Visualizing input data features.

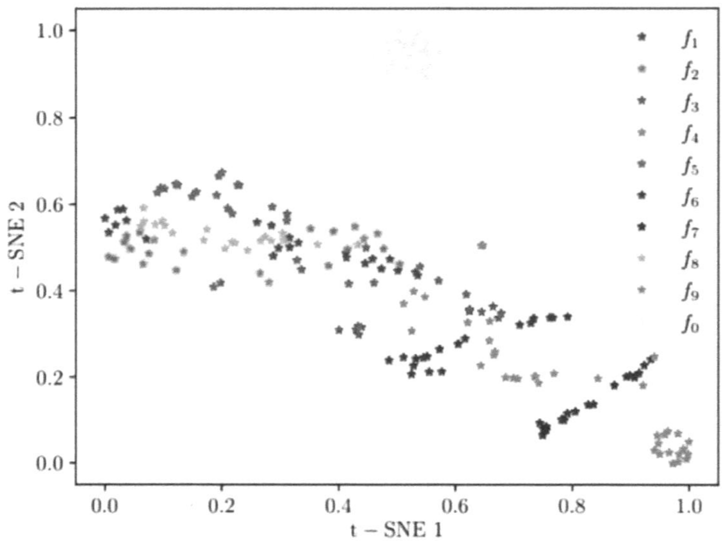

Figure 6.5 The FC1 fully linked layer's feature visualization.

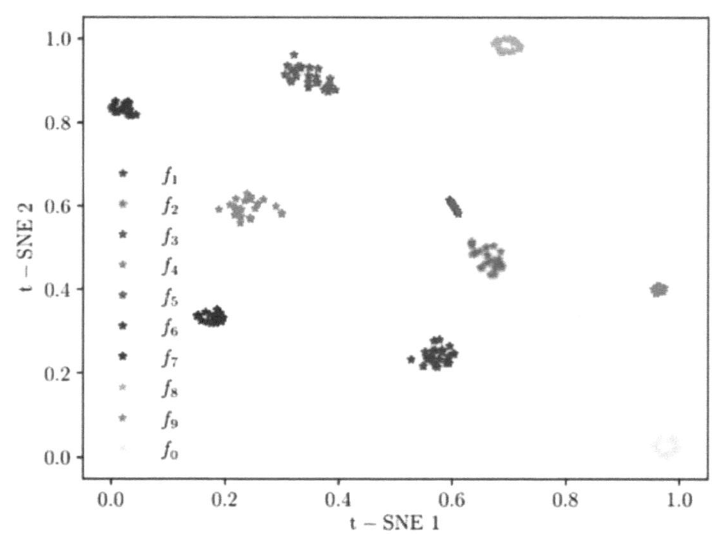

Figure 6.6 The FC2 fully linked layer's feature visualization.

the initial input data. In addition, compared to FC1, the layer that is completely linked to FC2 does a superior job at feature grouping. Lastly, Figure 6.7 shows a high concentration of traits belonging to the same class. When comparing the feature mapping outcomes in FC1 and FC2, the gap between the distributions of features

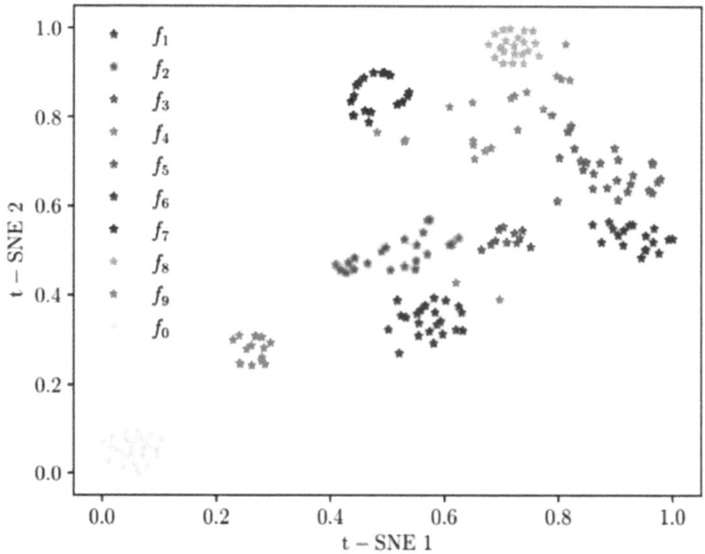

Figure 6.7 Fully connected layer FC_1^* feature visualization.

in $FisC_1^*$ is the biggest. Classifications of the dataset are executed effortlessly by the classifier, showcasing the outstanding results of the classification.
Compare the following methods for 1D-CNN, 2D-CNN, DBN, CWT+2D-CNN, and 1D-CNN+2D-CNN.

1. One-dimensional CNNs have the following layers: input, two 1D convolutional layers, two-layer pooling, completely linked, a softmax classifier, and output. The convolutional layer's input channels have kernel sizes of 6, 7, 4, and 7, whereas the pooling layer's output channels have kernel sizes of 7. Similar to the 1D-CNN, the remaining structural parameters in the suggested model are also unchanged.
2. Before being input into the 2D-CNN, the initial indicator data is transformed into a 33 × 33, n-dimensional matrix, since the dimensions of the 2D-CNN are 1025. This classifier is called the softmax. In every other way, the 2D-CNN in the proposed model is the same as the 2D-CNN.
3. CWT+2D-CNN chooses the Morlet cmor3-3 wavelet basis functional and uses CWT to extract time-frequency characteristics. A 2D-CNN with a softmax classifier is fed the CWT time-frequency picture. Similar to the 2D-CNN in the previous model, the remaining structural parameters are also included in the new one.
4. The network architecture of the DBN is [1025, 55, 25, 15], and it has three levels: an input layer, two layers that are hidden, and an output layer. A

mini-batch size of 9 is used, with 550 iterations, and the average learning rate is 0.06.

5. In terms of network architecture, 1D-CNN+2D-CNN is identical to TC-CNN. With TC-CNN, feature extraction is done using FFT and GST, which is different. The frequency spectrum and the time-frequency pictures are the inputs to the channels of two, correspondingly. The novel input signal and the 2D matrix turned into the exclusive indicator are the data inputs of the two channels of the 1D-CNN+2D-CNN, respectively.

Applying k-fold cross-validation helps prevent biased findings caused by randomly splitting training and testing datasets. The data are split into k equal-sized subgroups that are not mutually exclusive. Out of these subsets, k – 2 serves as training samples, while the other subset is utilized for testing. A training set and a validation set are created from the training samples. The outcome of the experiment is the average of the k precision and F1 rating outcomes obtained from an overall of k experiments. The study uses a ratio of 8:3:2 for the training dataset, validating set, and test set, with k = 15 and the sample set partitioned into 15 subsets.

DBN achieves the lowest values for precision and F1 ratings among various models, whereas approaches using layers of convolution and pooling are better capable of fault identification. The information richness of TC-CNN improves its fault diagnostic performance compared to the single-channel CNN model. For fault identification, 2D-CNN consistently performs better than 1D-CNN when compared to CWT+2DCNN and 1D-CNN. A two-channel model is also used by 1D-CNN+ 2D-CNN. Its fault diagnostic effect is superior compared to 1D-CNN but lower to 2D-CNN when compared to single-channel CNN. The reason for this is due to the signals being sent into both channels are authentic.

In a balanced dataset, over-fitting occurs because some extracted characteristics are irrelevant or redundant. The diagnostic findings of the various algorithms using balanced datasets presented earlier confirm that TC-CNN is better.

6.4.4 Findings from the Unbalanced Dataset for Fault Diagnosis

The preceding part covers the topic of TC-diagnostic CNN's performance on balanced datasets. To train traditional machine learning models with high performance, a huge number of samples is required. Nevertheless, data imbalance is prevalent in practice, and the quantity of inaccurate data is negligible. Consequently, rectifying the issue of defect identification in imbalanced datasets is crucial. Various percentages of normal and fault information are combined in this section. Results from experiments using datasets with varying proportions further prove that TC-CNN is stable for defect diagnosis. In the training set, there is a 3:1 split between normal and defective samples, as well as 6:1, 11:1, 21:1, 31:1, and 51:1, accordingly. There is always one normal data point for every fault data point in the test set.

The findings of the fault diagnosis may be seen in Figures 6.8 and 6.9. The fault detection capacity of these approaches is contingent upon the quantity of fault samples. For the first case, below are the six methods' diagnostic adequacy and F1 score: (100.00%, 2.00), (78.70%, 0.79), (92.71%, 0.93), (93.90%, 2.00), (74.70%, 0.75), and (91.60%, 0.92).

The suggested approach has a diagnostic accuracy of 97.80% and an F1 score of 0.98; the number of defective training samples is 1/50 of the regular training samples. Various additional approaches yielded the following results: (41.80%, 0.38), (47.30%, 0.45), (88.40%, 0.88), (38.10%, 0.33), and (52.80%, 0.49). Even when the ratio of healthy samples to defective ones approaches 50:1, TC-CNN maintains its outstanding fault identification capabilities. Although the F1 score drops by 0.04, the diagnosis precision is only 4.20% less than in example 1. However, as the fault number of samples is reduced, the performance of the existing approaches for fault identification drops dramatically. These approaches do a poor job of training a model to detect faults when the training dataset's fault data decreases with increasing imbalance rates. This led to poor precision and F1 scores since the majority of the test dataset's flawed samples could not be appropriately identified. The TC-CNN model's advantage becomes more apparent when the unbalanced ratio of the data distribution grows. To improve the ability to differentiate between fault samples, the

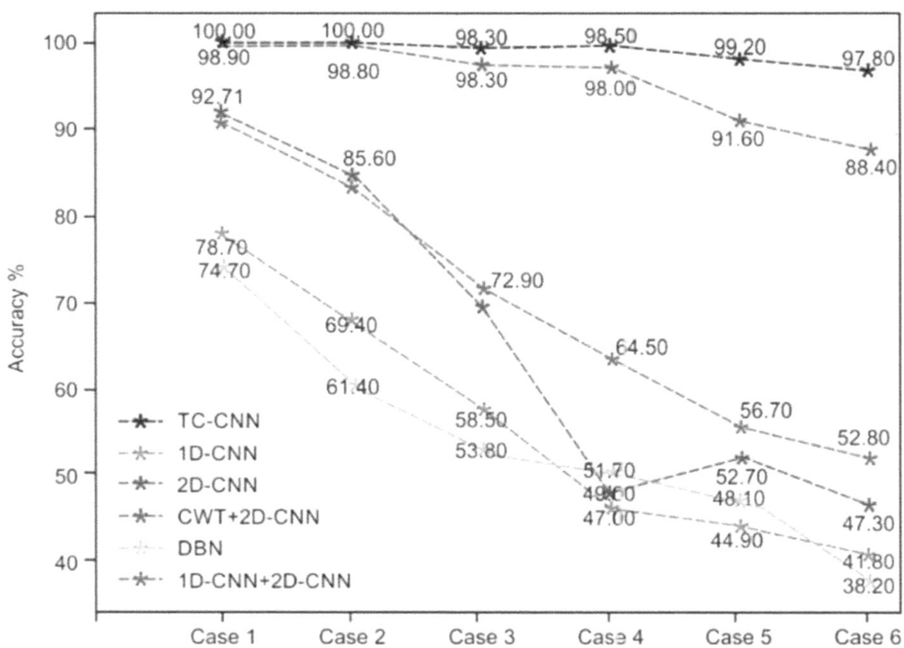

Figure 6.8 Comparing the precision of several approaches for six instances of imbalanced datasets.

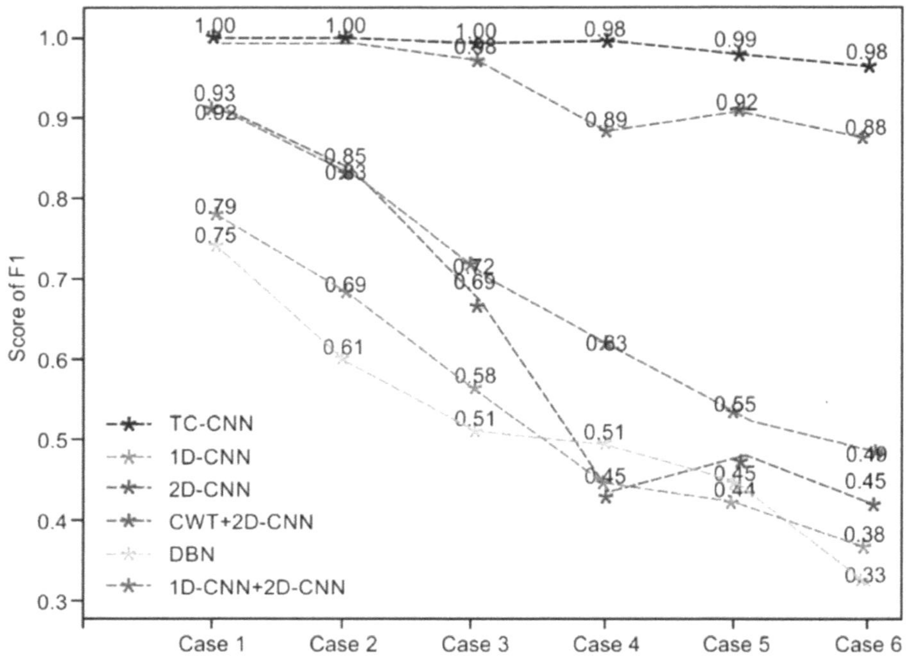

Figure 6.9 Different approaches' F1 scores for six situations of imbalanced datasets.

TC-CNN model incorporates fault feature information using FFT and GST. This allows for a deeper extraction of data features.

Compared to CWT+2D-CNN, TC-CNN does not significantly increase diagnostic performance when the imbalance in the ratio is minor, like 3:1. When the fault samples are balanced, the feature-rich advantage that TC-CNN has is not that significant. The fault diagnostic performance of TC-CNN remains relatively unchanged when the data unbalance ratio is raised from 6:1 to 51:1, in comparison to CWT+2D-CNN. The reason TC-CNN does more than CWT+2D CNN is that it also extracts time-frequency information. Complementing one another, enhancing fault characteristic data, and making fault samples simpler to recognize are the characteristics of varied dimensions. The fault diagnostic performance of 1D-CNN+2D-CNN is somewhat worse than that of 2D-CNN when the imbalance ratios are 3:1 and 6:1, respectively. However, 1D-CNN+2D-CNN outperforms 2D-CNN in fault identification as the imbalance ratio rises, suggesting that 2D-CNN is more vulnerable. The rationale for this is that, even with reasonably large training sets, 1D-CNN+2D-CNN has the potential to over-fit the model by extracting duplicate features. 1D CNN plus 2D-CNN performance improves for defect detection as data volume lowers because it can extract more usable features. By reducing the impact of data imbalance, FFT and GST allow TC-CNN to derive more fault information from fewer fault samples, making its fault diagnostic capabilities much greater than that of 1D-CNN+2D-CNN. Unfortunately, DBN isn't able to accurately portray fault information, and its training complexity rises sharply with decreasing training sample numbers.

Furthermore, the DBN model's performance is severely diminished due to its inadequate capacity to manage noise and other disruptive forces. Consequently, TC-CNN outperforms conventional approaches in defect identification, even though all models suffer from the extremely imbalanced dataset. When it comes to fault identification, the TC-CNN model performs well over a range of data imbalance ratios and is relatively unaffected by the absence of fault information.

6.5 Conclusions

To summarize the features of the suggested approach, the first step is to extract features from the frequency domain and the time-frequency domain using FFT and GST, respectively. This process increases the dimensionality of the features that are extracted, expands the error characteristics, and then enhances the mistake data. Additionally, a two-channel method is suggested that integrates 1D-CNN and 2D-CNN. This model would use the frequency spectral and time-frequency pictures as inputs. Next, to accomplish feature layer fusion, the training and learning processes include and combine the characteristics of distinct channels. The system outperforms numerous deep learning algorithms in fault detection and can correctly identify many kinds of faults, according to the experimental data. Also, even on very imbalanced datasets, the approach remains stable and resilient. This chapter's model input is a vibration signal, which is a restriction of the suggested technique. If better results are required when diagnosing faults, feature extraction using FFT and GST is the way to proceed. On the other hand, characteristics like current or sound waves could not be well-extracted using FFT and GST. It is essential to use appropriate feature extraction algorithms for sound or current signals. Hence, improving the TC-CNN model's fault diagnostic impact under imbalanced datasets by fusing several kinds of signal characteristics and researching more efficient feature extraction techniques would mostly include future efforts. Additionally, in future studies, GAN may be integrated with the suggested technique to get more accurate diagnostic outcomes.

References

1. Rahat, Irfan Sadiq, et al. "A step towards automated haematology: DL models for blood cell detection and classification." *EAI Endorsed Transactions on Pervasive Health and Technology* 10 (2024).

2. Kumar, K. Sunil Ratna, et al. "Iot and data mining techniques to detect and regulating of solar power systems." 2023 International Conference on Inventive Computation Technologies (ICICT). IEEE (2023).
3. Zhao, Chen, et al. "Fault diagnosis method for rolling mill multi-row bearings based on AMVMD-MC1DCNN under unbalanced dataset." *Sensors* 21.16 (2021): 5494.
4. Wang, Baiyang, et al. "A bearing fault diagnosis method based on spectrum map information fusion and convolutional neural network." *Processes* 10.7 (2022): 1426.
5. Liu, Xiaolin, Jiani Lu, and Zhuo Li. "Multi-scale fusion attention convolutional neural network for fault diagnosis of aero-engine rolling bearing." *IEEE Sensors Journal* (2023).
6. Xu, Ming, et al. "Bearing-fault diagnosis with signal-to-RGB image mapping and multichannel multiscale convolutional neural network." *Entropy* 24.11 (2022): 1569.
7. Zhang, Yanfei, et al. "Bearing non-uniform loading condition monitoring based on dual-channel fusion improved DenseNet network." *Lubricants* 11.6 (2023): 251.
8. Gong, Wenfeng, et al. "A novel deep learning method for intelligent fault diagnosis of rotating machinery based on improved CNN-SVM and multichannel data fusion." *Sensors* 19.7 (2019): 1693.
9. Sun, Kun, Dongdong Liu, and Lingli Cui. "Rotating machinery fault diagnosis based on optimized Hilbert curve images and a novel bi-channel CNN with an attention mechanism." *Measurement Science and Technology* 34.12 (2023): 125022.
10. Yu, Tianzhuang, et al. "A rolling bearing fault diagnosis method based on a new data fusion mechanism and improved CNN." *Proceedings of the Institution of Mechanical Engineers, Part O: Journal of Risk and Reliability* 238.6 (2023): 1156–1169.
11. Hakim, Mohammed, et al. "Bearing fault diagnosis using lightweight and robust one-dimensional convolution neural network in the frequency domain." *Sensors* 22.15 (2022): 5793.
12. Yao, Dechen, et al. "Fault diagnosis of rolling bearings based on dynamic convolution and dual-channel feature fusion under variable working conditions." *Measurement Science and Technology* 35 (2024): 1–24.
13. Liu, Xiaoxin, et al. "A model of binaural auditory nerve oscillator network for bearing fault diagnosis by integrating two-channel vibration signals." *Nonlinear Dynamics* 111.5 (2023): 4779–4805.
14. Wang, Youming, and Gongqing Cao. "A multiscale convolution neural network for bearing fault diagnosis based on frequency division denoising under complex noise conditions." *Complex & Intelligent Systems* 9.4 (2023): 4263–4285.
15. Huo, Jiuyuan, et al. "Data augmentation fault diagnosis method based on residual mixed self-attention for rolling bearings under imbalanced samples." *IEEE Transactions on Instrumentation and Measurement* 72 (2023).

Chapter 7

The Implementation of Artificial Intelligence for Auto Gearbox Failure Detection

Ashish A. Mulajkar, G. Laxmaiah, Raju Kumar, and Shruti Bhargava Choubey

7.1 Introduction

Machines both new and improved have emerged as a result of industrialization and a paradigm change in the scientific and technological communities [1]. An important part of the spinning machinery that transfers power is the gearbox [2]. The precision of a gearbox is crucial due to the rising need for more precise machinery in recent times [3]. Gear faults including tooth broken bones, backlash, and alignment may cause bearings, shafts, and other rotating equipment to vibrate [4]. These faults may be identified via vibration signals, which provide a plethora of information on the device's dynamic activity [5]. Failure of such crucial components may increase maintenance costs and even kill people [6]. Thus, gearbox failures must be detected. Three key procedures are needed to diagnose a rotating equipment problem: testing the machine, discovering the cause, and predicting fault propagation [7]. Artificial intelligence (AI) has various industrial uses, including rotating machinery fault diagnosis [8]. Academia is interested in it.

DOI: 10.1201/9781003479062-7

This issue involves pattern recognition [9]. The pattern needed for fault diagnosis is hard to discover in vibrating signals due to their complexity and abundant information [10]. Therefore, any defect diagnostics system needs two components [11]. Defect detection using extraction of features and data analysis [12]. Decision trees may choose features [13]. Machine and deep learning systems can classify features [14]. While comparing SVM to other machine learning (ML) algorithms, it has consistently shown to be the most efficient in terms of training time and accuracy for feature categorization tasks [15].

A gearbox equipped with an accelerator and a microphone is used in the experimental setup to capture the necessary vibration and sound signals, respectively, for the data collection process. First, the data is utilized to extract statistical features, and then the decision tree is used to choose features. The next step is to sort the characteristics that were retrieved using various deep learning and ML models. It is challenging to establish the necessary pattern for defect detection in vibrating signals due to the vast amount of rich and complicated information such signals include. For that reason, there are two essential steps in every defect diagnostic system. Following data processing with the aid of various intelligence algorithms, the next step is to use the processed data for defect detection. Decision trees and other similar methods may be used for feature selection. Feature classification is accomplished by using deep learning or ML methods.

The following procedures made up the project's methodology:

1. The data was acquired using an experimental setup that included a gearbox equipped with an accelerometer and microphone.
2. The necessary statistical characteristics, including median, skewness, kurtosis, etc., were retrieved when data gathering was finished.
3. The features that were needed were chosen using a decision tree technique after feature extraction.
4. Feature classification was performed using a combination of deep learning and ML techniques. The approach that proved to be the most effective in diagnosing gearbox faults was chosen according to the classification accuracy.

7.2 Experimental Study

7.2.1 Experimental Setup

The experimental setup consists of a 0.5 hp direct current motor with variable speed, a synchronous gearbox with four different types of gears, and four distinct speed ranges (500, 750, 1000, and 1250 rpm). Additionally, an eddy current dynamometer is included in the arrangement. The gearbox drive is inputted by the motor power, and the speed of the motor may be adjusted from 0 to 1440 rpm using the control panels. The output shaft of the gearbox is connected to an eddy current dynamometer, which alters the load on the gearbox to simulate gearbox loading scenarios. The gearbox is designed to absorb vibrations more efficiently with the help of a flexible connection. An AE sensor from Physical Acoustics and a piezo-electric accelerometer are directly adhered to the gearbox's smooth surface to detect vibrations. A microphone is placed next to the casing of the gearbox to obtain sound signals using a free field array.

7.2.2 Method

With a total of 36 test conditions as shown in Figure 7.1, the study was carried out utilizing the following test settings: three loads, three input motor speeds, and two scenarios including gear and bearing faults.

The Electric Discharge Grinding process was used to simulate defective bearing conditions by creating a fracture in the bearing. The simulation had two bearings and two gearbox sets. One of the gear sets had no faults, while the other had outer race difficulties and chipped gear. First, the gearbox is tested with all good gears (GGs) mounted and a load of 0-Nm at 500 rpm. The setup is made to operate for a certain amount of time to reach a steady state. The signal-conditioning unit is used to gather data from the microphone and accelerometer after it has established steady-state working conditions. This involves the use of an Analog-to-Digital Converter (ADC) and a charge amplifier to convey the signal. Samples were collected every second for each experiment, which lasted seconds. A Universal Serial Bus (USB) connection is used to input digital signals into the computer. The software SO analyzer is used for signal-conditioning and storage in the machine's secondary memory. The data on vibration and sound are acquired independently using the 8-channel data gathering system.

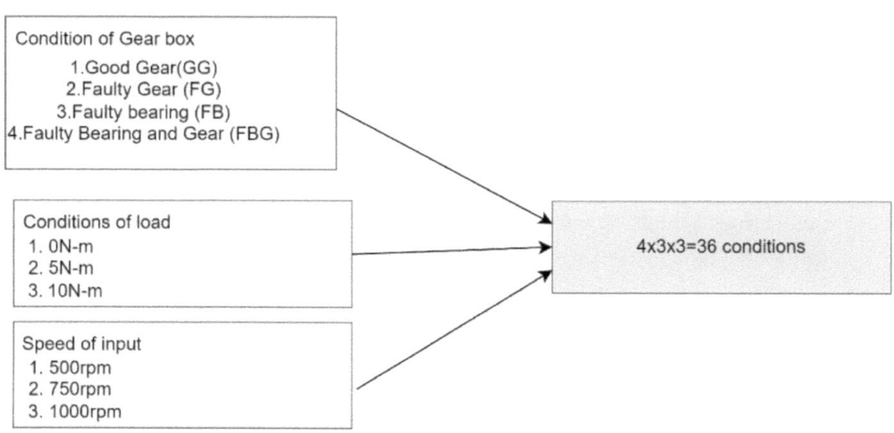

Figure 7.1 A total of 36 test conditions.

7.3 Extracting and Classifying Features

Now, every column's characteristics such as total, variance, mean, maximum, minimum, skewness, kurtosis, and standard deviation can be found using the raw data utilizing statistics in Python. Every gear goes through the same procedure, although at various speeds (500, 750, and 1000 rpm). Out of eight characteristics, choose the six that are most important using a decision tree. Several methods, including decision trees, artificial neural networks (ANNs), support vector machines (SVMs), and deep neural networks, have classification models built using these properties.

7.4 Machine Learning

As a branch of AI, ML enables computers to acquire new knowledge and skills automatically, without human intervention, via practice. When dealing with complicated tasks, standard algorithms fail, leading to the adoption of ML. ML has made it much simpler to examine massive datasets. This research makes use of decision trees, SVMs, and ANNs as ML techniques.

7.4.1 Decision Tree

It's a visual way to see all the potential outcomes of an issue or selection of options given certain parameters. The classification and regression issues may be addressed with the use of decision trees, a supervised learning approach.

The entropy and Gini index are the usual building blocks for the tree's split.

- Gini Index: Using the allocation of labeling in the subset, the Gini impurity metric determines the frequency with which a randomly chosen element in a set will be incorrectly identified.

$$G = \sum_{i=1}^{c} p_i * (1 - p_i) \tag{7.1}$$

- Data Accumulation (Entropy): At each step of the tree's development, it determines which characteristic to split on. Use that function for dividing more effectively as your information gain increases.

$$\sum_{i=1}^{c} - p_i * (p_i) \tag{7.2}$$

This T-structure classifier uses internal nodes to represent characteristics, branches to signify decision roots, and leaf nodes to show the result. The two main

Table 7.1 Classification Precision (DT)

Gear RPM	*Classification Accuracy (%)*
1_500	86
1_750	88.4
1_1000	78.2
2_500	88.14
2_750	87.80
2_1000	89.10
3_500	96
3_750	99.40
3_1000	100
4_500	97.8
4_750	96.8
4_1000	98
Mean	92.49

nodes in a decision tree are the leaf node and the decision node. Several branches emanate from the decision node, which is used for decision-making, and the leaf nodes represent the outcomes of these choices. When all input values are subtracted from the number of properly predicted values, the resulting ratio is the classification accuracy.

The decision tree approach was used to find the classification accuracy for various gears under varied settings; the results are shown in Table 7.1, with an average accuracy of 92.49%.

7.4.2 Support Vector Machine (SVM)

A popular supervised learning technique for solving regression and classification issues. To help build the hyperplane, SVM chooses the extreme points. The method termed SVM is connected with the concept of support vectors, which are the extreme points that help construct the hyperplane (Figure 7.2).

Finding the most effective boundary, in this case, a hyperplane, is one of the primary aims of the SVM algorithm, which aims to partition the space of n dimensions into groups for future data point placement. Three distinct kernels are used by SVMs:

- Linear
- Polynomial
- Ratio of Basis Function

Given the nature of the issue as a multi-class classification, the RBF kernel was used.

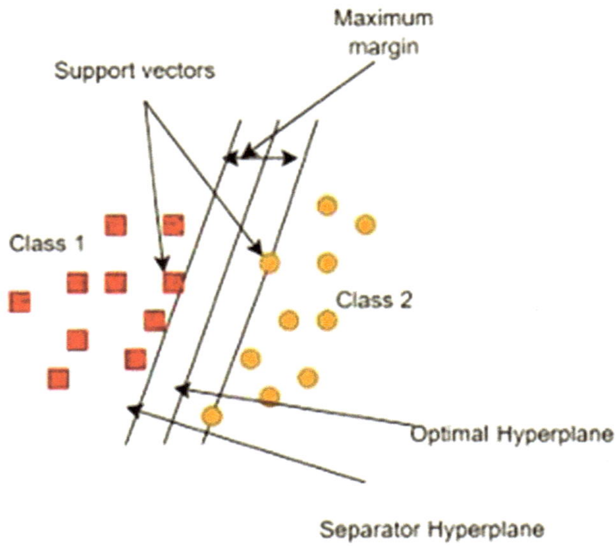

Figure 7.2 Support vector machine.

Procedures comprising SVM:

1. Data collection: The data logger provided raw data, which would be converted to CSV format after collection.
2. Data analysis for exploration: Analyze the data visually and statistically here. The data has no outliers and a Gaussian distribution.
3. Feature extraction and feature selection are part of data pre-processing.
   - Feature extraction: Extract useful information from raw data using statistical characteristics such as mode, kurtosis, standard deviation, root mean square, median, standard error, and variance.
   - Feature selection: One additional tree classifier may be used in this project to determine which feature is more important for understanding the product.
4. Splitting the data into train and test sets: Using the best parameters, divide the whole dataset into train and test sets.
5. Training the model: The SVM classifiers are used to fit the input and output features, and then hyperparameter tweaking is used to discover the optimal parameters for the model's accuracy.
6. This research used the RBF kernel because of its superior performance in classifying difficulties.

With a mean accuracy of 93.65%, the SVM method (RBF kernel) was used to determine the classification accuracy for various gears under various situations (see Table 7.2).

Table 7.2 Classification Precision (SVM)

Gear RPM	*Classification Accuracy (%)*
1_500	88
1_750	88.4
1_1000	80.36
2_500	94.09
2_750	87
2_1000	88.3
3_500	96.4
3_750	99
3_1000	98
4_500	99
4_750	98
4_1000	94.6
Mean	93.65

All of the diagonal elements in Figure 7.3 reflect the properly categorized data in the confusion matrices for the SVM algorithm. A confusion matrix is the end product of feature categorization. A confusion matrix, which is an M × M matrix with M being the number of targeted classes, is used to assess the efficacy of a classification model. The matrix evaluates the expected and actual values by comparing them with the ML model. The anticipated fault classes are shown in each column, whereas the actual fault classes are arranged in each row. The model's accuracy in classifying data points as faulty bearing (FB) is shown by the quantity of data points in the first row of the first column. The number of instances that should have been labeled defective gear (FB) but were instead mislabeled FBs is represented by the second component in the second column of the first row.

7.4.3 Artificial Neural Network (ANN)

An artificial neural network is sometimes known as a "neural net" since the term "neural" describes a neuron and "network" describes a structure that resembles a graph. These programmers mimic the behavior of the human nervous system to find solutions to any issue. Three layers constitute a typical neural network. An input layer, an output layer, and a hidden layer make up the layers. When working with an ANN, "learning" means adjusting the relative importance of the links between nodes. The architecture and computational techniques used to train a neural network influence its learning capabilities. Figure 7.4 shows a feedforward ANN network that accepts features as input, applies weights to them, and then passes them on to the hidden layer.

GG 0 Nm	GG 5 Nm	GG 10 Nm	GG 15 Nm	FB 0 Nm	FB 5 Nm	FB 10 Nm	FB 15 Nm	FG 0 Nm	FG 5 Nm	FG 10 Nm	FG 15 Nm	FGFB 0 Nm	FGFB 5 Nm	FGFB 10 Nm	FGFB 15 Nm	classified as
90	10	0	0	0	0	0	0	0	0	0	0	0	0	0	0	GG 0 Nm
13	84	3	0	0	0	0	0	0	0	0	0	0	0	0	0	GG 5 Nm
0	2	11	45	0	0	0	0	0	0	0	0	0	0	0	0	GG 10 Nm
0	2	27	69	0	0	0	0	0	0	0	2	0	0	0	0	GG 15 Nm
0	0	0	0	90	10	0	0	0	0	0	0	0	0	0	0	FB 0 Nm
0	0	0	0	15	71	14	0	0	0	0	0	0	0	0	0	FB 5 Nm
0	0	0	0	0	7	91	0	0	0	0	0	1	1	0	0	FB 10 Nm
0	0	0	0	0	0	0	100	0	0	0	0	0	0	0	0	FB 15 Nm
0	0	0	0	0	0	0	0	100	0	0	0	0	0	0	0	FG 0 Nm
0	0	0	0	0	0	0	0	1	99	0	0	0	0	0	0	FG 5 Nm
0	0	0	0	0	0	0	0	0	1	87	12	0	0	0	0	FG 10 Nm
0	0	0	0	0	0	0	0	0	0	8	92	0	0	0	0	FG 15 Nm
0	0	0	0	0	0	0	0	0	0	0	0	96	4	0	0	FGFB 0 Nm
0	0	0	0	0	0	0	0	0	0	0	0	7	72	18	3	FGFB 5 Nm
0	0	0	0	0	0	0	1	0	0	0	0	0	25	51	23	FGFB 10 Nm
0	0	0	0	0	0	0	2	0	0	0	0	0	5	18	75	FGFB 15 Nm

Figure 7.3 Confusion matrices.

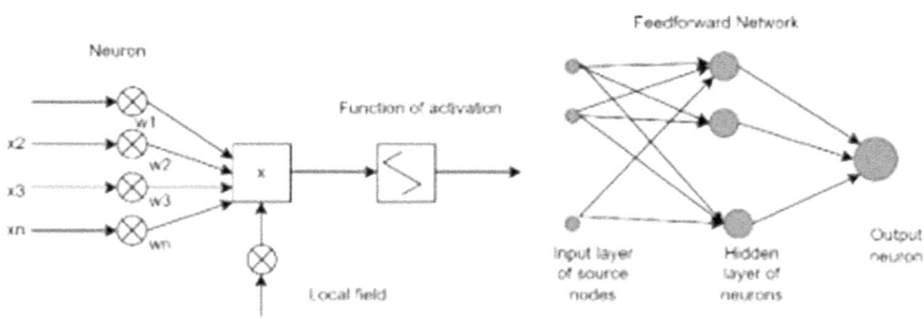

Figure 7.4 Artificial neural network.

In the hidden layer, the final product is the feature that has all the features converted to weights and then given a bias. After that, the y-value is subjected to the activation function.

Following that, it is sent to the output layer along with certain weights. This setup is a feedforward network or forward propagation method.

Table 7.3 Classification Precision (ANN)

Gear RPM	*Classification Precision (%)*
1_500	81.6
1_750	81.7
1_1000	74.3
2_500	85
2_750	82.8
2_1000	83.4
3_500	89
3_750	95.3
3_1000	98.3
4_500	94.3
4_750	94.2
4_1000	91.2
Mean	87.34

The model's evaluation: The information is provided on the discrepancy between the initial value (y) and the anticipated value $\left(\breve{y}\right)$. Backpropagation and weight adjustments will be made by the model in the event of an extremely significant loss.

The average accuracy for various gears and situations as determined by an artificial neural network categorization is 87.34%, as shown in Table 7.3.

7.5 Deep Learning

Deep learning is a subfield of ML that uses several layers to derive more complex characteristics from simpler ones.

7.5.1 Deep Neural Networks

In this specific kind of artificial neural network, the input and output layers are separated by a large number of hidden layers. This is an example of a one-way data flow from an input layer to an output layer.

7.5.2 The Function of Activation

The output of a neural network may be better determined with the use of an activation function. The significance of each neuron's input to the model's forecast determines when the network should be activated or not. This kind of function is related to each neuron in the network.

7.5.3 The Study's Activation Functions

7.5.3.1 The Function of Sigmoid Activation

A sigmoid function returns an integer between zero and one. The value is considered zero if it's less than half a half, and one otherwise. Therefore, it is the binary classification function that is employed the most.

7.5.3.2 Function of the tanh

The function in question is a hyperbolic tangent. The sigmoid and tanh functions have somewhat distinct curves. The sigmoid function ranges from 0 to 1, whereas the tanh function ranges from –1 to 1. This is the main difference between the two functions.

7.5.3.3 Rectified Linear Unit Function (ReLU)

The most preferred option because of its many advantages over the sigmoid and tanh functions.

With positive input, gradient saturation is not an issue. The computation time is much quicker due to the absence of non-linear relationships. For both forward and backward propagation, it beats the sigmoid and tanh functions (the latter two need more time to compute the exponent).

7.5.3.4 The SOFTMAX Function

When dealing with multi-class classification difficulties, it is often used at the output layer.

A deep neural network (Rectified Linear Unit Function (ReLU) Activation function) was used to find the classification accuracy for various gears under varied settings; the results are shown in Table 7.4, with an average accuracy of 92.3%.

7.6 Web Application

7.6.1 Web Intro

No one has tried to completely automate the fault categorization process yet, according to the study that has been conducted. The quick detection of defects is crucial to maintaining the machine's health, which is why automatic fault classification systems are expected to be in high demand shortly. We have developed a model to detect the errors using ML.

Table 7.4 Average Accuracy (DNN-Using ReLU)

Gear RPM	*Classification Precision (%)*
1_500	85.6
1_750	86.5
1_1000	79.6
2_500	93
2_750	84.6
2_1000	88.6
3_500	95
3_750	96.5
3_1000	98
4_500	96
4_750	100
4_1000	97
Mean	92.3

7.6.2 The Method for Making the Web App

Create a JavaScript file. Make a SVM model and save it as a .py file using the sci-kit learn, pandas, and NumPy libraries. Rename the .py file to "pickle" format. In a file named app.py, build a web app using the Flask framework. Import request and Flask libraries into the app.py file, then open the .pl file and display the template. Bring up the Local Host and execute the app.py file. The HEROKU a cloud-based should be used to deploy the model.

7.6.3 Developing the Web App

The SVM model will provide a data frame containing the predicted faults for the provided data when the predicting option is selected. The value count's function displays the total number of faults and the frequency with which each fault has occurred. The model will display the associated problem on the website if the fault occurs in the provided data.

7.7 Results

At 500, 750, and 1000 rpm, and under 0, 5, 10, and 15 N loading conditions, time-domain data was collected using the accelerometer and microphone mounted on the gearbox. To streamline the process of defect identification, the collected data was

Table 7.5 Classification Accuracy

Speed of Motor	*Gear*	*Algorithm*					
		DT (%)	*SVM (%)*	*ANN (%)*	*DNN*		
					Relu (%)	*tanh (%)*	*sigmoid (%)*
500	1	86	87	89.6	87.5	87	87
	2	89.13	94.2	85	85.3	94.2	97
	3	98	95.8	89	97.3	97.8	93
	4	98.3	96.0	96.3	92.6	96.4	86
750	1	89.3	86	84.3	93.8	86.7	94.9
	2	86.4	87	87.2	89.2	88	94.1
	3	96.3	96.2	84.9	75.9	93.2	94.7
	4	97.5	86.8	93.6	82.3	87.8	96.8
1000	1	76.9	99.5	81.6	99.0	95.5	88.9
	2	89.4	98.2	76.4	86.3	99.2	85.9
	3	86.9	95.1	94.7	92.6	96.1	79.7
	4	96	97.6	92.8	85.4	91.6	89.1

transformed into statistical characteristics using publicly accessible Python statistical algorithms, GG, Faulty Gear and Faulty Bearing (FGFB), Faulty Gear (FG), and FB are 4 of the 16 classes of the issue, and there are 4 loading scenarios. The feature selection process makes use of an additional tree classifier because of its resilience. To create the most effective classification model for vibrations and microphone data, this classifier was used to choose the six best features out of eight.

When looking at vibration data from various gears under various situations, Table 7.5 shows the classification accuracy discovered using various ML and deep learning techniques. As a result, SVM outperforms the competition.

Table 7.6 displays a comparison of the findings from several deep learning and ML algorithms regarding the categorization accuracy of vibration data from various gears under various scenarios. Compared to other algorithms, SVM performs better.

7.8 Conclusion

Gearbox failure detection using ML techniques is a major focus of condition monitoring research. The acquired signals were used to extract statistical data for several classifications. The additional tree classifier method begins with eight

Table 7.6 Categorization Accuracy of Vibration Data

Speed of Motor	*Gear*	*Algorithm*					
		DT (%)	*SVM (%)*	*ANN (%)*	*DNN*		
					Relu (%)	*tanh (%)*	*sigmoid (%)*
500	1	87	87.9	89.6	87.5	87	96.8
	2	87.13	94.2	85	85.3	94.2	87
	3	98	95.8	89	97.3	97.6	92
	4	99.3	96.0	96.3	92.6	96.4	86.9
750	1	82.3	86	84.3	93.8	86.7	97.9
	2	85.4	87	87.2	89.2	88	93.1
	3	98.3	96.2	84.9	75.9	93.2	97.7
	4	99.5	86.8	93.6	82.3	87.8	92.8
1000	1	72.9	99.5	81.6	99.0	95.0	89.5
	2	83.4	98.2	76.4	86.3	99.2	83.8
	3	82.9	95.1	94.7	92.6	96.1	71.5
	4	99	97.6	92.8	85.4	91.6	86.1

characteristics and selects the six that are most useful. ANN, SVM, Decision Tree, and DNN are used to train and classify the system once it is given the characteristics that were chosen. When compared to other algorithms run on the same data, the SVM comes out on top. Using the Flask framework, they built a web app that uses SVMs to automatically generate predictions based on data inputted in comma-separated values format (.csv).

References

1. Rahat, Irfan Sadiq, et al. "A step towards automated haematology: DL models for blood cell detection and classification." *EAI Endorsed Transactions on Pervasive Health and Technology* 10 (2024): 1–11.
2. Kumar, K. Sunil Ratna, et al. "Iot and data mining techniques to detect and regulating of solar power system." *2023 International Conference on Inventive Computation Technologies (ICICT)*. IEEE (2023).
3. AlShorman, Omar, et al. "A review of artificial intelligence methods for condition monitoring and fault diagnosis of rolling element bearings for induction motor." *Shock and Vibration* 2020.1 (2020): 8843759.

4. Brito, Lucas C., et al. "An explainable artificial intelligence approach for unsupervised fault detection and diagnosis in rotating machinery." *Mechanical Systems and Signal Processing* 163 (2022): 108105.
5. Cerrada, Mariela, et al. "AutoML for feature selection and model tuning applied to fault severity diagnosis in spur gearboxes." *Mathematical and Computational Applications* 27.1 (2022): 6.
6. Lang, Wangjie, et al. "Artificial intelligence-based technique for fault detection and diagnosis of EV motors: A review." *IEEE Transactions on Transportation Electrification* 8.1 (2021): 384–406.
7. Ahmed, Ryan, et al. "Automotive internal-combustion-engine fault detection and classification using artificial neural network techniques." *IEEE Transactions on Vehicular Technology* 64.1 (2014): 21–33.
8. Chen, Peng, et al. "An automatic speed adaption neural network model for planetary gearbox fault diagnosis." *Measurement* 171 (2021): 108784.
9. Nasiri, Sara, Mohammad Reza Khosravani, and Kerstin Weinberg. "Fracture mechanics and mechanical fault detection by artificial intelligence methods: A review." *Engineering Failure Analysis* 81 (2017): 270–293.
10. Saxena, Vimal, Nilendu Kar Chowdhury, and S. Devendiran. "Assessment of gearbox fault detection using vibration signal analysis and acoustic emission technique." *Journal of Mechanical and Civil Engineering* 7 (2013): 52–60.
11. Samanta, Biswanath. "Gear fault detection using artificial neural networks and support vector machines with genetic algorithms." *Mechanical Systems and Signal Processing* 18.3 (2004): 625–644.
12. Basangar, Santosh, and B. N. Tripathi. "Literature review on fault detection of equipment using machine learning techniques." *2020 International Conference on Computation, Automation and Knowledge Management (ICCAKM)*. IEEE (2020).
13. Nath, Aneesh G., Sandeep S. Udmale, and Sanjay Kumar Singh. "Role of artificial intelligence in rotor fault diagnosis: A comprehensive review." *Artificial Intelligence Review* 54 (2021): 2609–2668.
14. Desavale, Ramchandra Ganapati, Prakash M. Jadhav, and Nagaraj V. Dharwadkar. "Dynamic response analysis of gearbox to improve fault detection using empirical mode decomposition and artificial neural network techniques." *ASCE-ASME Journal of Risk and Uncertainty in Engineering Systems, Part B: Mechanical Engineering* 7.3 (2021): 031007.
15. Wang, Zirui, Jun Wang, and Youren Wang. "An intelligent diagnosis scheme based on generative adversarial learning deep neural networks and its application to planetary gearbox fault pattern recognition." *Neurocomputing* 310 (2018): 213–222.

Chapter 8

Evolutionary Algorithms to Optimise Deep Learning Model for Water Industry Forecasts

Ravi H. Gedam, Sonia H. Bajaj, Hemlata Rahul Kosrae, Vishakha Akhara, and Girish A. Garg Hate

8.1 Introduction

To manage urban water resources in a way that benefits both people and the environment, wastewater treatment plants (WWTPs) are an essential piece of infrastructure. Water resource recovery facilities (WRRFs) are a new evolution of these facilities that aim to sustainably collect water, energy (such as biogas), and nutrients for reuse [1–3]. On the other hand, WRRF plant-wide operations are very vulnerable to underperformance since sewage conditions fluctuate so randomly and often. Worst yet, integrated sewage processes are acting in an ever more unpredictable way due to complicated biochemical and chemical interactions; this is all because WRRFs are very dynamic and nonlinear systems. Because of this randomness, it is crucial to improve overall plant performance by forecasting important metrics like biogas generation in anaerobic digestion and effluent quality in activated sludge. Because of this, it is extremely desirable to model wastewater processes to regulate, predict, and monitor plant-wide operations [4–7].

DOI: 10.1201/9781003479062-8

A lot of work goes into WRRFs using computers for things like design, optimisation, and analysis. When doing research or design, many people turn to IWA mechanistic models or their commercial counterparts [8–11]. The foundation of these mechanical models is the conservation of mass, and they are augmented by chemical and biochemical rate equations that, although not entirely empirical, approximate aggregate biological functional groupings. Measurements taken in the field and the lab were used to create and verify these. When it comes to designing and controlling WRRFs, biochemical modelling of processes is an essential tool for determining feasibility and contingency. Mechanistic models have their uses, but they have limitations when it comes to studying complex systems in data-rich environments or without governing mechanisms [12–14]. Mechanistic models are great for designing and interpreting behaviour, but they can't capture all behaviours.

Since data-driven models are often reserved for evaluations of more extensive systems, they have seen little application in wastewater treatment. But with the advent of modern data-driven approaches like machine learning (ML), they have become a hotspot for data mining, optimisation, forecasting, and modelling in the last decade, all intending to provide insights to bolster water treatment systems. More opportunities to use ML approaches for systems analysis have arisen due to the availability of massive volumes of sensor data. Although it is still in its infancy, deep learning (DL) is showing enormous promise across several sectors, and it is quickly becoming one of the most popular areas of artificial intelligence [15–18]. To predict future states, DL approaches are outperforming other empirical techniques. Regression analysis using deep learning (DL) models includes Recurrent Neural Networks (RNNs), Long Short-Term Memory Networks (LSTMs), Gated Recurrent Units (GRUs), 1-Dimensional Convolutional Neural Networks (1D-CNNs), and Bidirectional Long Short-Term Memory Networks (Bi-LSTMs), where Bi-LSTMs stands for bidirectional long short-term memory networks. In terms of effluent concentration prediction and nutrient removal efficiency in wastewater treatment, the LSTM-based model has shown superior performance [19–21]. Among RNNs, bi-LSTMs are unique in that they can simultaneously retrieve future information states and recall past data inputs. When compared to RNNs, Bi-LSTMs perform better and are more suitable for forecasting time-series data with fundamental time constants that are both long- and short-term. This makes them ideal for use in wastewater treatment processes, where traditional empirical models struggle due to the large range of time constants involved, which may be anywhere from minutes (for dissolving gas) to week (for sludge composition). Improving effluent quality prediction, DL models predicting controls, soft sensors to supplement or even replace hard sensors, fault detection and diagnosis, and optimising removing pollutants from wastewater are just a few of the recent successes in the water industry's use of DL models to a variety of problems. The water business may lack the sophisticated expertise and computing resources needed to deploy ML models. Consequently, DL model setup may be computationally demanding and time-consuming [22,23].

Standard phases in an ML pipeline include data processing, feature selection (FS) or variable collection, training, validation, and exhibiting using knowledge algorithms, and evaluation. Finding the most important aspects of the given dataset is made easier with the aid of FS, which deletes superfluous and duplicated features. The prediction performance of the model with the original features is retained. How big the neural networks (NNs) input layer is heavily dependent on how well FS functions. There are a lot of moving parts in wastewater treatment procedures, but the number of factors that might significantly affect the expected result is quite small. So, while mining data from wastewater treatment, it is crucial to choose the most significant characteristics. There has been a lot of focus on DL method testing, while FS has received less attention in the water sector. Filter methods, wrapper methods, and embedded methods are common FS techniques. Metrics such as mutual information and correlation are the only criteria used by filter algorithms to choose features. To forecast a response variable, these approaches often examine the link between each characteristic and the target variable, rather than relying on findings from any model. A Pearson correlation coefficient (PCC) example would be an FS metric that primarily relies on I/O correlation and employs a threshold to exclude information. The linear relationships between features and targets cannot be detected by PCC, even though it is a quick and easy FS approach. In contrast, principal component analysis (PCA) is a statistical method for reducing dimensionality, visualising data, and extracting and selecting features. By superimposing orthogonal data of high-dimensional onto a main subspace, a linear distance of lower dimensional, it iteratively maximises the data variations. The following are some of the many uses for PCA: merging features, determining which characteristics are most important, and removing or merging associated features. When it comes to using ML, neither PCC nor PCA approaches evaluate the features' utility. Using wrapper and embedding techniques has become necessary due to the necessity to manage FS and training. While learning algorithms in embedding methods conduct FS by rating and score feature variables according to their relevance, in wrapper methods they choose the most suitable subset of features. Because of this, computationally, PCA and basic correlation algorithms are more efficient than wrapper and embedding approaches. One of the most well-known and extensively used search-embedded techniques for optimisation, including FS, is the genetic algorithms (GAs), an evolutionary computing approach based on genetic processes and natural selection. By using the notion of natural selection, GA allowed a random starting population of people, or candidate solutions, to develop towards a superior solution. The three main components of a basic GA are as follows: (1) the operator tasked with selecting the most promising members of the current population to carry the species forward; (2) the crossover operator, which uses a chromosome recombination procedure to create offspring from the chosen individuals; and (3) the operator for mutations, which randomly changes one or more genes within every newly produced human using probabilistic functions. Towards ensure that overall fitness does not decline over time, elitism may be used in conjunction with the other three operators to ensure that only the highest-scoring parents are passed on to future generations. After applying the choice, mutation, and crossover operators, a fitness function value is obtained for each individual in the original population and each subsequent generation. This produces another generation that aspires to be better than the one before it. While GAs have seen some process optimisation applications in the water sector, their usage in FS has been very limited. Training and evaluating DL models using a systematic FS approach, on the other hand, may increase performance and make the models easier to understand. To avoid adding superfluous features or variables to the model's complexity, it is crucial to first identify which aspects are most significant.

Hyperparameter optimisation (HPO) and NAS (neural architecture search) are crucial activities in constructing DL models that require investigating the vast potential state space of network architecture. Those duties were frequently established by hand otherwise by experience, which may lead to a model that is both unnecessarily complicated and time-consuming. Learning rates, batch size, momentum, optimisers, and other NN model parameters are known as hyperparameters. These parameters stay constant while the model is trained. An important aspect of the optimising model design is exploring the dimensionality of the hidden layers by finding the number of nodes in each one. The real I/O structure fixes the total amount of nodes in the two levels of input/output (I/O); however, you may choose the number of layers that are hidden and node per internal node. If you want an acceptable model that doesn't need too much structure and model modification iteration, you must mix HPO with NAS. Nevertheless, practitioners face challenges when trying to choose the most appropriate approach to solve a given problem due to the lack of adequate information for method selection. The number of layers and neurons may be changed as needed. Heuristic techniques, such as GAs, are among the many global and local searching and optimisation methodologies that have been created to circumvent human tweaking. Data-driven model performance has been enhanced by using GAs for hyperparameter and/or NAS optimisation. Nevertheless, it should be remembered that GAs, similar to other optimisation and search algorithms, do not always provide a global optimum solution in the absence of an analytical answer to the optimisation issue. Notwithstanding their shortcomings, HPO, automating ML, and neuro-evolution are just a few of the difficult areas where GAs have achieved remarkable progress. Optimisation of sludge activation systems with multiple objectives, weight optimisation, and FS and maximisation of biogas production are some of its restricted applications in the water sector. There may be room for more research and development into gas as a potential solution to optimisation issues in the water business. This chapter presents a clear and comprehensive approach to building strong deep-learning models that can forecast biogas output and effluent concentrations in wastewater treatment systems. It covers both the obstacles and the possibilities in this field. This study aims to accomplish two specific goals: first, to find the most effective groups of characteristics that improve DL model performance by investigating optimal feature choice using different methods, for example, PCC, PCA, or GA when each feature is distinct. The second objective is to use GA to choose features and build model optimisation during DL model creation, as well as HPO. This study uses GAs and other evolutionary computing methods as a result of the rising demand for ML methods; automated ML is being used to alleviate the burden of building models for the water industry. Also, learning boosts performance and model interpretability compared to standard tuning procedures that are considered state-of-the-art by combining GAs with FS, HPO, and NAS. Hybrid modelling, which integrates DL with mechanistic models, is another innovative aspect of this research. Due to differences in sampling frequency (e.g., 15 minutes versus 7 days) in the raw plant data, the mechanistic model may be used for data augmentation.

8.2 Methodology

8.2.1 Overview of the Procedure

An extensive WRRF in a particular nation provided the data needed for the development of models and optimisation. Activated sludge systems are used to remove organic waste, nitrogen, and phosphate at this plant. Anaerobic digestion is used to treat both the primary and activated sludge. Primary settlers, air-dissolved flotation, anaerobic digesters, secondary settlers, dewatering systems, and bioreactors operate as part of a five-stage process. The plant has six primary components, collectively referred to as Barden-pho. Final settling tanks enable the solids, combined with spirits, to settle, resulting in effluent that is free of debris. This is then prepared for ultimate release to the ocean's outfall with dispersion. As returning activated sludge, the bioreactors receive the settled solids. To prepare it for the anaerobic digesters, the bioreactors' active sludge is thickened. The operating temperature of these digesters is close to 36°C and has a solid retention duration of around 21 days. The digesters can mix gases and recirculate liquids. The floating headspace of the anaerobic digestion systems is used to extract biogas, which is then sent to generator engines to produce power. There was a great deal of basic time constant variation among the selected variables (for details, see below).

8.2.2 Summary of the Data Collection

Datasets were sourced from two different locations. This includes both online and offline measurements, as well as the Supervisory Control and Data Acquisition (SCADA) system. Water, sludge, biogas, stream, and airflow rates, together with 15-minute average data, were sent to the sludge activation system using the SCADA system. In contrast, measurements and records of compositional analysis were kept independently as part of the regular offline plant monitoring and not in the SCADA system. Anaerobic digesters, bioreactors, main effluent, effluent, bioreactor composite samples, and reject water were all part of the offline measurements taken once a week. Chemical oxygen demand (COD), the analyses of the influent and effluent included measurements of total solids, volatile substances, overall suspended particles, NH_4-N, NO_3^-, TKN, PO_4-P, and overall phosphorus are the elements in question. Alkaline content, solubility of calcium (Ca), magnesium (Mg), sodium (Na), and potassium (K) were measured. To address the data imbalance between the 15-minute SCADA data and the weekly offline data, augmentation was performed using verified mechanical models of the WWTP during the investigation. After being calibrated and confirmed using long-term operational data, the mechanistic model accurately reproduced the real system's variables, kinetics, processes, and operating conditions. The primary 20 characteristics included in this analysis were culled from the plant's most frequently monitored variables. Consistent with the SCADA data, the simulated data used a sample period of 15 minutes. As a byproduct of the anaerobic digester, biogas is a goal along with overall COD, PO_4-P, and NH_4-N within the effluent. The creation of a balanced and comprehensive dataset improved the accuracy of the projections.

8.2.3 An Optimisation Pipeline Utilising Deep Learning and GA Presented

Here are the four main steps of the ideal DL/GA optimisation pipeline: data preparation, FS, HPO, and NAS utilising genetic algorithmic approaches. Using a new dataset, test the improved model's prediction accuracy. Making use of all features AF, PCC, PCA, and GA, the method sought to identify the most effective qualities for the prediction task. After that, GA was used in conjunction with the selected characteristics to improve the Bi-LSTM model's NNs and hyperparameter settings. The objective was to predict biogas production and other critical effluent factors, such as nutrient concentrations. To make sure the optimised model can handle fresh data with ease and accuracy, its performance is evaluated by running it on a distinct dataset that hasn't been seen before (Figure 8.1).

8.2.3.1 Preparation of Data

As mentioned in Section 8.2.3.3, pre-treated data utilising automated scaling and choosing features approaches to efficiently extract the information important to DL model building. Applying Z-score normalisation, which entails centring each input over the means reduced in size to have a variation from the average of 1 to have their weights equal, helped standardise the data. In every dataset, there were samples used for training, validation, and testing (prediction). As part of the processing data, 65% of the dataset is allocated for training, 17% for model validation and hyperparameter tweaking (e.g., weights and biases), and 21% for objective assessment of the optimised NN model. To prevent overfitting and guarantee that

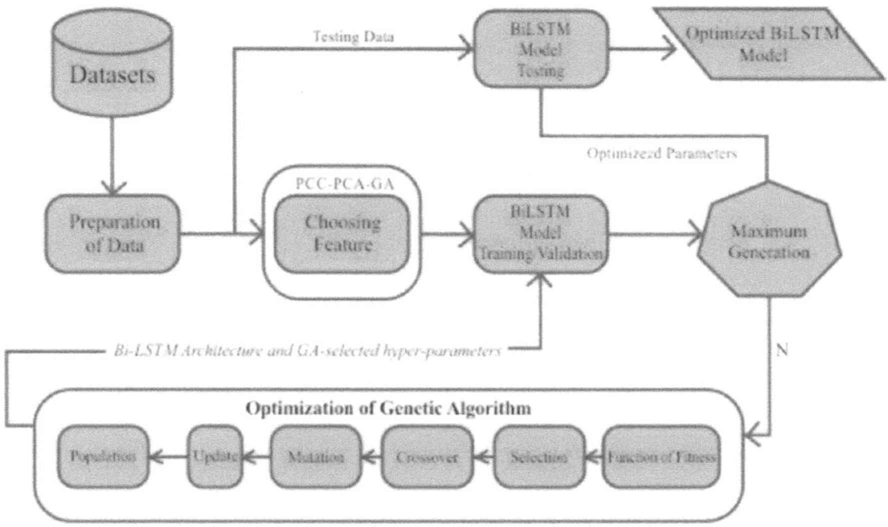

Figure 8.1 A proposed machine learning pipeline that uses genetic algorithms for feature selection.

the model can generalise, it is necessary to train the model on a big enough dataset and then verify and test it on separate datasets.

8.2.3.2 Algorithm Machine Learning

The optimisation process was tested using the learning algorithm used in the Bi-LSTM model. The GA employs a fitness function calculation as a mean squared error (MSE) during Bi-LSTM network training.

8.2.3.3 Choosing of Features

Finding the optimal subset of characteristics was the primary goal of the FS technique. A total of four FS methods AF, PCA, GA, and PCC were tested. FS for PCC, PCA, and AF was carried out separately from the selection of the DL model. The use of a GA for choosing features was done simultaneously with the selection of DL models for GA.

a. The PCC: The complete matrix of I/O correlations was created. Features having correlation coefficients below – 0.5 or above 0.5 were taken into consideration since they were linked to expected outputs. Also, traits that were pre-selected and shown to be significantly linked (with an absolute correlation criterion of 0.10–1) were deemed to have strong pairwise interdependence. To keep things simple, just trained on any one of those two variables.

b. PCA-based PCA: Only the principal component that accounted for 90% of the overall variance was kept after the data was transformed into the principal component space.

   The explained variation ratio to recreate the input variables went along with it. To minimise the reconstruction error, the dominant variable was determined as the one with the biggest loading on each principal component.

c. GAs: The learning algorithm and the input datasets were sifted through using GAs to find the best features. A value of seed was used to initialise the generator of casual numbers and guarantee repeatability. The procedure began by generating a starting population of 100 distinct solutions. Twenty genes made up each chromosome; these genes influenced the desired outputs, which were the biogas training datasets and effluent quality. Each characteristic in the dataset was represented by an item in a fixed-length binary string that the chromosomes were a part of. Following the application of genetic operators such as choosing, crossover, and mutation to generate new populations, every individual in the original population had their fitness value determined using the Bi-LSTM networks as the fitness function. For every generation after that, this procedure was carried out again. The researcher employs a two-tournament size selection strategy, as well as a 2-point crossover and flip-bit mutation method. In addition, there was a system of elitism that persisted for a full ten generations, with the top three persons from each generation being handed down. The evaluation datasets were used to evaluate the Bi-LSTM network's prediction skills after it had been trained on the training data. A lesser value for the mean square error (MSE) cost function, which was utilised to evaluate the individuals, indicated a better solution.

8.2.3.4 GA Improves Hyperparameters and Searches for Bi-LSTM Architectures

An optimisation algorithm called GA was used to fine-tune the neural architecture and hyperparameters, which include information batch quantity, sampling intervals, rate of learning, optimiser of network, and functions for activation. Every parameter was treated as a variable throughout the search. Table 8.1 shows many kinds of hyperparameters and model architecture that were represented by a floating-point-based chromosome throughout the GA optimisation process. A random seed was used to select the starting population's genes within the chromosome. A chromosome of a specific length was used to mimic the size of the hidden layers to accommodate the conventional genetics controller in the GA approach. Adaptive Gradient Descent (SDG), Adaptive Moment Estimation (Adam), or Adaptive Gradient Algorithm (Adagra) are three possible optimisation algorithm settings. To do this, converted it to a float value between 0 and 2.98. After that, rounded the

Table 8.1 The Number of Buried Layer Neurons, Optimisation Method, Activation Function, Batch Size, and Sampling Interval Are All Parameters that Must Be Defined Are Some of the Parameters of a Bi-LSTM Network that May Be Optimised Using GA

Training Parameters	*Range/Type*	*Chromosome Representation*
Batch size Data sampling Interval (hour)	20–40 0.25–2	[20, 40] [0.25, 2]
Learning rate Number of neurons in Layer 1	0.0005–0.05 5–20	[0.0005 0.05] [5, 20]
Number of neurons in Lay†Table_Bodyer 2	2–15	[2, 15]
Number of neurons in Layer 3	–2 to 15	[–2, 15]
Number of neurons in Layer 4	–10 to 15	[–10, 15]
Optimization algorithm	Adam, SGD, Adagrad	[0, 2.99]
Activation function	Sigmoid, Tanh, ReLu	[0, 2.99]

float value that was created at random to get either 0 or 2. After that, the optimisation process swapped out Adam for zero, SDG for one, and Adagrad for two. To train the rectified linear unit (ReLU), sigmoid, and tanh activation functions, the same procedure was followed. The MSE, the value of the fitness function, was used for training. Through the use of reproduction, crossover, and mutation, all of the contained parameters were updated with each generation of the model. The top four contenders maintained their positions in the contest by appealing to elitism (Table 8.1).

Secret layers 4 and 5's chromosomal ranges began with a negative number so that the number of layers may be controlled. If the random seed for hidden layer 4 was a negative integer, both secret layers 4 and 5 were deleted. On the other hand, if the seed for the forth hidden layer was negative, just that layer was eliminated. Although the chromosome did not specify it, the optimal individual after calculation was used to calculate the number of layers that are hidden [3,5].

8.2.4 Testing of Premature Convergence

To prevent the GA from being trapped in a local minimum, researchers tested its convergence too soon. This was accomplished by varying the starting population size from 10 to 200, which is one of the genetic operators. In addition, for four expected target effluent NH_4-N, PO_4-P, COD, and flowrate of biogas, a total of 16 optimisation experiments were conducted with starting population sizes of 10, 50, 100, and 200, respectively.

8.3 Outcomes

8.3.1 Choosing of Selections

8.3.1.1 Choosing of Feature through PCC

The 22 characteristics, including targets, are shown in Figure 8.2 as PCCs are shown as a heatmap. Several data factors seemed to have modest correlations, according to these findings. For example, the main clarifier's overflow COD (0.54), underflow COD (0.44), influent total suspended particles (0.49), and effluent COD (0.54) were all positively connected. Correlation values of 0.5 and 0.44 for the COD in bioreactor 17 with the MLSS in mixed liquids showed a negative correlation with the COD in the effluent, respectively.

The effluent COD predicted target pool had 12 characteristics eliminated during ML optimisation because they were deemed uninteresting.

8.3.1.2 Choosing of Features with PCA

Figure 8.2 in the supplemental material shows that 11 main components were identified from the data, which accounted for 90% of the variation. Influential NH_5-N

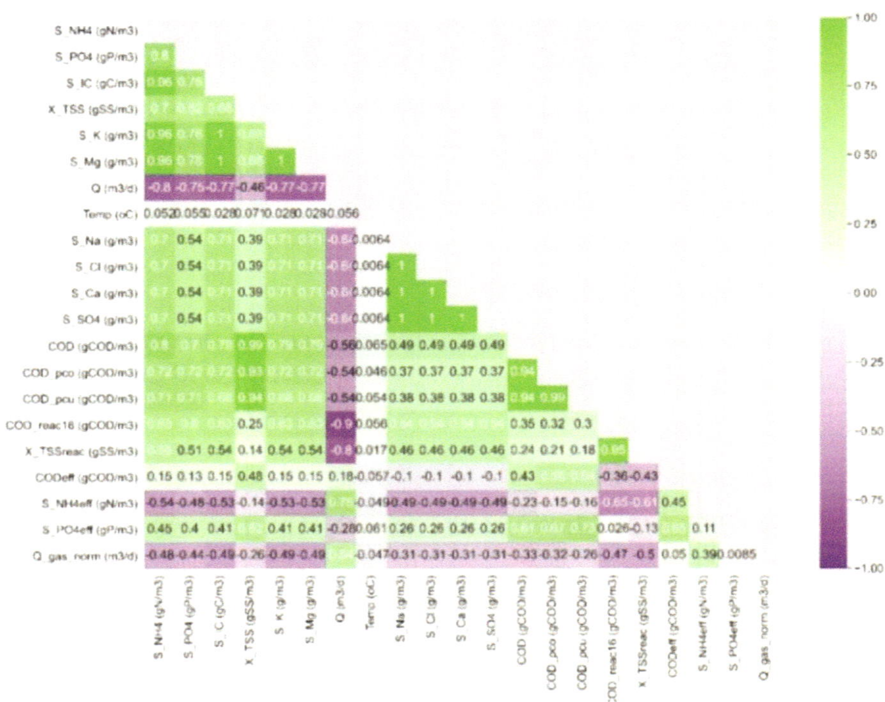

Figure 8.2 All the inputs used to choose important characteristics for target outputs were correlated using Pearson's coefficients.

was the first and most important component, explaining 57% of the total variation. After that came influent calcium (19% of the total), temperature (9%), and total dissolved solids (6%).

8.3.1.3 Choosing of Features with GA

Table 8.2 displays all of the characteristics that were chosen, as mentioned earlier, as well as those that were gained via optimisation utilising GA-based FS. To achieve FS for every target, a Bi-LSTM model was trained using default parameters. As an example, the following features were selected for the projected effluent NH_4-N target using the GA-based feature selection method: input NH_4-N, PO_4-P, total solids, flow rate (Q), input sodium, input sulphate, chemical oxygen demand in the first clarifier overflow, and chemical oxygen demand in the primary clarifier underflow.

Table 8.2 also displays the optimum characteristics chosen for the other projected objectives, including biogas flow rate, effluent COD, and effluent PO_4-P.

8.3.2 Merged NAS with HPO

To guarantee repeatability, the first GA used a seed to randomly create 100 solutions (chromosomes). When training the Bi-LSTM model, rely on the data recorded in each chromosome, which included the number of batches, hyperparameters, and NN architecture. This section solely presents the findings of PCC-GAO since fitness profiles from all 17 optimisation experiments were quite similar. Statistical analysis of genetic transfer for minimum and average fitness during eleven generations is shown in Figure 8.3 for the Bi-LSTM networks trained using PCC features. While the minimal fitness values remained constant from generation to generation, the average fitness dropped as the number of solutions increased, according to the data. The mean value was approaching the minimal fitness threshold criterion as early as the ninth generation. Typically, peak fitness was achieved before generation 11, and subsequent generations saw no further improvement. Additionally, these studies show that minimal fitness has been declining for generations. The minimal wellness in the original generation (population 0) is 0.0033 in the PCC-GAO scenario for the biogas anticipated goal, for instance, but it dropped to 0.0008 by the tenth generation. The wide variation in size of the starting population (100 chromosomes) is likely responsible for the discrepancy between the lowest and average fitness levels.

To test premature convergence concerning the number of hidden layers, Figure 8.4 displays the outcomes of GA optimisation with various starting populations (i.e., 11, 51, 101, 201). The number of concealed layers was shown to decrease as the original population size increased. The NH_4-N target had four hidden layers at the 11-person beginning population size and has three separate objectives for PO_4-P, COD, and biogas, respectively. The evolutionary method increased the number of hidden layers despite starting with a smaller population because of premature convergence, which caused it to settle on a local minimum instead of a global solution. Starting

Table 8.2 The Anticipated Objectives, Which Include Biogas Flowrate, Effluent COD, Effluent NH_4-N, and Effluent PO_4-P, Were Determined by Selecting Characteristics through PCC, PCA, or GA

Features	*Effluent NH4-N*			*Effluent PO4-P*			*Effluent COD*			*Biogas Flow Rate*		
	PCC	*PCA*	*GA*	*PCC*	*PCA*	*GA*	*PCC*	*PCA*	*GA*	*PCC*	*PCA*	*GA*
S_NH4 (gN.m^{-3})	✓	✓	✓	✓	✓	✓		✓		✓	✓	✓
S_PO4 (gP.m^{-3})	✓	✓		✓	✓			✓		✓	✓	
S_IC (gC.m^{-3})			✓									
X_TSS (gSS.m^{-3})		✓		✓	✓		✓	✓	✓		✓	
S_K (g.m^{-3})						✓				✓		
S_Mg (g.m^{-3})						✓						
Q(m^3.d^{-1})	✓		✓			✓				✓		
Temp (oC)		✓		✓	✓		✓	✓			✓	
S_Na (g.m^{-3})	✓		✓						✓			✓
S_Cl (g.m^{-3})	✓					✓		✓			✓	
S_Ca (g.m^{-3})	✓	✓			✓	✓		✓			✓	
S_SO4 (gS.m^{-3})	✓		✓						✓			
COD (gCOD.m^{-3})				✓		✓	✓					
COD_pco (gCOD.m^{-3})			✓	✓		✓	✓		✓			
COD_pcu (gCOD.m^{-3})			✓	✓		✓	✓		✓			
COD_reac 16 (gCOD.m^{-3})			✓	✓		✓	✓		✓			
X_TSS reac (gSS.m^{-3})	✓	✓	✓		✓			✓			✓	
Nh4eff (gN.m^{-3})					✓		✓	✓		✓	✓	
Po4eff (gP.m^{-3})	✓	✓					✓	✓		✓	✓	
CODeff (gCOD.m^{-3})	✓	✓	✓	✓	✓						✓	✓
Biogas flowrate (m3/d)		✓			✓							
Number of Features	11	9	10	8	9	7	7	9	7	6	9	5

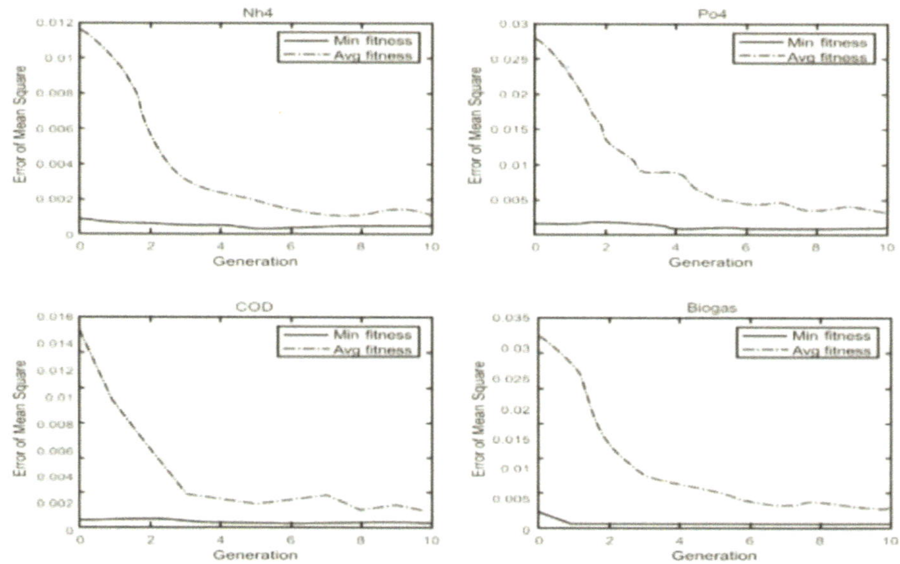

Figure 8.3 Evaluation of the PCC-GAO example across ten generations in terms of minimum versus average health value (MSE) for several effluent parameters, including biogas flowrate objectives, COD, and NH_4-N.

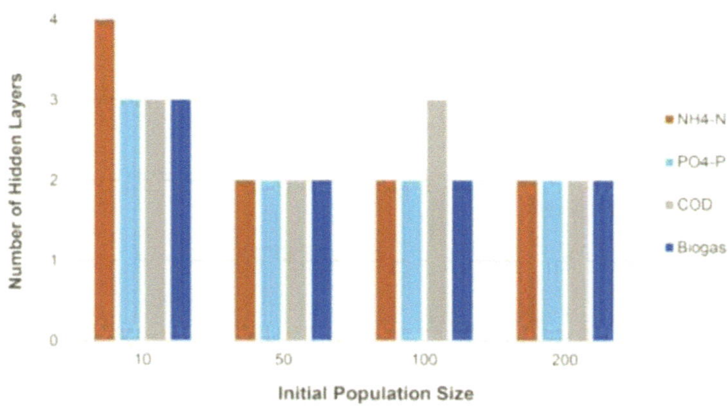

Figure 8.4 The optimal number of hidden layers derived from neural architecture in PCA-GAO genetic algorithms concerning the starting population size.

with a population range of 51–201, lower the total amount of hidden layers to 2. Based on these findings, it seems that keeping the starting population diverse (with at least 51 people) is important for GA optimisation to prevent early convergence but adding more individuals than 201 doesn't help much.

8.3.3 Optimisation of BI-LSTM Models for Prediction

Figure 8.5 shows the PCC-GAO case's validation and training losses summarised. The validation and training error (losses) of the Bi-LSTM method are shown in the figure. As the number of iterations rose, these losses reduced monotonically with some random oscillations. Eventually, they became stable, suggesting a satisfactory match. The loss function was created regarding the training dataset, and the Bi-SLTM model was trained by iteratively reducing it. Nevertheless, when the network started to overfit, the measured error about the validation set went from decreasing to slightly increasing. The researcher used an early stopping method to halt training at the lowermost fault relative to a dataset of validation to generate a Bi-LSTM model with good generalisation performance. This would explain why there is a

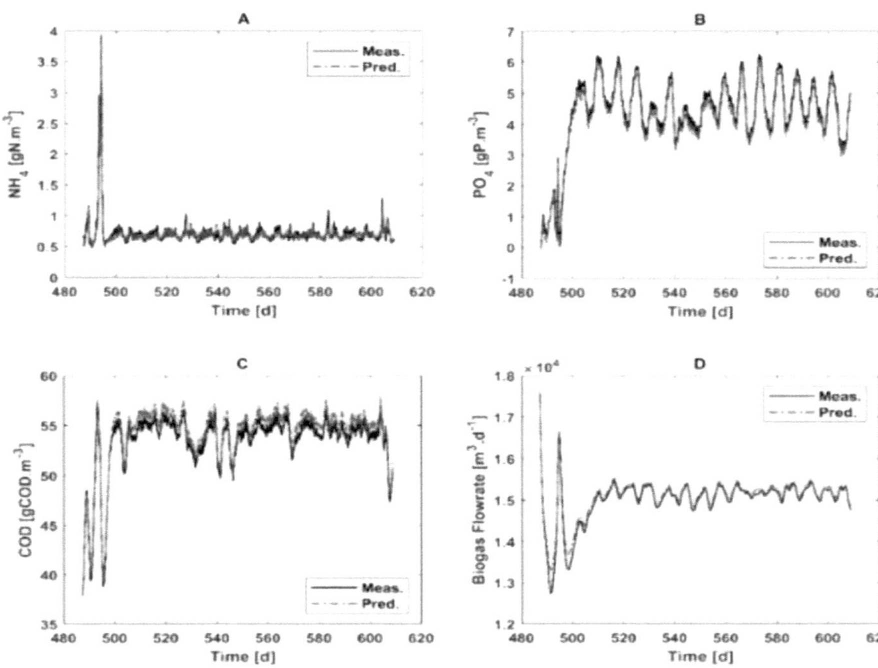

Figure 8.5 Results of PCC, hyperparameter, and architectural optimisation tests conducted on the Bi-LSTM models using evolutionary algorithms, shown graphically.

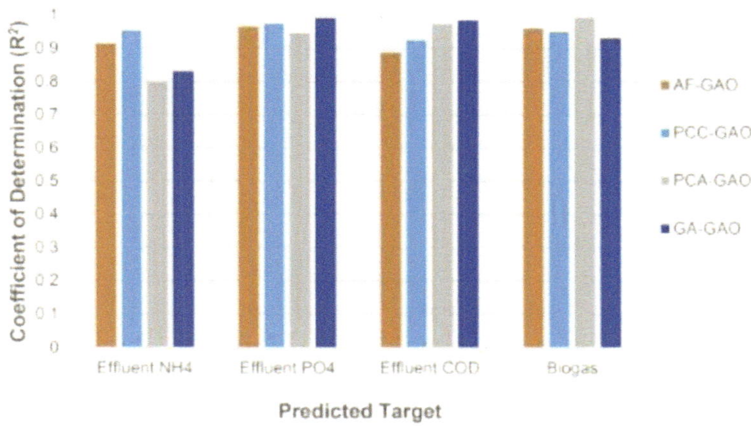

Figure 8.6 Evaluation of the four methods' Bi-LSTM models by calculating the coefficients of determination for the four projected goals.

significant difference in the number of iterations needed to train the models, with NH_4-N requiring 71 epochs and COD just 15 epochs.

In Figure 8.6, see how the observed dataset and the projected dataset, which were generated using characteristics chosen using PCC, look at the biogas flow rate, COD, NH_4-N, and PO_4-P in the effluent. All four goals are well-fitted by the PCC-GA-based Bi-LSTM models on the test dataset (ranging from 488 to 610 days). As seen in the supplementary material, these outcomes persisted across all of the other FS techniques as well.

Biogas flow rate, effluent NH_4-N, PO_4-P, and COD dataset comparisons using the NN model are shown in Table 8.3 and Figure 8.6, respectively. The Bi-LSTM model that used the PCC method to pick features performed better than the others, according to the optimisation findings. In contrast to the average R^2 values of around 0.93 for the PCA, GA, or AF technique, the biogas flowrate, effluent NH_4-N, effluent PO_4-P, and effluent COD all had R^2 values of 0.95, 0.98, 0.93, and 0.95, respectively.

Figure 8.7 displays the outcomes of optimising the Bi-LSTM model's neural architecture with various starting population sizes, as well as the model's performance with hyperparameters. While the model's performance was less than ideal with an average population size of 10, it improved noticeably with a population number of 50 or more, according to the data. For instance, there was a 400% improvement in the value of R^2 for effluent NH_4-N prediction. Nevertheless, the R^2 value decreased by 1% when the population size was increased from 50 to 200 and by 10% when the population size was increased from 50 to 100. It seems that 50 is the optimal amount for accomplishing all 4 of the targets based on these calculations. According to the results, a decrease in population variety might cause convergence to happen too soon.

Table 8.3 Data Synthesis of the Bi-LSTM Model's Predictive Abilities Next Optimisation of Its Architecture and Hyperparameters

Performance	*Effluent NH_rN (gN/m^3)*			*Effluent PO_rP (gP/m^3)*					*Effluent COD (gCOD/m^3)*				*Effluent Biogas (m^3/d)*			
	AF	*PCC*	*PCA*	*GA*	*AF*	*PCC*	*PCA*	*GA*	*AF*	*PCC*	*PCA*	*GA*	*AF*	*PCC*	*PCA*	*GA*
MAE	0.06	0.04	0.12	0.07	0.22	0.19	0.27	0.11	0.96	0.87	0.39	0.32	104	31	93	141
RMSE	0.08	0.06	0.13	0.11	0.23	0.19	0.29	0.11	1.10	0.89	0.50	0.39	107	121	39	146
R^2	0.91	0.95	0.80	0.83	0.97	0.98	0.95	0.99	0.89	0.93	0.97	0.99	0.96	0.95	0.99	0.93

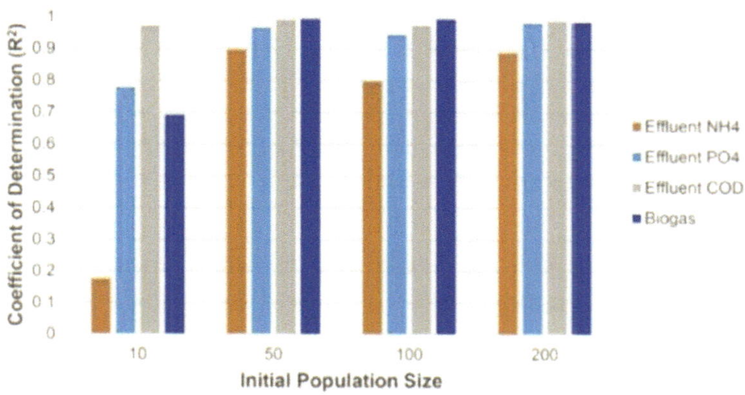

Figure 8.7 Results of testing Bi-LSTM models (i.e., coefficient of determination) as a function of the GA starting population.

8.4 Discussion

8.4.1 An Overview of the Offered Features

Analysing samples collected from wastewater activities for the important features, like solvable and particular matter COD, TSS, NH_4-N, PO_4-P, total phosphorus, can be costly and data-intensive, especially when using DL for data-driven modelling. This study's findings, however, show that just the most important variables should be included by a model using DL for estimating biogas output or effluent quality, which might lead to reduced analysis costs. Not only that but using FS to preprocess inputs may simplify NN topologies, which in turn can lead to models that are more interpretable and computationally efficient (Section 10.4.2). In line with previous research, the FS findings showed that FS approaches rely on specific data and do not apply to all cases. Effluent NH_4-N model performance was best achieved by PCC; effluent PO_4-P and COD models were best handled by GA; and biogas NN model performance was best achieved by PCA. The most time-consuming DL models used GA to optimise all 20 characteristics for every target (NH_4-N, PO_4-P, COD, and biogas flow rate) had the most unexpected outcome: Bi-LSTM models with respectable prediction efficiency. This could be because of the DL algorithm's built-in robust feature engineering. The NH_4-N and PO_4-P models did not perform better after using PCA, despite its widespread usage in the water sector for dimensionality reduction and data pre-processing. This is probably because the models were overfitted to data of training and failed to appropriately fit unknown information. Since this FS technique included learning the algorithm during optimisation, it was the most computationally costly, but it was also seen with GA.

Figure 8.7 shows that all FS approaches were successful in lowering the number of features needed for Bi-LSTM model training. However, each method performed optimally for a different objective. Incorporating FS approaches into GA or in-silico analyses allows for the identification of the best FS method, which is particularly useful before intense sampling operations. For the first feature screening, PCC or PCA might be used when there are several features.

8.4.2 Optimising Neural Architecture and Coupling Hyperparameters

Conventionally, hyperparameters are either left at their default settings or fine-tuned by hand. Nevertheless, as shown in Figures 8.5 and 8.6, the work implies that NN models' prediction accuracy may be enhanced by integrating GA-based HPO with NAS. The computational complexity lies in the combined search of neural architectures and hyperparameters since each of these topics has been examined separately. The ideal learning rate values, however, which ranged from 0.0014 to 0.0087, were in agreement with the values used in the majority of data science books and databases. The optimisation findings showed that, compared to traditional SDG and Adagra, Adam was the best stochastic optimisation approach. Among Adam's numerous advantages is his capacity to deal with very noisy issues with sparse gradients. Since Tanh performed better than ReLU (with PO_4-P) and Sigmoid, it was used as the activation function for all of the hidden layers. The optimisation findings also showed that deep models and training time are significantly affected by batch size. The recommended batch size ranged from 25 to 40, having an average of 32, according to the recommendation for a default batching size of 33 to offer rapid training iterations. The results of this study and the early convergence test suggest that a reasonably deep and narrow Bi-LSTM network might accomplish better learning, as shown in Figure 8.3. Even though the method is typically used to create DL models (see Table 8.1), it has the potential to produce shallow architectural designs that are easy to understand and work well for actual time deployment in environments with limited resources and little memory needs. This result agrees with other research that has shown fewer hidden layers in NNs that tend to have higher generalisation capabilities and less overfitting issues. Analytics tools that enable the assessment of different hidden layer architectures are essential for the successful evolution of data-driven forecasting methods in the water business. Previous studies have used DL models with 512 neurons for PO_4-P and TSS and trained them with up to 200 LSTM hidden elements.

8.4.3 Benefits and Drawbacks of the Research

Combining DL models with GAs has several potential advantages, one of which is an improvement in the model's predictive power. Shorter training periods are one advantage. The trained models are usually easier to understand since they have fewer hidden layers and neurons, which is an additional benefit. More importantly, DL models trained in this manner often outperform their trained counterparts when presented with novel input data or datasets. The study's usage of a computationally intensive DL model fitting with distinct solution candidates from a search space is a drawback of the study's utilisation of a single-threaded GA optimisation technique. Nevertheless, by executing several combinations of hyperparameters simultaneously on separate computers or processes, the computing demands of GA optimisation may be drastically reduced by parallelisation. Because numerous processes on a single computer may assess a population of N members simultaneously via multi-threading or inter-process communication, GA techniques are highly parallelisable.

8.5 Conclusions

According to the results, various FS strategies had varying degrees of success with each dataset. The PO_4-P along with the COD effluent model was better managed by GA (R^2 = 0.99) than the PCC approach; however, the NH_4-N modelling was best handled by PCC. However, with an R^2 value of 0.99, PCA emerged as the superior technique for the biogas NN model. To further automate the procedure, the research also suggests integrating FS techniques into the GA optimisation cycle. The GA-based method also optimised NN design and hyperparameters at the same time, leading to DL models that were both simple and very predictive, using lower secret layers and neurons. Due to the considerable variability of individual solutions, the GA pre-mature convergence test showed that a starting population size of 50 delivered the greatest outcomes across the expected objectives.

References

1. Rahat, Irfan Sadiq, et al. "A step towards automated haematology: DL models for blood cell detection and classification." *EAI Endorsed Transactions on Pervasive Health and Technology* 10 (2024): 1–17.Dheda, Dhruti, Ling Cheng, and Adnan M. Abu-Mahfouz. "Long short-term memory water quality predictive model discrepancy mitigation through genetic algorithm optimisation and ensemble modelling." *IEEE Access* 10 (2022): 24638–24658.
2. Kumar, K. Sunil Ratna, et al. "Iot and data mining techniques to detect and regulating of solar power system." *2023 International Conference on Inventive Computation Technologies (ICICT)*, IEEE, 2023. Al-Sahaf, Harith, et al. "A survey on evolutionary machine learning." *Journal of the Royal Society of New Zealand* 49.2 (2019): 205–228.
3. Darwish, Ashraf, Aboul Ella Hassanien, and Swagatam Das. "A survey of swarm and evolutionary computing approaches for deep learning." *Artificial Intelligence Review* 53.3 (2020): 1767–1812.

4. Alam, Gulzar, et al. "Applications of artificial intelligence in water treatment for optimization and automation of adsorption processes: Recent advances and prospects." *Chemical Engineering Journal* 427 (2022): 130011.
5. Ibrahim, Abdelhameed, et al. "Wind speed ensemble forecasting based on deep learning using an adaptive dynamic optimization algorithm." *IEEE Access* 9 (2021): 125787–125804.
6. Granata, Francesco, et al. "Machine learning algorithms for the forecasting of wastewater quality indicators." *Water* 9.2 (2017): 105.
7. Maier, Holger R., et al. "Introductory overview: Optimization using evolutionary algorithms and other metaheuristics." *Environmental Modelling & Software* 114 (2019): 195–213.
8. Ahmad, Zohaib, et al. "Enhancing time series forecasting with an optimized binary gravitational search algorithm for echo state networks." *IEEE Access* 11 (2023): 79466–79479.
9. Bouktif, Salah, et al. "Multi-sequence LSTM-RNN deep learning and metaheuristics for electric load forecasting." *Energies* 13.2 (2020): 391.
10. Liu, Shuangyin, et al. "A hybrid approach of support vector regression with genetic algorithm optimization for aquaculture water quality prediction." *Mathematical and Computer Modelling* 58.3–4 (2013): 458–465.
11. Yu, Jin-Won, et al. "Water quality forecasting based on data decomposition, fuzzy clustering and deep learning neural network." *Environmental Pollution* 303 (2022): 119136.
12. Tikhamarine, Yazid, et al. "Improving artificial intelligence models accuracy for monthly streamflow forecasting using grey Wolf optimization (GWO) algorithm." *Journal of Hydrology* 582 (2020): 124435.
13. Bui, Dieu Tien, et al. "Novel hybrid evolutionary algorithms for spatial prediction of floods." *Scientific Reports* 8.1 (2018): 15364.
14. Far, Saeed Banaeian, Azadeh Imani Rad, and Maryam Rajabzadeh Asaar. "Blockchain and its derived technologies shape the future generation of digital businesses: A focus on decentralized finance and the Metaverse." *Data Science and Management* 6.3 (2023): 183–197.
15. Kumar, Rohit, Bramah Hazela, and Trilok Suthar. "Crack on brick wall detection by computer vision using machine learning." *2022 6th International Conference on Electronics, Communication and Aerospace Technology*, IEEE, 2022.
16. Sit, Muhammed, et al. "A comprehensive review of deep learning applications in hydrology and water resources." *Water Science and Technology* 82.12 (2020): 2635–2670.
17. Deng, Tianan, Kwok-Wing Chau, and Huan-Feng Duan. "Machine learning based marine water quality prediction for coastal hydro-environment management." *Journal of Environmental Management* 284 (2021): 112051.
18. Pulido-Calvo, Inmaculada, and Juan Carlos Gutiérrez-Estrada. "Improved irrigation water demand forecasting using a soft-computing hybrid model." *Biosystems Engineering* 102.2 (2009): 202–218.
19. Fu, Minglei, et al. "Deep learning data-intelligence model based on adjusted forecasting window scale: application in daily streamflow simulation." *Ieee Access* 8 (2020): 32632–32651.

20. Jahandideh-Tehrani, Mahsa, Omid Bozorg-Haddad, and Hugo A. Loáiciga. "Application of non-animal–inspired evolutionary algorithms to reservoir operation: an overview." *Environmental Monitoring and Assessment* 191 (2019): 1–21.
21. Zhu, Mengyuan, et al. "A review of the application of machine learning in water quality evaluation." *Eco-Environment & Health* 1.2 (2022): 107–116.
22. Ravi, Vadlamani, Dadabada Pradeepkumar, and Kalyanmoy Deb. "Financial time series prediction using hybrids of chaos theory, multi-layer perceptron and multi-objective evolutionary algorithms." *Swarm and Evolutionary Computation* 36 (2017): 136–149.
23. El Bilali, Ali, Abdeslam Taleb, and Youssef Brouziyne. "Groundwater quality forecasting using machine learning algorithms for irrigation purposes." *Agricultural Water Management* 245 (2021): 10662.

Chapter 9

Artificial Intelligence Anomaly Detection and Root Cause Analysis

Vishal Gangadhar Puranik, T. Chithrakumar, E. Naresh, and Shruti Bhargava Choubey

9.1 Introduction

With the advent of 5G, new, highly bandwidth- and latency-sensitive use cases are becoming possible [1]. The proliferation of 5G's associated frequency bands and radio access technologies is adding complexity to radio access networks (RANs) [2]. Additional difficulties arise in the following areas while optimizing and managing a commercial multiple Radio Access Technology (RAT) network [3]. The first is that parameter setup and network architecture become more complex and adaptable with the advent of network slicing technologies [4]. The second is that quality of service (QoS) is becoming more granular, which means that a major challenge is to distribute network resources appropriately in response to the varied needs of services [5]. Thirdly, there is a growing need for more network-wide coordination due to the increasing frequency of interoperation across various networks [6]. The need to decrease the overall cost of operation and maintenance (OAM) due to massive human labour necessitates an immediate improvement in efficiency, which brings us to the fourth point [7]. Hence, to enhance the efficacy of network administration and operation, smarter system automation is necessary to substitute the conventional approaches [8].

 DOI: 10.1201/9781003479062-9

Key components of network OAM automation are anomaly detection (AD) and root cause analysis (RCA) [9]. As the first stage of OAM, anomaly detection seeks to identify the specific instances and locations of RAN performance reduction [10]. After AD, the RCA procedure suggests solutions by determining the origin of the discovered abnormalities [11].

Detection rules based on thresholds are the mainstay of conventional OAM systems for automating AD and RCA [12]. The complete process of managing anomalies is tied together by rules, which record maintenance actions and generate tickets [13]. The two current aspects of rule definition are QoS and quality of experience [14]. Key quality indicators (KQIs) that assess the experience directly from the consumers are the QoE indicators [15]. In contrast, QoS indicators are described as key performance indicators (KPIs) that assess the efficiency of a network [14–19]. To identify user experience abnormalities, KQIs are more significant; yet, KQIs do not provide much about what is causing the RAN user experience to degrade. To identify network-side abnormalities, KPIs are required [20].

The following are some of the current difficulties encountered by the conventional rule-based system:

- **Complex KPI Anomaly Is Challenging to Model:**

The KPI criteria are often unified for all cells without considering adequate case details, and they are confined to a narrow range of KPIs due to human labour constraints. As a result, complex anomalies are challenging to model using the principles. But RAN anomalies are complicated [21–23]. For instance, the issue with the air interface might arise from factors such as landscape, inappropriate configurations, user equipment, interference, and the growing number of users resulting from social actions. Some reasons may exhibit similar abnormal patterns on KPIs [24]. First, the accuracy of KPI anomaly detection is low because of the difficulty. Second, analysis becomes more difficult and automation level drops because the KPI rules typically fact to a general source. Third, the same root cause can trigger different KPI rules, which is problematic because it leads to duplication.

- **Challenging KQI Anomaly to Manage:**

KQI anomalies offer minimal information about RAN, making it difficult to treat them. One approach may be to link KPI-detected RAN abnormalities with KQI anomalies. Few KQI-KPI anomaly correlation systems are now commonly used in commercial networks.

A complex network optimization and management system that incorporates artificial intelligence (AI) is necessary to overcome these obstacles. Data centre hardware, including servers and routers, has undergone certain AD methods based on AI. More difficulties arise for the AD of RAN. For the RAN situation, AD procedures must be ubiquitous and computationally cheap, which implies that they should be able to handle a high number of cells and input characteristics, in addition to the complexity of anomaly sources. The absence of labelled data is another issue with the present state of AD/RCA verification. Despite the massive amounts of data generated from the RAN, point-wise labelling them is a challenging and costly process since RAN anomalies are complicated.

An AI-based framework, a new Alzheimer's disease (AD) procedure called SODA, and an auto-labelled cell AD data called Atomicity, Consistency, Isolation, and Durability (ACID) are presented in this research. More precise AD and a two-tiered RCA with a list of critical aberrant parameters and an explanation that a person can understand are both provided by the framework. In field testing, the suggested AI-based architecture was partially validated. When compared to the state-of-the-art (SOTA) method, the suggested approach SODA significantly decreases the time cost while marginally sacrificing performance. A KPI dataset called ACID is auto-labelled using KQI tickets.

This chapter's organization continues below. We explain the solution framework in Section 9.2. Section 9.3 describes AD algorithms. Section 9.4 covers the dataset ACID and SODA performance, while Section 9.5 presents the closed-loop validation based on O-RAN planning for inside scenarios and conclusions in Section 9.6.

9.2 Framework

9.2.1 General Structure

Figure 9.1 shows the overall structure. A two-phase RCA system, KPI anomaly detection, and KQI anomaly detection are all parts of the system. A list of the most out-of-the-ordinary KPIs is provided by phase I RCA, and the causes of each anomaly in a way that humans can understand are given by phase II RCA. Although severe KPI anomalies may potentially activate the system, KQI anomalies are the primary triggers. For instance, in the event of a KQI anomaly, which might serve as a trigger, there will be a correlation with any discovered KPI anomalies that occurred within the same period. After that, the KPI AD algorithm sorts a list of abnormal KPIs and gives an estimate of how anomalous each KPI is for each KQI anomaly; this is called phase I RCA. The second phase of RCA involves sending the abnormalities to a human-readable source. Finally, the report is prepared and sent on an as-needed basis, containing the AD results, phase I RCA, and phase II RCA. We present each component in the following subsections for closed-loop verification or OAM engineering.

9.2.2 Detecting KQI Anomalies

In the commercial network, KPIs are more common than KQIs, which are established from the user experience's point of view. Consequently, KQI rules based on thresholds are sufficient for modelling the impact of abnormalities on user experience. It is

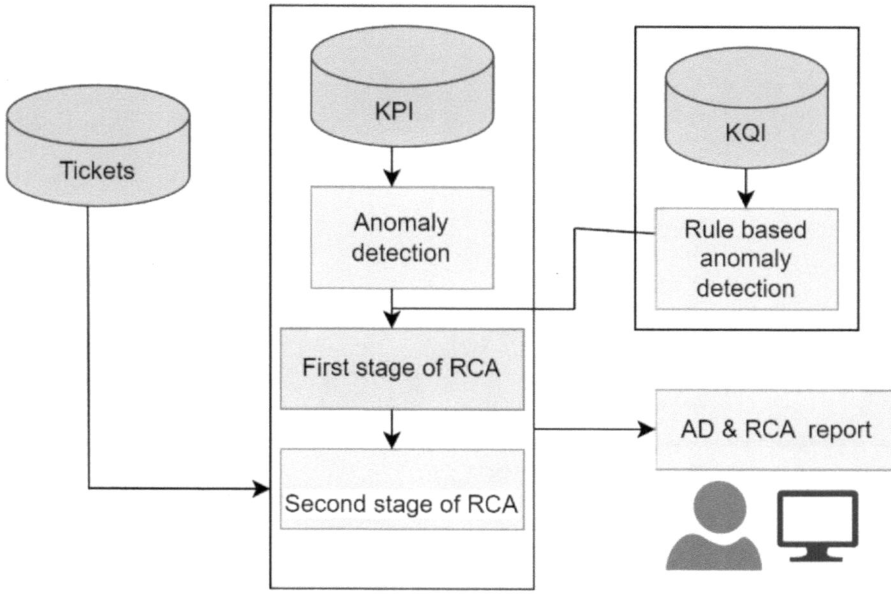

Figure 9.1 Overall structure.

not immediately clear that KQI rules that are based on thresholds have drawbacks. Therefore, the KQI anomaly detection continues to use threshold-based rules.

9.2.3 Phase I RCA and KPI Anomaly Detection

Full modelling of KPI anomalies is challenging when using threshold-based methods. Use the SODA AD approach, which is based on variational autoencoders (VAEs), instead of rules. As a bonus, SODA can provide feature-wise anomaly scores in addition to detecting anomaly points; these scores might serve as the outcome of phase I RCA. SODA receives a series of KPIs as input and returns their respective anomaly ratings. Every KPI sequence stands for a cell's condition at a certain moment and a brief interval just before that time. The first step of KPI RCA involves using the feature-wise anomaly ratings to create sorted lists of aberrant KPIs. The individual scores for each characteristic are combined to get the total score for each KPI order. The aggregate scores of each KPI order are used to determine whether the cell state, as indicated by the KPI order, is irregular based on a predetermined quantile edge.

9.2.4 Phase II Data Labelling/RCA

To produce root cause statements that a person can understand, labelled data is necessary for phase II RCA. A ticket-based labelling element handles the challenging

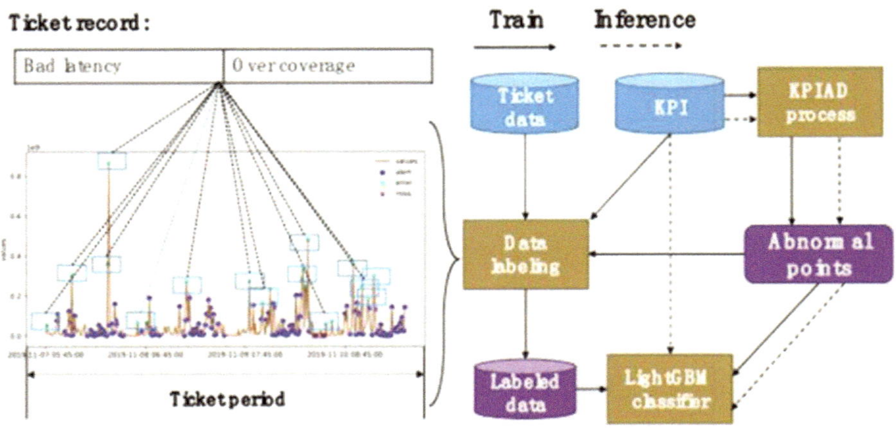

Figure 9.2 Procedure for phase II RCA.

element of phase II RCA, which is labelling. Figure 9.2 depicts the procedure for phase II RCA. During the training phase, the tickets and KPIs are both seen as data inputs. The KPI AD identifies the out-of-the-ordinary points. Each cell is labelled with a ticket record for any aberrant points that occur during the same period as the ticket. The ticket record contains many types of cause information, which are labelled the structured first-level resolution and second-level source information. The comprehensive data on the selected ticket and label in the assessment. Once the data has been labelled, the phase II RCA is approached as a standard classification technique, with an increasing tree used as the classifier. The inference phase is similar to the training phase but does not include the labelling process.

The F1 score for the second phase of the RCA may be seen in Figure 9.3. Next, they gather the KPI data for each cell in the trial area by filtering their valid KQI tickets. KPI data and KQI tickets were then used for labelling. A test set and training set are created from the tagged data. Two methods exist for splitting datasets: first, randomly selecting a training set from all the data, and second, selecting an appropriate date to divide the dataset. Since the cells in both the training and test sets are distributed randomly, the classifier may be prone to learning the difference between cells rather than the causes, leading to an extremely high F1 score. Since the cells in the testing and training datasets are drastically dissimilar in this scenario, the fact that accuracy dropped on the split-by-date dataset provided some evidence of this.

9.3 Algorithm Proposed

SODA is suggested in this study. OmniAnomaly, an unsupervised recurrent VAE-based approach, served as inspiration for SODA, a multivariate AD algorithm.

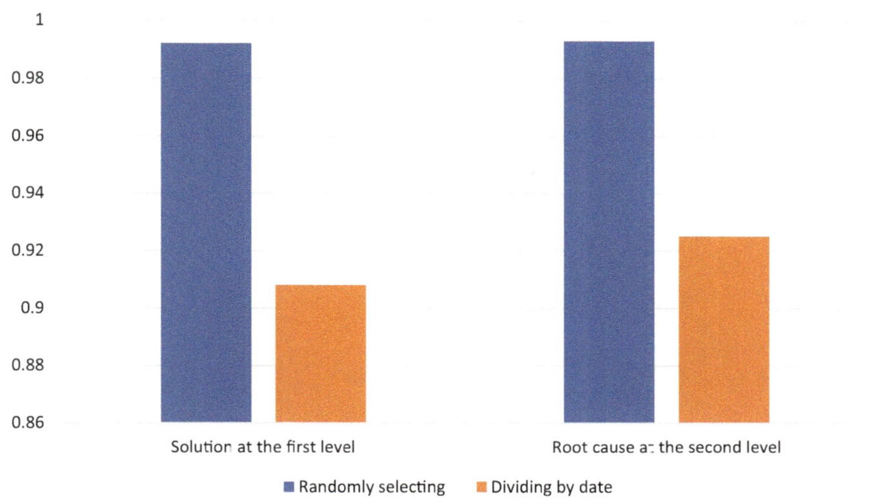

Figure 9.3 Second phase of the RCA.

OmniAnomaly's recurrent nature makes it exceedingly computationally expensive to train and infer from multidimensional sequential data, which is its biggest drawback. The significant computation time cost hinders its adoption in large-scale cell applications. SODA goes OmniAnomaly's recurrent structure in favour of a convolutional layer to process sequential input more efficiently, meeting the processing speed requirements of commercial networks.

Figure 9.4 depicts the SODA structure. Consisting of a convolutional neural network (CNN) encoder, decoder, and hidden variable z, it typically adheres to the conventional VAE architecture. The standard Gaussian distribution is assumed to be the prior distribution of z, and it is represented as $p_\theta(z)$. Let X represent the input data. The encoder network approximates the posterior distribution $p(x \mid z)$, while the decoder network approximates the posterior distribution p(z|x). The approximated posterior distributions are denoted as $p_\theta(x \mid z)$ *and* $q_\phi(z \mid x)$, with p-net representing the decoder network and q-net representing the encoder network. θ *and* ϕ represent the limits of the corresponding q-net and p-net, respectively. Consider the input information as 2D with a single station and a size of $T \times N$; each input to the model consists of a time sequence of KPIs from a cell, where T is the length of the order and N is the number of KPIs.

The SODA encoder and decoder consist of convolutional and then trans-convolutional layers. The CNN-encoder performs a downsample on the input. Subsequently, the two subsequent linear layers will generate the values μ_{z_0} *and* σ_{z_0}, which represent the mean and the standard deviation of a Gaussian distribution z_0, respectively. The input z_0 is then converted to z_K using a planar normalizing layer;

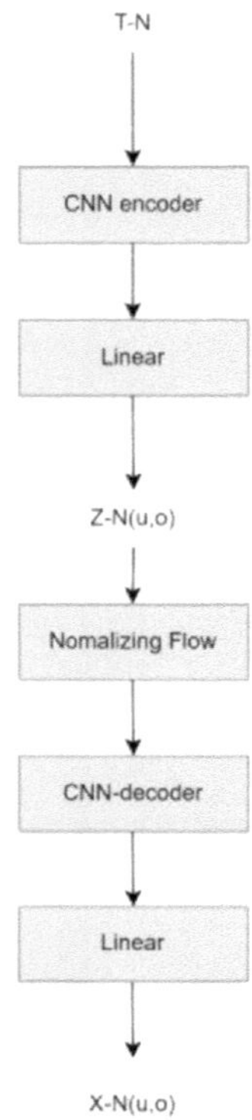

Figure 9.4 SODA structure.

it accommodates a potentially more intricate posterior distribution $q_\phi(z \mid x)$. Here, K represents the number of planar flows. Identify z as the l values and use section and timer z_K. The CNN-decoder uses $z^{(l)}$ as its input and employs a difficult-to-perform form of up-sampling on $z^{(l)}$, restoring it to the original size of the model input. The last phase involves the use of two linear layers to compute the standard deviation $\sigma_{\check{x}}$ and mean $\mu_{\check{x}}$ for the posterior distribution $p_\theta(\check{x}_T \mid z)$. This distribution is considered to be a diagonal Gaussian distribution, where flexible $\check{x}_T$ represents the renovation of the final component of the input system.

Train the network using the Stochastic Gradient Variational Bayes estimator (SGVB) and maximize the evidence lower bound (ELBO). The estimated ELBO for input $x^{(i)}$ might be expressed in the following equation using Monte Carlo estimation:

$$L^a\left(\theta,\phi,x^{(i)}\right)\sum_{l=0}^{L} - p_\theta(x_T^{(i)} \mid z^{(i)}) log\ log\ p_\theta\left(z^{(l)}\right) log\ log\ p_\theta\left(z^{(l)}\right) + log\ log\ q_\phi\ (z^{(l)} \mid x^{(i)}) \tag{9.1}$$

where $z^{(i)}$ is the lth sample and L is the total amount of samples derived from $q_\phi(z \mid x)$.

The conditional probability $p_\theta(\breve{x}_T \mid z)$ is $p_\theta(\breve{x}_T \mid z) = \sum_{j=0}^{N} p_\theta(x_T^j \mid z)$, where x_T^j represents the jth feature of the Tth input element. $p_\theta(\breve{x}_T \mid z)$ represents the anomalous score of the Tth element in the input sequence, with each KPI being considered separately. An increase in the probability of $\breve{x}_T$ reconstruction is shown by a greater $p_\theta(\breve{x}_T \mid z)$. Normal patterns are easier to learn since they appear more often and exhibit less unpredictability. Therefore, $\breve{x}_T$ with a higher probability of reconstruction is more likely to follow a normal pattern.

9.4 Algorithm and Dataset Results

To evaluate SODA on actual cellular datasets, this research builds an auto-labelled cell KPI dataset named ACID. Compared to the earlier auto-labelled root cause dataset, it is evident that ACID is an auto-labelled AD dataset. ACID is obtained from a vast number of cells, including a wide range of situations in the business network. The quantity of Radio Resource Control (RRC) networks, uplink/downlink throughput, success rate of handover, downlink Medium Access Control (MAC) block error rate, and more than a hundred other performance measures are all part of ACID and are used by OAM engineers. The chemical formula for ACID is as follows:

- Desensitized KPI data, encompassing all cells, identifiable with desensitized cell ID and time and date.
- The uniquely developed KQI rules automatically produce desensitized KPI data and anomaly labels to cells that have been filtered using KQI criteria. For “normal KQI” the label is 0, whereas for “abnormal KQI” it is 1.

SODA, the suggested AD algorithm, is evaluated on the Surface-Mount Device (SMD) dataset and the second half of the ACID dataset. Isolation Forest (IF) and the recurring technique, Omni Anomaly, are both utilized in the process of anomaly detection. In reality, anomalies often develop over time and occur continually, creating a section of anomalies. So, to determine false or true on tested datasets, use a point-adjust approach.

When evaluating the actual anomalous group, if any abnormality is observed, the section is considered to be appropriately recognized and all data inside it is considered to be correct. The metrics used are the highest F1 scores. To get the same outcome for SODA and IF on SMD, it is recommended to use the OmniAnomaly on SMD rather than engaging in training or seeking the ideal F1 score on each cell. Train and search for the ideal edge on every mechanism so that their behaviour is

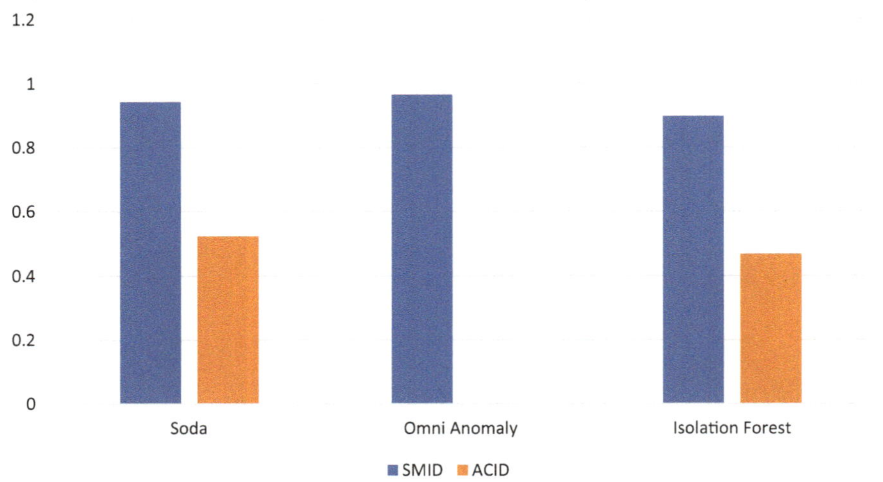

Figure 9.5 Outcomes of the comparison.

consistent with ACID. Rather, search for the greatest F1 score once across the whole dataset and train the system to better match the circumstances.

The outcomes of the comparison are shown in Figure 9.5. Compared to the recurrent approach, SODA performs somewhat worse on the SMD dataset. However, it outperforms the IF algorithm on both the SMD and ACID datasets.

Figure 9.6 displays a comparison of the three methods' computational efficiency on the SMD dataset. Time spent computing during training and inference is a measure of computational efficiency. OmniAnomaly and SODA are both epochs that have been gained. IF is configured to make use of all available CPU cores for both training and inference. Compared to OmniAnomaly, the suggested approach is 32 times quicker during training and 15 times quicker during inference processing. Because there are fewer parameters and a much more parallel structure than in recurrent structure, this is the result. Although SODA is slower in training and inference than the ensemble method IF, it outperforms IF since it can handle sequential input. When compared to IF, SODA may also provide feature-wise anomaly scores. SODA can be used on large-scale data due to its cheap computation time. The best estimates put the number of cells in the testing area—the central business district of a provincial capital—at about a dozen. SODA can train cells with only one Graphics Processing Unit (GPU), whereas the recurrent method needs four GPUs to do the same amount of work.

9.5 Loop-Closure Verification

O-RAN, an acronym for open and clever RAN, is considered a crucial architecture for the integration of AI with RAN in both 5G and 4G systems. Hence, the

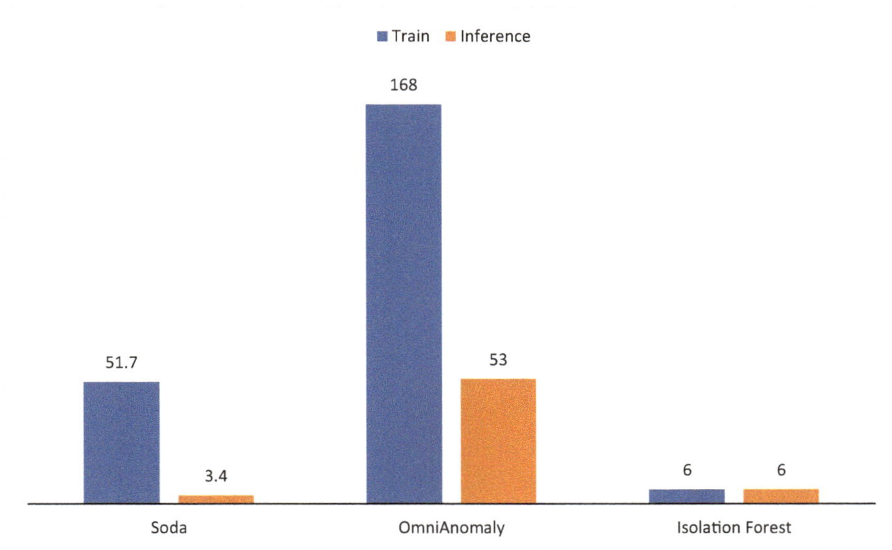

Figure 9.6 Comparison of the three methods of computational efficiency.

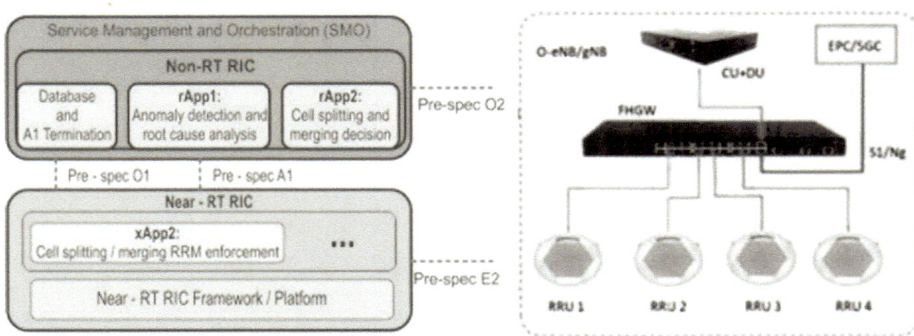

Figure 9.7 System design of the verification of the closed-loop application.

loop-closure verification, which incorporates the suggested anomaly detection (AD), RCA, and cell merging and splitting solution, to address capacity issues, has been successfully tested at many base station locations using the O-RAN architecture.

9.5.1 The Structure of the Network

Figure 9.7 illustrates the system design of the closed-loop applications authentication, which adheres to the stated O-RAN design. The intelligent component of the RAN comprises the Non-Real-Time RAN Intelligent Controller (NRT RIC) and the Non-Real-Time RAN Intelligent Controller (NRT RIC). The NRT RIC, which offers improved information for network management, is often installed on the same level as conventional network management systems. To provide smarter scheduling and control at a finer level, the nRT RIC is located nearer to the base classes. The new RIC can get wireless data reports and control capabilities in near-real-time over the E2 interface. The equipment being referred to is the indoor dispersed Pico bases, which are composed of a baseband unit, a fronthaul entry, and a remote receiver unit. The O1 interface is the operation and management link between the SMO system and the O-RAN system operations. Connecting the NRT RIC in the SMO to the nRT RIC for RAN optimization is the A1 edge. The field trial's proof-of-concept uses pre-specification versions of the O1, A1, and E2 interfaces.

This general-purpose server hosts both NRT RIC and nRT RIC. An application that is organized in the container type is NRT RIC RCA and anomaly detection. The anomaly report is generated by an application that uses KPI values gathered by the network management system at a 1-minute granularity to pinpoint the source of the problem. Through the graphical user interface, the anomaly report may be displayed for the OAM experts to review. Cell merging and splitting decision applications will be notified if anomaly detection and root cause investigation indicate a capacity concern. Collaborative efforts between NRT RIC, eNB/gNB, and NRT RIC are carried out during the Pico RRU decision and subsequent UE migration by the cell split or merge decision from NRT RIC. To make better decisions about cell splitting and merging, RCA and anomaly detection are used continuously to monitor performance. The design has confirmed the automated closed-loop for the capacity issue.

9.5.2 *The Outcome of the Verification*

The 4G network is used to confirm the performance in the first phase. The location is now undergoing the deployment of 5G equipment. The outcome demonstrates that the performance decrease was noticed by the AI-based anomaly detection system before the ticket threshold was triggered. The proper source for the capacity issue was revealed by the remedy. The number of tickets has been reduced with the aid of the closed-loop method. In the meantime, both the downlink rate and the efficacy rate of establishing RRC connections have been enhanced. There was absolutely no human error involved.

9.6 Conclusion

Automating networks relies heavily on the new AI-based AD and RCA methods. This study presents an AD algorithm called SODA and explains the AD and RCA architecture that was employed in field testing. When compared to the SOTA method, SODA has been shown to have significantly reduced computational cost while only experiencing a tiny decrease in performance. As a result, SODA may be used in a large-scale system. To assess the procedure's effectiveness on a dataset that involves big entities and various attributes, namely in an actual cellular situation, construct a dataset including over 10,000 cells. The KQI tickets are used to loosely label the dataset. Additionally, many O-RAN base station sites have validated the automated network closed-loop approach for capacity problems, which includes the planned RCA and AD. In the future, there are still certain parts that might need some improvement. Using weak-supervised techniques or a finer mapping process might enhance the present ticket-labelling strategy. Similarly, a new solution based on natural language processing (NLP) could improve the extraction of unformatted root cause information.

References

1. Rahat, Irfan Sadiq, et al. "A step towards automated haematology: DL models for blood cell detection and classification." *EAI Endorsed Transactions on Pervasive Health and Technology* 10 (2024): 1–14.
2. Kumar, T. Santhosh, et al. "Enhanced performance of solar PV array-based machine drives using zeta converter." *2020 IEEE international conference on advances and developments in electrical and electronics engineering (ICADEE)*. IEEE (2020).
3. Steenwinckel, Bram, et al. "FLAGS: A methodology for adaptive anomaly detection and root cause analysis on sensor data streams by fusing expert knowledge with machine learning." *Future Generation Computer Systems* 116 (2021): 30–48.
4. Abdelrahman, Osama, and Pantea Keikhosrokiani. "Assembly line anomaly detection and root cause analysis using machine learning." *IEEE Access* 8 (2020): 189661–189672.
5. Abdelkefi, Atef, Yuming Jiang, and Sachin Sharma. "SENATUS: An approach to joint traffic anomaly detection and root cause analysis." *2018 2nd cyber security in networking conference (CSNet)*. IEEE (2018).
6. Soldani, Jacopo, and Antonio Brogi. "Anomaly detection and failure root cause analysis in (micro) service-based cloud applications: A survey." *ACM Computing Surveys (CSUR)* 55.3 (2022): 1–39.
7. Mdini, Maha. Anomaly detection and root cause diagnosis in cellular networks. Diss. Ecole nationale supérieure Mines-Télécom Atlantique (2019).
8. Carletti, Mattia, et al. "Explainable machine learning in industry 4.0: Evaluating feature importance in anomaly detection to enable root cause analysis." *2019 IEEE international conference on systems, man and cybernetics (SMC)*. IEEE (2019).
9. Mobtahej, Pooyan, et al. "Anomaly detection by employing root cause analysis and machine learning-based approach using compressors time-series data." *IIE annual conference. Proceedings*. Institute of Industrial and Systems Engineers (IISE) (2022).
10. Brito, Lucas C., et al. "An explainable artificial intelligence approach for unsupervised fault detection and diagnosis in rotating machinery." *Mechanical Systems and Signal Processing* 163 (2022): 108105.
11. Dix, Marcel, et al. "Anomaly detection in the time-series data of industrial plants using neural network architectures." *2021 IEEE seventh international conference on big data computing service and applications (BigDataService)*. IEEE (2021).

12. Sauvanaud, Carla, et al. "Anomaly detection and root cause localization in virtual network functions." *2016 IEEE 27th international symposium on software reliability engineering (ISSRE)*. IEEE (2016).
13. Bogatinovski, Jasmin, et al. "Artificial intelligence for its operations (oops) workshop white paper." *arXiv preprint arXiv:2101.06054* (2021).
14. Zhang, Chuxu, et al. "A deep neural network for unsupervised anomaly detection and diagnosis in multivariate time series data." *Proceedings of the AAAI conference on artificial intelligence*. Vol. 33. No. 01 (2019).
15. Du, Min, et al. "Deeplog: Anomaly detection and diagnosis from system logs through deep learning." *Proceedings of the 2017 ACM SIGSAC conference on computer and communications security* (2017).
16. Papageorgiou, Konstantinos, et al. "A systematic review on machine learning methods for root cause analysis towards zero-defect manufacturing." *Frontiers in Manufacturing Technology* 2 (2022): 972712.
17. Du, Min, et al. "Deeplog: Anomaly detection and diagnosis from system logs through deep learning." *Proceedings of the 2017 ACM SIGSAC conference on computer and communications security* (2017).
18. Anello, Eugenia, et al. "Anomaly detection for the industrial internet of things: An unsupervised approach for fast root cause analysis." *2022 IEEE conference on control technology and applications (CCTA)*. IEEE (2022).
19. Thottan, Marina, and Chuanyi Ji. "Anomaly detection in IP networks." *IEEE Transactions on Signal Processing* 51.8 (2003): 2191–2204.
20. Huang, Zhengxing, et al. "On local anomaly detection and analysis for clinical pathways." *Artificial Intelligence in Medicine* 65.3 (2015): 167–177.
21. Nikula, Miro-Markus. "Machine learning-based anomaly detection and root cause analysis in mobile networks." (2024).
22. Böhmer, Kristof, and Stefanie Rinderle-Ma. "Mining association rules for anomaly detection in dynamic process runtime behavior and explaining the root cause to users." *Information Systems* 90 (2020): 101438.
23. Danti, Piero. "Application of artificial intelligence for anomaly detection in energy generation systems." (2024).
24. Liao, Yaohua, et al. "Research on line loss analysis and intelligent diagnosis of abnormal causes in distribution networks: Artificial intelligence-based method." *PeerJ Computer Science* 9 (2023): e1753.

Chapter 10

Artificial Intelligence and Internet of Things-Based Intelligent Scheduling for Load Distribution in Power Grids

Selvam Suganya, Mukesh Soni,
Sushilkumar Dangiya, Maher Ali Rusho, and
Sagar Dhanraj Pande

10.1 Introduction

In the distribution network, coordinated dispatch of generation, grid, load, and energy storage involves various interactions to enhance the power dynamic balance capability of the power system. This improves the economic, efficient, and secure operation of the grid, thereby maximizing the utilization of energy resources. With the growing severity of environmental pollution and energy crisis, new technologies such as renewable energy generation, electric vehicles, and microgrids have emerged. The development of these technologies has brought significant changes to traditional distribution networks [1]. Consequently, the operational architecture of distribution networks is facing substantial changes, and traditional dispatch mechanisms for

DOI: 10.1201/9781003479062-10

generation, grid, load, and storage may no longer be efficient. Therefore, researching new distribution network dispatch mechanisms is essential [2].

Traditional Distribution Network Architecture and Challenges: The traditional distribution network architecture has been impacted by the construction of the Ubiquitous Power Internet of Things (UPIoT) [3]. Simultaneously, new power electronic technologies like microgrids, distributed generation (renewable energy), electric vehicle charging facilities, and energy storage have been widely applied in distribution networks, challenging the passive one-way power supply mode. To address this, the concept of an active distribution network was designed. However, due to data isolation and limited perception capabilities, the effectiveness of coordinated dispatch in an active distribution network is limited [4].

UPIoT: UPIoT refers to the application of ubiquitous Internet of Things (IoT) technologies in the power grid. The rapid development of IoT in recent years has made UPIoT implementation possible. Ubiquitous IoT means the interconnectedness and interaction of information between any time, location, person, and object. UPIoT achieves the interconnection between various power system operations and human–machine interactions through advanced technologies such as edge computing, artificial intelligence, and blockchain. This enables the exploration of the latent value of data. UPIoT's foundation lies in information acquisition, emphasizing real-time interaction and connection between users and the power grid. It involves the real-time high-speed transmission, collection, and analysis of massive data to enhance the grid's intelligent processing and decision-support capabilities. Overall, UPIoT brings significant benefits to coordinated dispatch of generation, grid, load, and storage in distribution networks by breaking data barriers and involving users more in power scheduling. From a data flow perspective, it forms a natural distributed intelligent scheduling system [5,6].

Challenges and the Need for New Dispatch Mechanisms: Under the context of UPIoT construction, the existing coordinated dispatch mechanisms for generation, grid, load, and storage cannot meet the new requirements. For example, traditional coordinated dispatch does not permit the implementation of active management and control on the distribution network side to accomplish peak pruning and valley filling. Nevertheless, the elimination of data barriers facilitates enhanced interactivity in coordinated dispatch, thereby providing improved technological assistance in the development of novel dispatch mechanisms. The purpose of this chapter is to investigate a UPIoT-based distributed state-aware coordinated dispatch of generation, load, grid, and storage to reduce additional power intelligent scheduling costs, valley filling, peak trimming, and valley filling.

This approach aims to enhance the stability, security, and cost-effectiveness of electricity scheduling, which is both groundbreaking and significant. Under the UPIoT framework, the purpose of this chapter is to propose a new state-aware coordinated dispatch mechanism for generation, grid, demand, and storage. With the objectives of attaining peak pruning, valley filling, and cost reduction in power scheduling, this approach optimizes energy resource utilization and enhances power system balance.

Future Trends and Objectives of Research: In light of the present state of distribution networks and the UPIoT framework's massive changes in distribution network topology, the investigation of novel dispatch mechanisms is an unavoidable development. Before examining the implications of UPIoT on distribution networks, this chapter conducts an analysis of the present state of distribution networks in light of the objectives and prerequisites of UPIoT development and the optimization of critical technologies in the new distribution network architecture. Then, it proposes a distributed state-aware coordinated dispatch of generation, grid, load, and storage by analysing the distribution network system architecture, establishing a system model, and designing the dispatch mechanism. The objective is to achieve peak shaving, and valley filling, and reduce additional power intelligent scheduling expenses. Finally, the chapter presents a simulation experiment based on real data from the grid, which validates the effectiveness of the coordinated dispatch mechanism.

The contribution of this work lies in addressing the challenges posed by the evolving distribution network architecture under the influence of the UPIoT. The traditional passive one-way power supply mode is no longer sufficient to accommodate new technologies like renewable energy generation, electric vehicles, and microgrids. As a result, the concept of an active distribution network has been introduced, but its effectiveness is hindered by data isolation and limited perception capabilities.

10.2 Related Work

Currently, scholars have researched the interaction and coordination of generation and load. In Reference [7], a large-scale friendly interaction system between generation and load was experimentally studied. The chapter introduced various new power technologies such as dynamic frequency calculation to provide technical means and effective analysis methods for system debugging experiments, thereby optimizing the control of loads in the distribution network. Reference [8] focused on considering the emergence of new demands on the load side, specifically demand-side resource response compatibility, and proposed an optimized dispatch model for generation, grid, and storage. The case analysis results demonstrated that this model effectively reduces the overall system cost and enhances the safety and stability of system operation. The load control strategy adopted in this model is primarily based on price adjustments, which may be subject to some subjective influence from users. Additionally, the system cost referred only to the economic aspect, without considering energy losses or the integration of new energy sources. Therefore, further research could be conducted to address these aspects. Subsequent studies could consider unifying these two aspects by applying load reality technology to enhance the cost-effectiveness of grid operation. In the context of the Ubiquitous Power Internet of Things, the matching of user electricity consumption and electricity generation could also be further explored.

Regarding interactive research on energy systems, most studies focus on managing various energy sources, especially new energy, in the context of the Energy Internet to optimize energy utilization. Reference [9] considers a coordinated and optimized control structure for a flexible EC distribution network with multiple complementary energy sources. The chapter primarily focuses on the coordinated operation of multiple energy sources and considers the integration of new energy and various distributed generation devices. By employing a discrete-time consensus algorithm, the chapter achieves multi-objective optimal control to minimize the economic expenses of the entire grid. However, this chapter primarily considers multiple complementary energy sources from the source end, while integrating the state of energy storage devices to meet load demands without adjusting the loads themselves. Therefore, it is worth considering incorporating load adjustment into the coordinated control to further optimize grid operation.

Furthermore, due to the increasing demand for grid regional autonomy, many researchers have studied the distribution network dispatch architecture and mechanisms in the substation area. Some scholars have studied the energy management systems of area Energy Internet. Reference [10] proposed an effective energy management system and organized various mechanisms for multiple energy sources to complement each other through the Internet. Reference [11] conducted a characteristic analysis of various energy sources and loads and concluded that the information control system is the core of the Energy Internet. References [10,11] each proposed their system structures for area energy management but did not provide corresponding control strategies and plans. Nevertheless, they still offer a foundational architecture for researching coordinated dispatch of generation, grid, load, and storage under the context of the Ubiquitous Power Internet of Things. Reference [11] provided an overview of the cooperative control methods for an energy system in coastal areas with specific loads (desalination loads). It summarized and looked ahead at relevant research on generation–grid–load systems and explained the interaction between generation, grid, and load systems. It also summarized the methods for coordinating control and multi-time scale energy intelligent scheduling in coastal multi-source and multi-load systems, providing direction for defining and optimizing dispatch models for generation–grid–load systems [11–13]. The above research studies and overviews on microgrid system load scheduling, regional autonomy methods, and technologies provide valuable references for constructing the system framework of distribution networks [14,15].

There is another group of studies that focus on optimization dispatch algorithms or models based on specific objectives. In Reference [16], the perspective of investment operators is taken into account, proposing a basic method for coordinated optimization and operation of generation, grid, load, and storage in the Energy Internet.

This method involves four steps: basic condition analysis, system planning, system operation, and overall evaluation [17]. However, the chapter does not provide a specific and comprehensive solution, and its optimization objectives mainly consider economic costs while neglecting the impact of energy losses [18,19]. Reference [20] constructs an active distribution network optimization dispatch model, which is a multi-objective optimization model aiming to minimize network losses, maximize user satisfaction, and optimize renewable energy utilization [21]. This study narrows down the search range for the optimal value of the objective function as it does not consider optimizing power flow by changing switch connections [22]. Additionally, it overlooks the characteristics of unstable and fluctuating renewable energy output, treating renewable energy as a controllable discrete variable. Various research studies address different optimization objectives in different regions [23–26].

In summary, the aforementioned literature explores various aspects of coordinated output from the source end, demand-side adjustments of flexible loads, microgrid system architecture, and coordinated optimization dispatch models for generation, grid, load, and storage. These studies have provided some directions for solving specific issues and also laid the groundwork for subsequent research specifically:

1. Different dispatch architectures should be adopted based on the characteristic electricity consumption and generation of different regions. The real-time and accurate control of coordinated optimization results in generation, grid, load, and storage also relies on the dispatch architecture. Thus, further research on the coordinated dispatch architecture is still highly meaningful.
2. Considering the different optimization objectives for regions with distinct electricity consumption and generation characteristics, such as economic costs, energy losses, and renewable energy integration, there are still many optimization and dispatch methods worth studying or improving.
3. Current research on the interaction and dispatch of various components in generation, grid, load, and storage, such as generation to generation, generation to the grid, grid to load, and generation to load, mainly focuses on one or two of these aspects. Future research should aim to integrate these aspects to establish a more comprehensive coordinated dispatch model for generation–grid–load–storage.

Existing works have not considered the background of the UPIoT construction. However, the new scenarios of dispatch mechanisms can be inspired by the existing research. Therefore, starting from the third point mentioned above, this chapter designs a comprehensive, UPIoT-based, distributed state-aware coordinated dispatch mechanism for generation, grid, load, and storage.

10.3 Design of the Coordinated Dispatch Mechanism for Generation, Grid, Load, and Storage

To design a coordinated dispatch mechanism for generation, grid, load, and storage under the context of the UPIoT, this section first analyses the distribution network architecture under UPIoT to construct a reasonable distribution network system. Then, with the objectives of peak shaving and valley filling, as well as reducing additional power intelligent scheduling expenses, mathematical models are developed for the elements in the system. Finally, referring to existing optimization dispatch methods [27–30], an optimization dispatch is designed to meet the current scenario's requirements; incorporating UPIoT demands to achieve a coordinated dispatch mechanism for generation, grid, load, and storage.

10.3.1 Analysis of the Active Distribution Network under UPIoT

As shown in Figure 10.1, the components and their connections in the distribution system under UPIoT can be roughly illustrated as follows: generation, grid, load, and storage are represented as separate entities, with the source indicating distributed power sources, including traditional power plants and new energy sources, responsible for primary power supply in the distribution network system. Energy flow represents the electricity flow among generation, grid, load, and storage. According to the definition of UPIoT, data barriers are broken, and all data is highly shared. The information flow in the figure demonstrates the high-level coordination among

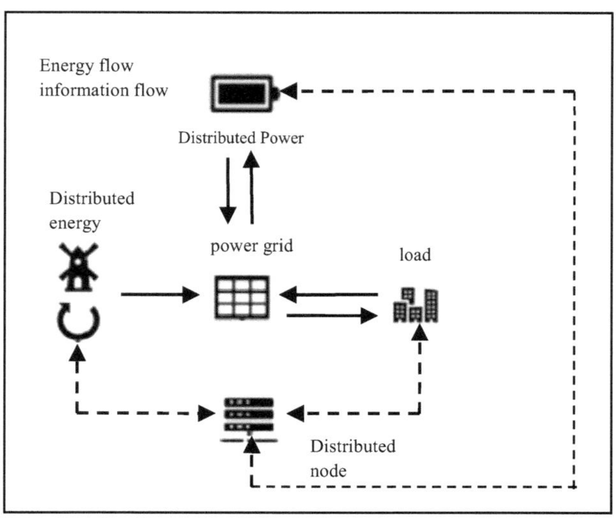

Figure 10.1 Illustration of the distribution system under UPIoT.

generation, grid, load, and storage under UPIoT. The distribution node aggregates data generated by all distributed power devices, and based on upper-level application requirements and grid security and stability demands, it carries out coordinated dispatch.

According to the definition of intelligent perception nodes in UPIoT, the nodes possess both data acquisition capabilities and the ability to adjust voltage, current, and power based on upper-level instructions. Therefore, the coordinated dispatch mechanism designed in this chapter is a platform layer operating within a distributed power system under UPIoT. It utilizes data acquired from the UPIoT perception layer and combines it with other applications, such as new energy generation forecasting and orderly electric vehicle charging and discharging management, to optimize and dispatch the current power resources. The objective is to achieve peak shaving, and valley filling, and reduce additional power intelligent scheduling losses.

10.3.1.1 Significance and Methods of Peak Shaving and Valley Filling

Peak shaving and valley filling are measures to adjust electricity consumption loads. The purpose is to decrease electricity consumption during peak load periods and fill the load gap during low periods, thereby reducing the peak-to-valley difference in electricity consumption on the power grid. The significance of these measures lies in achieving a balance between power generation and consumption, improving load rates, reducing investment in power generation units, and ultimately ensuring efficient utilization of energy resources.

Currently, there are two main methods adopted in the power grid. One approach is to implement policies at the user end (load side), such as time-of-use pricing and incentivizing load shifting, to encourage users to voluntarily shift their electricity consumption away from peak periods. However, this method has its limitations in terms of effectiveness. The other approach involves coordinated dispatch of flexible loads and energy storage devices. During peak periods, the devices discharge, and during low-load periods, they charge. This method is commonly used in active distribution networks.

Under the construction of UPIoT, the breaking of data barriers allows both methods, i.e., policy-based load shifting and coordination of flexible and controllable loads, to work together. The dispatch mechanism designed in this chapter adopts this approach, where both flexible and controllable loads participate in coordinated dispatch. On the load side, based on policy measures, users' electricity consumption intentions are obtained. The distribution node then receives real-time user intentions from the cloud, calculates the load's controllable range, and subsequently performs coordinated dispatch in combination with other controllable variables. In theory, this approach is more effective than separately adopting the two traditional methods.

10.3.1.2 Methods to Reduce Additional Power Intelligent Scheduling Losses

In the distribution network, losses mainly consist of line losses and network losses. Line losses are primarily determined by the distribution network's network structure and supply voltage level, representing losses generated in the lines during the generation, transmission, and consumption stages. Network losses, on the other hand, encompass the total losses resulting from line losses and losses incurred by various substations and compensation devices in the entire power grid. The purpose of power intelligent scheduling is to balance loads and ensure the stable and secure operation of the power grid.

Imbalanced three-phase loads can lead to increased line and network losses. Therefore, during the intelligent scheduling process, it is necessary to adjust the power output of generators to achieve load-voltage balance. The losses generated during this adjustment process are termed additional losses, which are incurred by the transformer and compensator at each node in the distribution network during the power and voltage adjustments. As a result, the larger the difference in power and voltage adjustments during scheduling, the greater the additional losses. Traditional methods to reduce additional power intelligent scheduling losses mainly involve research and development improvements to existing distribution and power equipment or considering the characteristics of various equipment when designing the distribution network architecture; both approaches are costly and challenging to implement. Hence, this chapter leverages the advantages of UPIoT's perception layer and network layer to address the issue of power intelligent scheduling losses by minimizing the power adjustment values during the intelligent scheduling process.

10.3.2 Modelling of the Coordinated Dispatch System under UPIoT

The intelligent scheduling architecture proposed in this chapter adopts the centralized architecture suggested in Reference [31], where all data information is aggregated and processed at a central distribution node. For the sake of research convenience, the scope of intelligent scheduling is specific to a distribution area, as shown in Figure 10.2, which represents a simplified ring-shaped distribution network.
Now, let's define all nodes in the distribution network:

Distributed Generation: This represents the collection of all distributed energy sources, including new energy generation nodes such as photovoltaic, wind power, and other controllable generation nodes like fuel cells in the system. This chapter considers photovoltaic and wind power generation nodes, utilizing full power tracking to meet the demand for new energy consumption. The power values of other types of generation nodes are adjusted based on optimization results and combined with distributed

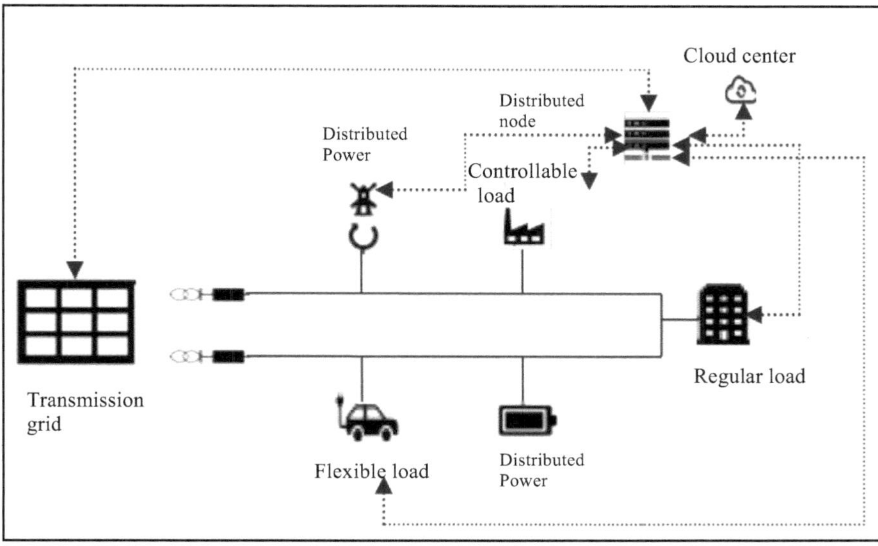

Figure 10.2 Simplified distribution network architecture.

generation output predictions to provide power adjustment ranges for the next intelligent scheduling cycle.

Loads: Loads are divided into two categories: controllable loads, whose power can be adjusted within a certain range, and conventional loads, such as regular users or factories in urban areas that have autonomous power consumption. In the system, controllable loads are monitored to obtain their current power values and minimum power limits. Meanwhile, predictions for conventional loads affected by electricity prices are obtained (considering active interactions between users and the power grid under UPIoT). Conventional loads act as the main follow-up variable.

Distributed Energy Storage: A collection of multiple distributed energy storage nodes, where the total capacity of energy storage devices significantly impacts the effectiveness of peak shaving and valley filling. This chapter assumes that the distribution area adopts an optimal energy storage device addressing and sizing strategy.

Flexible Loads: These refer to loads that can be charged and discharged. The chapter provides an example of charging station devices equipped with electric vehicles. During scheduling, charging and discharging plans, along with power adjustment ranges for each intelligent scheduling cycle, are obtained in collaboration with user intentions (obtained from relevant UPIoT business applications).

Transmission Grid: This is the main power supply component in the distribution area. The main monitoring data in the power grid are active power, reactive power, voltage, and current. After thorough investigation and research, this chapter selects active power during grid operation as the variable of interest for adjustments.

The intelligent scheduling objectives are to reduce additional power intelligent scheduling losses and perform peak shaving and valley filling. Both objectives are positively correlated. When there is minimal fluctuation in grid output, resulting in no significant peak-to-valley differences, the corresponding additional losses are small, as the losses mainly arise from active power adjustments at substations and power line transmission losses. Line losses depend on active power magnitude and are unavoidable, while the losses induced by active power adjustments at substations increase with the magnitude of adjustments. The primary manifestation of load peak and valley lies in substantial fluctuations in active power at loads within the distribution network. Hence, selecting active power as the variable object for the mathematical model is reasonable. The research objective of this chapter is to minimize variations in active power throughout the entire distribution network.

10.3.2.1 Mathematical Models

In this section, mathematical modelling descriptions will be provided for all the mentioned objects. For ease of description, all power values mentioned later refer to active power, and the active power values obtained during intelligent scheduling from each node are valid, i.e., power values after deducting self-loss and network adjustment loss from the generated power or load consumption. This convention ensures that the data processed during intelligent scheduling is actual power values within the distribution network, making adjustments more precise. Existing research has shown that intelligent sensing devices under UPIoT can perform these operations [35, 36]. Below are definitions for the mathematical models appearing in the collaborative intelligent scheduling of sources, grids, and storage.

1. EG = {Type, Q_{EG}min, Q_{EG}max, Q_{EG}}: Used to describe distributed generation. Here, Type = 0 or 1. When Type = 0, it represents clean energy generation nodes such as photovoltaic or wind power, and Q_{EG} denotes the forecasted output for the next intelligent scheduling cycle. When Type = 1, it represents other distributed generation nodes like fuel cells with adjustable power. Q_{EG}max and Q_{EG}min indicate the upper and lower power limits for adjustment.
2. DES = {ToC, ε, R, Q_{char}max, Q_{dis}max}: Used to describe distributed energy storage. ToC takes values of −1, 0, or 1, representing the current state of the energy storage device. ToC = −1 indicates that the node can only charge, ToC = 0 indicates it can both charge and discharge, and ToC = 1 indicates it can only discharge. ε = R/Rc represents the remaining energy capacity of the node,

with Rc being the full energy capacity. Q_{cha}max and Q_{dis}max are the power limits during charging and discharging, respectively.

3. TC = {State, Q_{TC}min, Q_{TC}max}: Used to describe flexible loads. It is assumed that all flexible load SC represents a collection of charging stations equipped with electric vehicles (in practice, they are divided into regions, with one region corresponding to one collection). The state takes values of –1, 0, or 1, representing the willingness for the next intelligent scheduling cycle. State = –1 indicates that the user represented by the node can only charge during this period, with Q_{TC}max and Q_{TC}min denoting the upper and lower limits of the charging power they are willing to accept. State = 1 indicates that the vehicle owner is willing to participate in discharging and has sufficient battery capacity. Q_{TC}min and Q_{TC}max represent the adjustable power range during discharging. Charging stations that cannot participate in discharging but are willing to shift charging time are represented with State = 0. Q_{TC}min and Q_{TC}max indicate the adjustable power range during charging. These stations' electric vehicles do not have sufficient capacity for discharging but are open to shifting their charging time, which plays an important role in peak shaving and valley filling, reducing overall load during peak periods and increasing overall load during off-peak periods.
4. DL = {Q_D, $Q_D$min} is used to describe controllable loads. These nodes represent loads whose power can be adjusted, such as central air conditioning. Here, QD represents the current operating power level, and $Q_D$min represents the minimum limit of adjustable power.
5. Load = {Q_{Load}} is used to describe conventional loads, which refers to the total power consumption of all conventional loads. In the intelligent scheduling process, this value is the load forecast for the next intelligent scheduling cycle [31–33].
6. Grid = {Q_G} is used to describe the total power supply from the higher-level grid or other active distribution networks in the region.
7. The total line losses in the distribution network are denoted by Q_{Loss}.

10.3.2.2 Objective Function

The purpose of this intelligent scheduling mechanism is to achieve peak shaving and valley filling while reducing additional power intelligent scheduling losses. These two objectives are positively correlated. Below are several assumptions made for the application scenario of this intelligent scheduling mechanism.

Firstly, the intelligent scheduling architecture employs a centralized control approach, and the intelligent scheduling operates in a specific distribution region where the elements mentioned in Section 10.3.2.1 may be partially or entirely present.

Secondly, intelligent scheduling divides one day into equal-length intelligent scheduling periods. Within each intelligent scheduling period, all components have their operational values and forecasts (requiring collaboration between nodes and the cloud centre). This means that each node reports accurate measurements and an effective power adjustment range derived from predictive techniques [2] or power output plans. Based on this data, the distribution nodes optimize the system and send back the intelligent scheduling results for appropriate adjustments by each node, with the intelligent scheduling taking place before the next operational period.

During the intelligent scheduling process, the power output PG from the grid (cloud centre) and Q_{EG} from EG are summed up to represent the total load. The objective of this intelligent scheduling mechanism is to minimize the fluctuations in the total load, i.e., $Q_G + Q_{EG}$, in the distribution network. Therefore, the entire collaborative intelligent scheduling process becomes an integration of quadratic programming problems with mixed constraints and resource-intelligent scheduling problems.

To achieve the minimum power intelligent scheduling losses and peak shaving objectives, this study considers minimizing the total power fluctuation in the distribution network for each intelligent scheduling period. This can be mathematically expressed as follows:

$$\min(Q)^2 = \min\left(Q_{\text{new}} - Q_{\text{now}}\right)^2 \tag{10.1}$$

Among them, Q is the value of $Q_G + Q_{EG}$, and ΔQ represents the active power difference between the new cycle and the current cycle distribution network during dispatching.

10.3.2.3 Constraints

(1) Grid Power Balance and Power Flow Constraints: The condition for the safe and stable operation of the entire distribution network is that the generated power is equal to the consumed power, that is, energy conservation, that is:

$$Q_{DES} + Q_{SC} + Q_{DL} + Q_{\text{Load}} + Q_{\text{Loss}} = Q_{EG} + Q_G \tag{10.2}$$

Among them, PG is the access power of the grid; Q_{DL} is the total value of energy connected to all nodes of the controllable load part; Q_{Load} is the total value of the conventional load; Q_{Loss} is the sum of the power loss of the entire distribution network line; Q_{DES} is the sum of the power of all energy storage nodes, the positive value represents the charging power, and the negative value means the discharge power.

(2) EG Output Range: In the EG part, for each EG_i, the upper limit of i is the total number of EG nodes. EG Type=0 means that the node is a new energy generation node such as a photovoltaic or wind turbine. Q_{EG_i} is the predicted output value of the next dispatch cycle, which is a constant item.

$$Q_{EG_i}\min \leqslant Q_{EG_i} \leqslant Q_{EG_i}\max \tag{10.3}$$

(3) DES States and Output Range: In the DES part, for each DES_i where i is within the limit of the total number of DES nodes, obtain each DES node's charging and discharging states. The effective power output range in the discharging state is given by

$$-Q_{\text{dis}}\max \leqslant Q_{DES_i} \leqslant 0 \tag{10.4}$$

The effective power consumption range in the charging state is given by

$$0 \leqslant Q_{DES_i} \leqslant Q_{\text{char}}\max \tag{10.5}$$

For nodes that can be both charged and discharged, their power adjustment range is given by

$$-Q_{\text{dis}}\max \leqslant Q_{DES_i} \leqslant Q_{\text{char}}\max \tag{10.6}$$

(4) TC Current Capacity and State: For each TC_i where i is within the limit of the total number of TC nodes, their power output or consumption range in the charging and discharging states is defined as follows:

$$Q_{T_i}\min \leqslant Q_{T_i} \leqslant Q_{T_i}\max \tag{10.7}$$

The charging power range is calculated based on the charging schedule time provided by upper-layer applications. The minimum discharge power value is typically set to 0, meaning only the maximum discharging power is limited.

(5) Controllable Load Range Constraint: The DL part obtains the power load situation for all controllable loads. Assuming there are a total of n nodes, the constraint is as follows:

$$Q_{DL} = \sum_{i=0}^{n} Q_{D_i} \tag{10.8}$$

$$Q_{D_i}\min \leqslant Q_{D_i} \tag{10.9}$$

where i is within the limit of the total number of DL nodes.

Finally, it should be noted that the loss value for the next intelligent scheduling period is calculated based on the theoretical method using the maximum load for that period, resulting in Q_{Loss}. In an active distribution network, when the power fluctuates within a small range, the network will autonomously adjust the power supply and demand relationship, ensuring that the total power consumption equals the total power supply.

10.3.3 Design of the Source–Grid–Load–Storage Cooperative Intelligent Scheduling Mechanism

The source–grid–load—storage cooperative intelligent scheduling mechanism designed in this chapter can be divided into two main parts: optimization and adjustment strategies. Section 10.3.2 has already introduced the optimization problem, and now we will provide a detailed explanation of the intelligent scheduling mechanism's process and power adjustment strategy.

10.3.3.1 Intelligent Scheduling Process

The entire intelligent scheduling mechanism can be divided into five steps, and the complete operational process is illustrated in Figure 10.3:

Step 1: Data Acquisition
The intelligent scheduling occurs before the next operational cycle, during which all nodes in the area report their data to the distribution node. Distributed power generation nodes report power output predictions and adjustment ranges. Distributed energy storage nodes report their energy storage status and power adjustment ranges for the next cycle. Flexible load nodes report their charging and discharging plans and adjustment ranges for the next cycle. Controllable load nodes report their lower adjustment limits and current power values. Each feeder node calculates the predicted load for the next cycle and reports the corresponding line loss power. Load predictions for the next cycle are obtained from the power grid centre.

Step 2: Data Processing
The first task is to organize the data obtained in Step 1 and calculate a constant value $D = Q_{Loss} + Q_{Load} - Q_{G}now - Q_{EG}now$. Additionally, the value ranges for Q_{EG}, Q_{DES}, Q_{SC}, and Q_{CL} are obtained, where positive values represent power consumption and negative values represent power output. For distributed power generation, the lower adjustment limit is set as the maximum output power of all clean energy sources, and the upper limit is the sum of the maximum adjustment powers of all

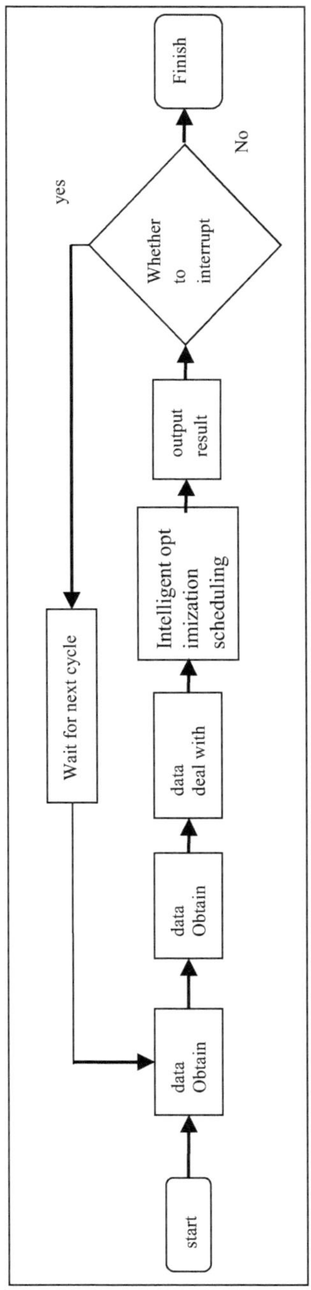

Figure 10.3 Source–grid–load–storage cooperative intelligent scheduling process flowchart.

nodes. For distributed energy storage, the lower adjustment limit is the sum of the lower adjustment limits of all ToC = 0, –1 nodes (maximum discharging power), and the upper limit is the sum of the upper adjustment limits of all ToC = 0, 1 nodes (maximum charging power). For flexible load nodes, the lower adjustment limit is the sum of the minimum power values (positive values) of all State = –1 nodes and the minimum adjustment power values (negative values) of all State = 1 nodes, which means considering discharging after fulfilling all charging intentions in the region. The upper adjustment limit is the sum of the maximum power values of all State = 0, 1 nodes, allowing the effective transfer of part of the load from the peak periods to off-peak periods. For controllable loads, the formulas were already given in Section 10.3.2.2. Data preprocessing prepares for the next optimization task.

Step 3: Optimization Intelligent Scheduling
Based on the value ranges obtained in Step 2, the optimization can be viewed as follows:

$$\begin{cases} \min f(x) = min\left(Q_G \text{ new} + Q_{EG} \text{ new} - Q_G \text{ now} - Q_{EG} \text{ now}\right)^2 \\ \text{s. t. } \min v \leqslant g_i(x) \leqslant maxv, i = 0,1,2,3 \\ h(x) = 0 \end{cases} \tag{10.10}$$

where minv ≤ $g_i(x)$ ≤ maxv, i = 0, 1, 2, 3 represent the value ranges for Q_{EG}, Q_{DES}, Q_{SC}, Q_{DL}, which are the results obtained in Step 2. h(x) = 0 represents the power balance constraint, and f(x) is the objective function. Therefore, the optimization process is a quadratic programming problem with mixed constraints. Various heuristic methods such as simulated annealing, genetic algorithms, ant colony algorithms, and particle swarm optimization can be used to solve this problem, and the choice depends on specific needs.

Step 4: Strategy Formulation
After the optimization scheduling, the total power values for each part are obtained. The next step involves distributing and adjusting the power for each part according to the optimization results. Different power adjustment strategies are formulated for each part, which will be explained in Section 10.3.3.2.

Step 5: Output Results
The final power adjustment values for each node, based on the strategy, are transmitted to the respective nodes, and all the results are recorded in the cloud centre to support other business processes. The intelligent scheduling mechanism will run according to this process in each intelligent scheduling cycle until manually stopped or interrupted due to fault handling.

10.3.3.2 Power Adjustment Strategies

After obtaining the optimized results, it is necessary to adjust each node's power. The chapter follows the principle of fair allocation and formulates power distribution strategies for EG, DES, SC, and DL. The overall goal is to achieve a fair distribution of load or supply power among all nodes. Fair power distribution benefits power supply equipment by prolonging its lifespan, and for user loads, equitable load adjustment leads to better overall user satisfaction. The study of power adjustment strategies can be considered a separate research topic. The focus of this chapter is on the overall intelligent scheduling mechanism. In practice, different power distribution strategies can be applied to different parts based on regional characteristics to achieve better results. For example, a minimum equipment principle can be applied to power distribution for EG. The power adjustment order for EG, DES, SC, and DL is not specified, and their power adjustments do not affect each other. The power adjustments for each part are asynchronous, based on the optimization results, and only require the calculation of power adjustment values for each node before execution.

(1) **EG Part:**
Assuming there are a total of n EG nodes, where m (n ≥ m) are new energy nodes ($EG_{Type} = 0$), which operate at their maximum power, and the remaining n – m non-new energy nodes have a power of

$$\frac{\left(Q_{EG} - \Sigma\, EG_{Type=0}\right) * \left(Q_{EG_{Type-1}j}\max - Q_{EG_{Type-1}j}\min\right)}{\Sigma\left(Q_{EG_{Type-1}i}\max - Q_{EG_{Type-1}i}\min\right)} \tag{10.11}$$

Among them, j ranges from 1 to n – m. Equation (10.11) represents the power distribution based on the power adjustment capabilities.

(2) **For the DES Part:**
First, the optimization result of PDES is evaluated. If $Q_{DES} > 0$, it requires charging for energy storage, which might be during a low-load period. The value of $Q_{DES} - \Sigma DES_{Toc} = -1$ max is checked. If it is less than 0, all the energy storage devices that can only be charged are charged with average power. If it is greater than 0, all ToC = –1 nodes are charged at maximum power, and the remaining part is proportionally distributed among the ToC = 0 nodes based on their ε values. The smaller the ε value, the higher the charging power for the nodes. If $Q_{DES} \leq 0$, it requires discharging for energy storage, which might be during a peak load period. The value of $|Q_{DES}| - \Sigma DES_{Toc} = 1$ max is checked. If it is less than 0, all the energy storage devices that can only be discharged are set to discharge with average power. If it is greater than 0, all ToC = 1 nodes are discharged at maximum power, and the remaining part

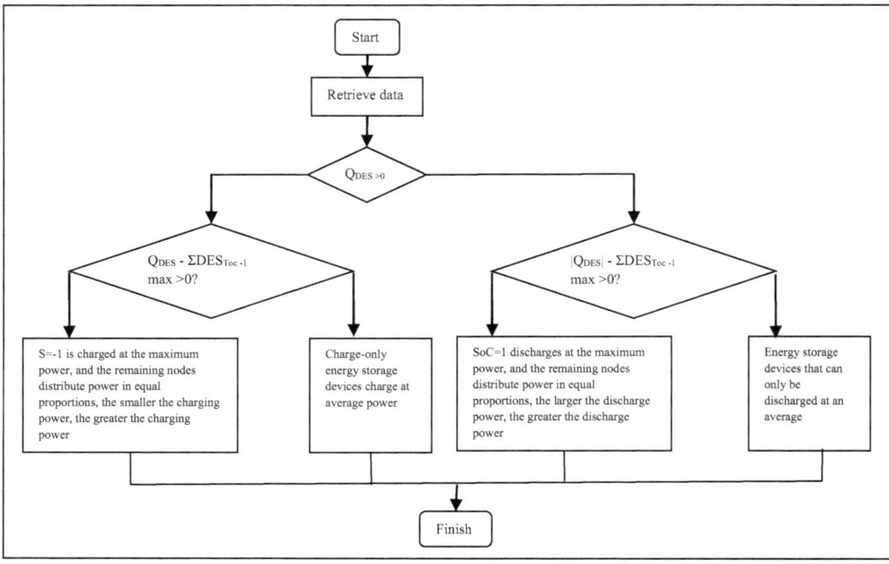

Figure 10.4 Simplified diagram of distribution network architecture.

is proportionally distributed among the ToC = 0 nodes based on their ε values. The larger the ε value, the higher the discharge power for the nodes (Figure 10.4).

(3) The SC Part:

First, the optimization result Q_{SC} is evaluated. If $Q_{SC} > 0$, the SC part needs to charge and consume power. The value of $Q_SC - \Sigma SC_{State} = -1$ min is checked. If it is greater than 0, it means some charging stations do not need to discharge, so $Q_{SC} - \Sigma SC_{State} = -1$ max is checked. If it is less than 0, then the power is evenly distributed among all the State = −1 charging stations. If it is greater than 0, δ is used to represent the difference, indicating that more charging stations are needed to consume the excess power. In this case, the charging power is sequentially allocated to the State = 0 charging stations until it is fully distributed. If $Q_{SC} - \Sigma SC_{State} = -1$ min is less than 0, it means that State = 1 nodes need to release some energy to meet the minimum demand of State = −1 nodes. The lacking power is evenly distributed among the State = 1 charging stations for discharge, while State = −1 nodes are charged at the minimum power. When $Q_{SC} \leq 0$, the SC part needs to be discharged. The discharge power should be the remaining value after satisfying the minimum charging power of State = 1 node. State = −1 nodes are charged at the minimum power, and for each State = 1 node (assuming there is j of them), the power is allocated as shown in the following equation:

$$\frac{\sum SC_{\text{State}} = -1\min + \left|Q_{SC}\right|}{j} \tag{10.12}$$

(4) **For the DL Part:**
The power distribution in Q_{DL} adopts an average allocation method, aiming to improve overall user satisfaction. User satisfaction in intelligent scheduling has been studied in some literature. Since this chapter does not focus on this particular issue, the conclusions from existing research are applied [8]. The load power value for each node is controlled as shown in Equation (10.13). Assuming there are n controllable load nodes, the range of i in Equation (10.13) is from 1 to n.

$$\frac{Q_{DL}{}^{*}\left(Q_{Di} - QP_{Di}\min\right)}{\Sigma\left(Q_{Di} - Q_{Di}\min\right)} \tag{10.13}$$

Thus, the entire optimization intelligent scheduling mechanism design is completed. Next, a brief analysis of the effectiveness of the proposed intelligent scheduling mechanism is presented. Firstly, using energy storage devices can fully accommodate distributed power sources, i.e., new energy sources. Traditional scheduling methods passively adjust the power grid supply load by tracking changes in demand. As a result, the total load on the power grid is stabilized to some extent through the mutual adjustments between the two. At the same time, the addition of controllable loads and flexible loads widens the range of adjustable power, which, combined with the optimization objectives, effectively achieves peak shaving, and valley filling, and reduces additional power intelligent scheduling losses. A more detailed analysis is provided in Section 10.4.

10.4 Simulation Experiment

The architecture shown in Figure 10.2 is used as the distribution network structure, and the active power of the source, network, load, and storage parts within one day is the monitoring object. The day is divided into 96 scheduling and operating periods at 15-minute intervals, with predictions of conventional load for the day. The predicted output plan for distributed power generation is also set for distributed storage and flexible load devices. To verify the effectiveness of the scheduling mechanism, actual conforming situations are used as reference benchmarks. At the same time, two scheduling methods are used in the experiment: one is the scheduling model designed in this chapter, and the other is a scheduling method that does not consider flexible loads and controllable loads. In this method, no control is applied to the load side (before the construction of the ubiquitous power Internet, data barriers prevented load participation in scheduling), allowing for a comprehensive comparison of the collaborative scheduling mechanism and the single interactive scheduling mechanism. The results obtained from these two scheduling methods are compared with the actual load conditions.

The data collected in the experiment includes the optimization results after each scheduling period, which are used to calculate the active power adjustment values between adjacent scheduling periods, as well as the curves of active power changes for the source, network, load, and storage parts in the distribution network throughout the day. The curve of the total active power change for the distribution network throughout the day is also recorded.

10.4.1 Experimental Data

The actual load data used in the experiment is the variation of the actual load in a certain region's power system throughout the day. This chapter uses a typical 24-hour load curve for a summer day, as shown in Figure 10.5.

In the peak shaving and valley filling problem of power scheduling, specific data values are not crucial, as long as the load data's curve shape is generally consistent, following the pattern of a typical daily load curve. The data range may vary depending on the region. Therefore, based on a typical daily load curve, this chapter uses a total load range of 7000 kW–12000 kW. In the curve graph, the x-axis represents the 24 hours of a day, and the y-axis represents the power values in kilowatts (kW). All subsequent curve graphs in the chapter will have the same format. In the experiment, the data from this curve is used to represent the actual load situation in the distribution network without using any regulation means such as energy storage or flexible loads.

(1) Load Forecast Input:

The scheduling mechanism in this chapter assumes that it operates before the next operating period. Therefore, the scheduling needs to obtain the load forecast for the next scheduling period. In previous scheduling mechanisms, this method was not considered due to limitations in private networks and artificial intelligence technology accuracy. However, load forecasting technology has become relatively mature, and combined with the context of UPIoT, short-term load forecasting accuracy is sufficient, making the operation of the scheduling mechanism feasible. In this chapter, the load forecasting error is assumed to be between 3% and 5%, which

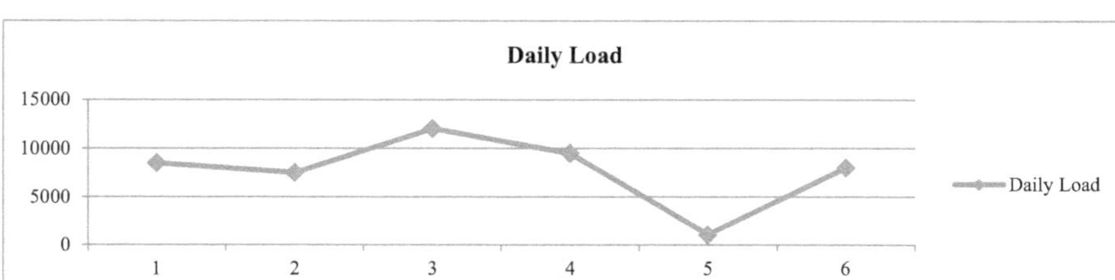

Figure 10.5 Typical 24-hour load curve for a power system in summer.

is already considered a relatively large error range according to surveys. Figure 10.6 shows the comparison between the load forecast and the actual load data used in the chapter.

(2) Distributed Power Data:

In the proposed distributed power sources, a unified management approach is used for clean new energy sources and other controllable distributed power sources. Considering that most distribution areas use clean new energy generation and draw power from the higher-level grid, this chapter only considers new energy generation nodes in the experiment. Based on the actual curve as a template and given appropriate power ranges, the chapter simulates the daily output power forecast for clean energy sources. In the experiment, the photovoltaic and wind power generation for each scheduling period is shown in Figure 10.7, with solid lines representing photovoltaic power generation and dashed lines representing wind power generation.

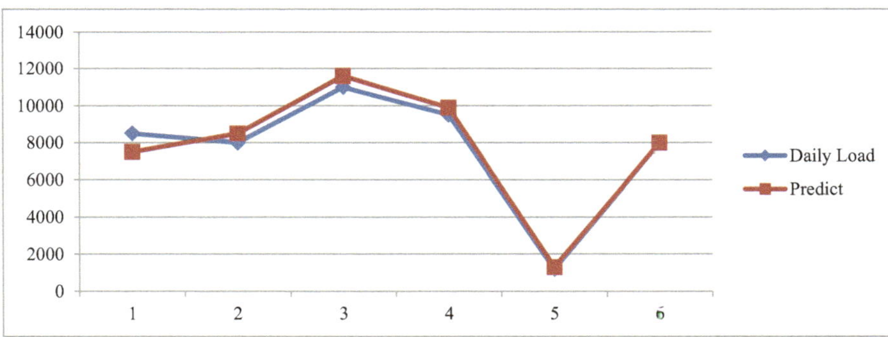

Figure 10.6 Comparison of load forecast and actual load curve.

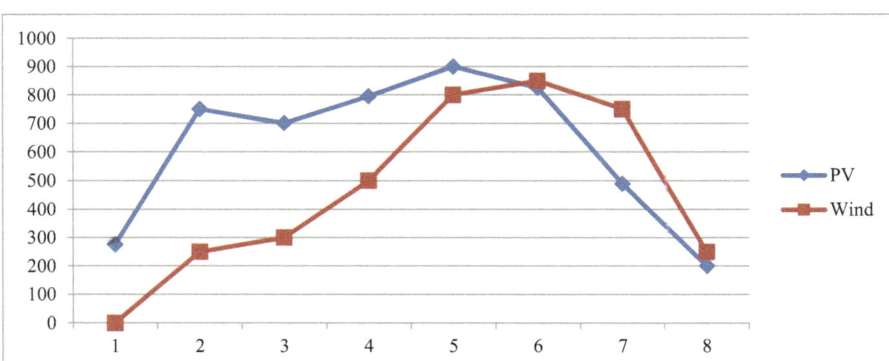

Figure 10.7 Photovoltaic and wind power generation for each scheduling period.

Table 10.1 Initial State of Distributed Energy Storage Nodes

Variable	*Qc/kW*	ε/%	Q_{min}/(kW/h)	Q_{max}/(kW/h)
DES0	250	50	–500	500
DES1	250	50	–500	500
DES2	250	50	–500	500

(3) Setting of Energy Storage, Flexible Load, and Controllable Load:
For distributed energy storage nodes, an initial energy level is given, and power limits are set. In the experiment, the power values are adjusted based on the scheduling mechanism, and their states are updated accordingly. This chapter uses three distributed energy storage nodes in the experiment, all with consistent initial states. Table 10.1 shows the initial energy status and the maximum charging and discharging power for these nodes. The maximum charging and discharging power correspond to the energy storage device capacity, which is determined through capacity planning for energy storage [5]. As explained in Section 10.3, the values presented here are obtained through multiple experiments and represent an optimal setting.

As for the setting of flexible load and controllable load, their power values are separated from the actual load during the experiment. The flexible load considers user participation, so in the simulation, it is assumed that each node reports data based on an orderly charging algorithm for electric vehicles at charging stations. This means that the power distribution at charging stations in the experiment follows a power distribution statistically obtained from daily life [31–34]. For instance, most electric vehicle charging is scheduled during the early morning, while during the day, most nodes are willing to shift their charging times. At night, a random function is used to generate charging rate ranges. The control range of the controllable load is generated randomly within a total control unchanged principle for each period. A monitoring thread is employed in the simulation programme to ensure that the load changes while adhering to the principle of total control unchanged. Specific data can be observed in the experimental result graphs.

10.4.2 Experimental Results and Analysis

This section uses two indicators to evaluate the effectiveness of the scheduling mechanism.

1. The magnitude of the change in the total load of the distribution network between adjacent periods (Δ). Assuming the current scheduling time is it, the total load in the distribution network at this time is Q_{t_i}, and the total load at a time t_{i+1}, after the scheduling is $Q_{t_{i+1}}$. Thus, $\Delta = Q_{t_{i+1}} - Q_{t_i}$; the smaller Δ, the lower the additional losses.

2. Overall trend of the total load. The effect of peak shaving and valley filling can be visually observed by simulating the total load curve after running for one day. Meanwhile, the average value of Δ can also reflect the peak and valley characteristics of the load.

10.4.2.1 Additional Losses and Analysis

As shown in Figure 10.8, the variations of Δ values in each period under the three scheduling modes are different and have distinct gaps. In Figure 10.8, the solid line represents the variation of Δ values under the coordinated scheduling mechanism proposed in this chapter, and the discrete points represent the variation of Δ values under the comparative mechanism, which only considers energy storage in a single coordinated scheduling mode, and the dashed line represents the variation of Δ values without adopting any coordinated scheduling mechanism. It is evident from Figure 10.8 that the Δ values under the coordinated scheduling designed in this chapter are consistently smaller than the other two scheduling modes in almost every period, and in many operating periods, they are significantly smaller than the others.

Table 10.2 presents the average load fluctuation in each period obtained using the three scheduling modes in the experiment. When using the proposed scheduling mechanism in this chapter, the overall average load adjustment in the distribution network is reduced by 50% compared to no coordinated scheduling, and it is also reduced by 40% compared to the existing mechanism that considers only energy

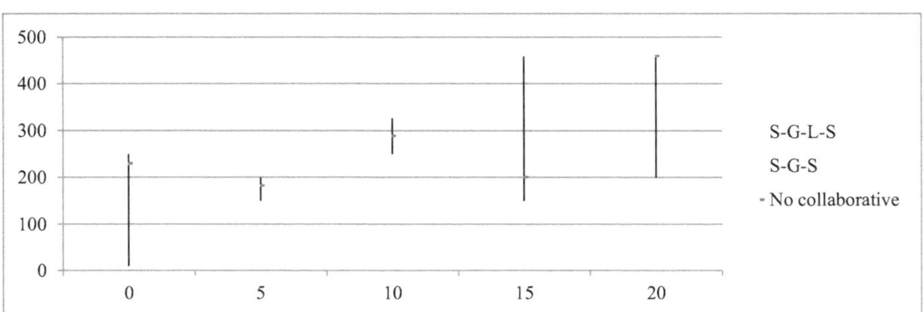

Figure 10.8 Variation of Δ values.

Table 10.2 Average Load Adjustment Variation in the Total Load

Scheduling Mode	*Mean Adjust Value/kW*
Collaborative scheduling is not used	242.58
Source–grid–load–storage collaborative scheduling	114.61
Collaborative scheduling without SC and DL	186.92

storage for coordinated scheduling. This indicates that the average losses generated during the power adjustment process by the scheduling mechanism proposed in this chapter are much lower than the other two methods. The data in Table 10.2 demonstrates the effectiveness of coordinated scheduling in reducing additional losses. Therefore, the experimental results show that the coordinated scheduling mechanism of source–network–load–storage can effectively reduce additional losses in power scheduling.

Although single-coordinated scheduling also has some effects, in many scheduling periods, its Δ values are much larger than the actual situation. This is because the adjustment capacity of energy storage devices is limited in the coordinated scheduling process. In the scheduling process, to smooth the fluctuations in load, priority is given to using energy storage for adjustment. However, once there is a sudden surge in load, the adjustment capacity of energy storage devices becomes insufficient. If the energy storage devices are already fully charged or discharged due to previous adjustments, they lose their adjustment capacity, resulting in larger adjustment values in individual periods compared to no coordinated scheduling. Moreover, through repeated experiments, it is found that arranging the energy storage capacity properly can improve the scheduling performance. The specific situation depends on the output of renewable energy and load characteristics in the actual region. This experimental result validates the correctness of the necessity to make reasonable distribution network planning based on the coordination objectives for the corresponding region [35, 36]. The source network load storage coordinated scheduling can adjust more elements, the adjustment ability is greatly strengthened, and there is a certain self-adaptive ability (flexible load part), so this situation is avoided.

10.4.2.2 Overall Load Change and Analysis

The effect of load levelling is mainly reflected in the overall load variation during the day. The variations in the grid's total load under the three scheduling modes are shown in Figure 10.8. In Figure 10.9, the solid line represents the coordinated scheduling proposed in this chapter for source–network–load–storage, the discrete points represent single coordinated scheduling, and the dashed line represents scheduling without coordination. By observing the figure, it can be seen that under the coordinated scheduling, the total load will maintain a certain smoothness during the variation process, especially when using the coordinated scheduling mechanism proposed in this chapter, the overall variation forms a smooth curve, except for two time periods of load surge and sharp decline. Therefore, it can be concluded that the coordinated scheduling mechanism can effectively smooth the peak-valley difference, greatly reducing the average adjustment in each period, achieving the purpose of load levelling, and the more comprehensive the coordination, the better the effect.

From the perspective of total load, the total load remains constant under single coordinated scheduling because the energy status of energy storage devices is set to be consistent from the beginning to the end of the experiment, ensuring energy

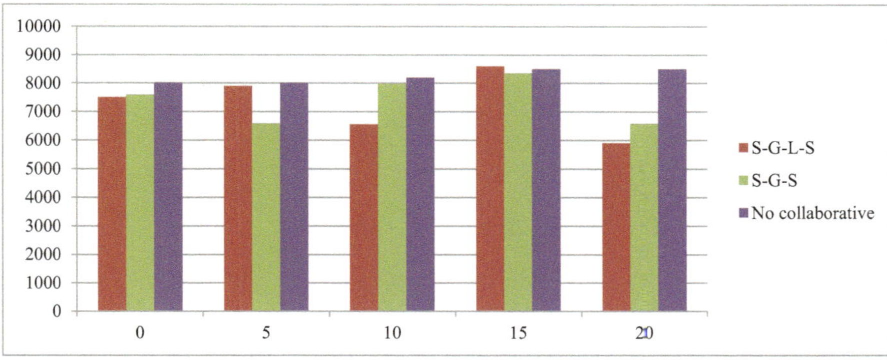

Figure 10.9 Total load variation.

conservation. However, under comprehensive coordinated scheduling, it is found that the total load decreases slightly within a day. This is due to the random setting of SC and DL components during the simulation experiment, which has some irrationality, resulting in a reduction of active consumption (i.e., a controllable load is basically in a minimum energy consumption state). Nevertheless, overall, the principle of load conservation is followed. This result provides direction and assistance for studying the interaction mechanism on the load side, namely, how to increase the overall adjustable load while meeting user demands.

10.4.2.3 Analysis of Operation of Source–Network–Load–Storage Nodes

Figure 10.10 presents the power variation of all nodes during the scheduling process. Figure 10.10(a) and (b), respectively, shows the power variation of all nodes under single coordinated scheduling and the comprehensive coordinated scheduling proposed in this chapter. Each curve corresponds to the load variation of a category of nodes during the scheduling process.

Power variation of all nodes under single coordinated scheduling mode Figure 10.10(b) Power variation of all nodes under comprehensive coordinated scheduling mode Figure 10.10(c) Power variation of controllable nodes under comprehensive coordinated scheduling mode

From Figure 10.10 it can be observed that coordinated scheduling effectively reduces the power output of the power backbone network. On one hand, this is achieved by adopting the maximum power strategy for new energy generation, which is used in both scheduling methods, hence the consistency of the DG curve in Figure 10.10(a) and (b). On the other hand, coordinated scheduling utilizes energy storage devices, controllable loads, and flexible loads to effectively shift the load to appropriate periods, achieving peak-valley levelling. Moreover, from Figure 10.10(b), it can be further observed that comprehensive coordinated scheduling not

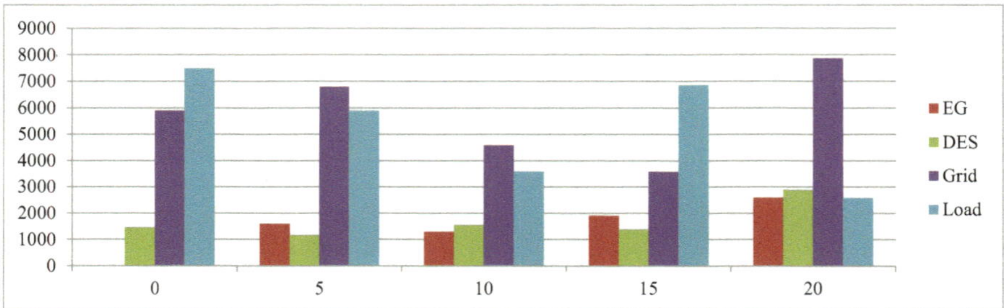

(a) Power change curves of all nodes in a single cooperative scheduling mode

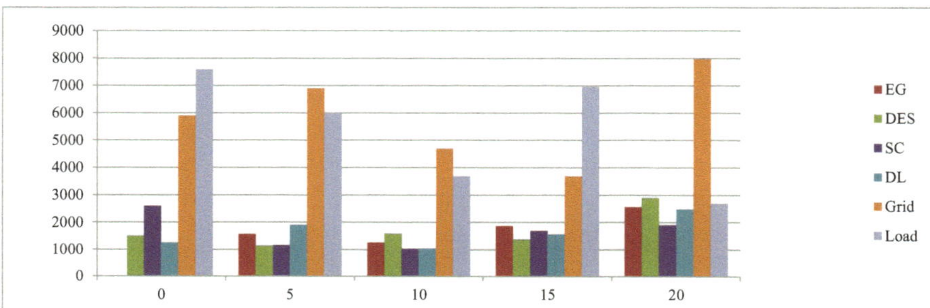

(b) Power change curves of all nodes in the comprehensive cooperative scheduling mode

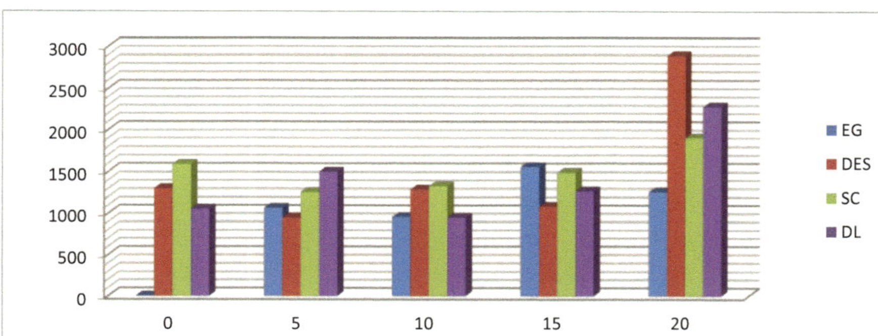

(c) Curve diagram of controllable node power variation under the comprehensive cooperative scheduling mode

Figure 10.10 Power variation of all nodes during the scheduling process. (a) Power change curves of all nodes in a single cooperative scheduling mode; (b) power change curves of all nodes in the comprehensive cooperative scheduling mode; and (c) curve diagram of controllable node power variation under the comprehensive cooperative scheduling mode.

only reduces the power transmission of the power backbone network but also makes the transmission variation smoother. This helps to reduce the energy loss caused by frequent high-power adjustments during transmission, further emphasizing the role of source–network–load–storage coordinated scheduling in reducing additional power dispatch losses. Figure 10(c) presents the power variation of nodes other than the total load and power backbone network output from Figure 10.10(b). It can be seen that all nodes participate in coordinated scheduling and play a role, with their power adjustment being frequent and within a considerable range.

Considering the losses at individual feeder nodes, frequent power variations may not be beneficial for prolonging equipment lifespan, and some energy loss occurs during this process. However, compared to the entire distribution network, the volume of these losses is relatively small. In actual practice, the power grid adopts a hierarchical centralized management mode, and the total losses from all nodes within each layer can still be significant and should not be underestimated. Therefore, clustering management of feeder nodes in coordinated scheduling could be explored as a direction to minimize losses within the clusters.

10.5 Conclusion

Under the construction of UPIoT, active distribution networks have gained access to more information and bandwidth resources, enabling a more comprehensive source–network–load–storage coordinated intelligent scheduling mechanism. However, the development of UPIoT also brings forth a series of new challenges to the distribution network, such as expanding external business and changes in internal business topology. Consequently, in this scenario, the source–network–load–storage coordinated scheduling mechanism needs to meet the diverse application requirements of external business while also accommodating the changes in internal business processing. Existing source–network–load–storage optimization dispatch methods may become less efficient in this context.

In response to these new changes and demands, this chapter focuses on three main aspects of work:

1. Conducted research on source–network–load–storage coordinated intelligent scheduling technology and summarized several research directions under the context of UPIoT construction.
2. Based on the existing source–network–load–storage interactive intelligent scheduling, thorough research considering the new characteristics and future trends of distribution networks, especially in the context of utilizing new technologies. The aim was to achieve the requirements of stability, economy, and safety for distribution networks while satisfying the demands of UPIoT construction. Ultimately, this chapter proposed a source–network–load–storage coordinated intelligent scheduling mechanism oriented towards UPIoT.

3. Designed and conducted simulation experiments, and the results validated that the proposed source–network–load–storage coordinated intelligent scheduling mechanism not only meets the requirements of secure and stable dispatch and basic functionality of new energy consumption capacity but also achieves a certain level of peak shaving and reduces additional losses.

At the same time, there are some limitations in this study, such as only considering active power regulation while neglecting reactive power regulation. In certain distribution areas, the regulation of reactive power is more important. Future work can be directed towards three aspects:

1. Investigate the combination of active and reactive power optimization in coordinated intelligent scheduling, considering both aspects in subsequent research.
2. Explore the influence of individual interactive adjustments on the overall coordinated intelligent scheduling mechanism. Researchers have studied such problems, for example, combining other research areas like user behaviour under time-of-use tariffs, and individually studying user-side demands. In future research, besides studying more comprehensive source–network–load–storage coordinated intelligent scheduling mechanisms, focusing on specific interactive mechanisms within it remains meaningful.
3. The intelligent scheduling assumed certain requirements for terminal nodes. Subsequent research can incorporate edge intelligence technology to further investigate the collaborative interaction mechanisms of these nodes.

References

[1] K. Sunil Ratna Kumar, et al. IoT and data mining techniques to detect and regulating of solar power system, 2023 International Conference on Inventive Computation Technologies (ICICT), IEEE, 2023.

[2] Irfan Sadiq Rahat, et al. A step towards automated haematology: DL models for blood cell detection and classification. EAI Endorsed Transactions on Pervasive Health and Technology, Volume 10 2024, 1–30.

[3] Beneyaz Ara Begum, Satyanarayana V. Nandury, Data aggregation protocols for WSN and IoT applications – A comprehensive survey, Journal of King Saud University – Computer and Information Sciences, Volume 35, Issue 2, 2023, 651–681, ISSN 1319-1578, https://doi.org/10.1016/j.jksuci.2023.01.008

[4] Elham Fazel, Hamid Esmaeili Najafabadi, Mohammad Rezaei, Henry Leung, Unlocking the power of mist computing through clustering techniques in IoT networks, Internet of Things, Volume 22, 2023, 100710, ISSN 2542-6605, https://doi.org/10.1016/j.iot.2023.100710

[5] Abdullah Ayub Khan, Asif Ali Laghari, Mamoon Rashid, Hang Li, Abdul Rehman Javed, Thippa Reddy Gadekallu, Artificial intelligence and blockchain technology

for secure smart grid and power distribution Automation: A state-of-the-art review, Sustainable Energy Technologies and Assessments, Volume 57, 2023, 103282, ISSN 2213-1388, https://doi.org/10.1016/j.seta.2023.103282

[6] Rachneet Kaur, Clara Schaye, Kevin Thompson, Daniel C. Yee, Rachel Zilz, Ravi S. Sreenivas, Richard B. Sowers, Machine learning and price-based load scheduling for an optimal IoT control in the smart and frugal home, Energy and AI, Volume 3, 2021, 100042, ISSN 2666-5468, https://doi.org/10.1016/j.egyai.2020.100042

[7] Poornima M.R., Vimala H.S., Shreyas J., Holistic survey on energy-aware routing techniques for IoT applications, Journal of Network and Computer Applications, Volume 213, 2023, 103584, ISSN 1084-8045, https://doi.org/10.1016/j.jnca.2023.103584

[8] N. Renugadevi, S. Saravanan, C.M. Naga Sudha, IoT-based smart energy grid for sustainable cities, Materials Today: Proceedings, Volume 81, Part 2, 2023, 98–104, ISSN 2214-7853, https://doi.org/10.1016/j.matpr.2021.02.270

[9] Zhang Xiaoyi, Wang Dongling, Zhang Yuming, Karthik Bala Manokaran, A. Benny Antony, IoT driven framework based efficient green energy management in smart cities using multi-objective distributed dispatching algorithm, Environmental Impact Assessment Review, Volume 88, 2021, 106567, ISSN 0195-9255, https://doi.org/10.1016/j.eiar.2021.106567

[10] B.T. Geetha, P. Santhosh Kumar, B. Sathya Bama, S. Neelakandan, Chiranjit Dutta, D. Vijendra Babu, Green energy aware and cluster-based communication for future load prediction in IoT, Sustainable Energy Technologies and Assessments, Volume 52, Part C, 2022, 102244, ISSN 2213-1388, https://doi.org/10.1016/j.seta.2022.102244

[11] Lakshmi Kanthan Narayanan, Priyanga Subbiah, S.A. Sahaaya Arul Mary, R. Rengaraj Alias Muralidharan, Iswarya Gururajan, Ranjani Sampathkumar, Arun Prasad Baskaran, Chapter 4 – A smart and efficient IoT-AI and ML-based multi-functional system for multilevel power distribution management, Editor(s): Sanjeevikumar Padmanaban, Jens Bo Holm-Nielsen, Kayal Padmanandam, Rajesh Kumar Dhanaraj, Balamurugan Balusamy, Smart Energy and Electric Power Systems, Elsevier, 2023, Pages 49–63, ISBN 9780323916646, https://doi.org/10.1016/B978-0-323-91664-6.00008-5

[12] Faiza Qayyum, Harun Jamil, Faisal Jamil, Shabir Ahmad, Do-Hyeun Kim, IoT-orchestration based nanogrid energy management system and optimal time-aware scheduling for efficient energy usage in nanogrid, International Journal of Electrical Power & Energy Systems, Volume 142, Part A, 2022, 108292, ISSN 0142-0615, https://doi.org/10.1016/j.ijepes.2022.108292

[13] Priya R. Krishnan, Josephkutty Jacob, An IOT based efficient energy management in smart grid using DHOCSA technique, Sustainable Cities and Society, Volume 79, 2022, 103727, ISSN 2210-6707, https://doi.org/10.1016/j.scs.2022.103727

[14] Nawin Ra, Aritra Ghosh, Ankur Bhattacharjee, IoT-based smart energy management for solar vanadium redox flow battery powered switchable building glazing satisfying the HVAC system of EV charging stations, Energy Conversion and Management, Volume 281, 2023, 116851, ISSN 0196-8904, https://doi.org/10.1016/j.enconman.2023.116851

[15] Madani Abdu Alomar, An IOT-based smart grid system for advanced cooperative transmission and communication, Physical Communication, Volume 58, 2023, 102069, ISSN 1874-4907, https://doi.org/10.1016/j.phycom.2023.102069

[16] Sami Ben Slama, Prosumer in smart grids based on intelligent edge computing: A review on artificial intelligence scheduling techniques, Ain Shams Engineering Journal, Volume 13, Issue 1, 2022, 101504, ISSN 2090-4479, https://doi.org/10.1016/j.asej.2021.05.018

[17] Shikha Mathur, Anshuman Kalla, Gürkan Gür, Manoj Kumar Bohra, Madhusanka Liyanage, A survey on role of blockchain for IoT: Applications and technical aspects, Computer Networks, Volume 227, 2023, 109726, ISSN 1389-1286, https://doi.org/10.1016/j.comnet.2023.109726

[18] Mustapha Oudani, Anass Sebbar, Karim Zkik, Imad El Harraki, Amine Belhadi, Green blockchain-based IoT for the secured supply chain of hazardous materials, Computers & Industrial Engineering, Volume 175, 2023, 108814, ISSN 0360-8352, https://doi.org/10.1016/j.cie.2022.108814

[19] Prakash Pawar, Panduranga Vittal K., Design and development of advanced smart energy management system integrated with IoT framework in a smart grid environment, Journal of Energy Storage, Volume 25, 2019, 100846, ISSN 2352-152X, https://doi.org/10.1016/j.est.2019.100846

[20] Khalid M Alhamed, Celestine Iwendi, Ashit Kumar Dutta, Badr Almutairi, Hisham Alsaghier, Sultan Almotairi, Building construction based on video surveillance and deep reinforcement learning using smart grid power system, Computers and Electrical Engineering, Volume 103, 2022, 108273, ISSN 0045-7906, https://doi.org/10.1016/j.compeleceng.2022.108273

[21] Ananya Banerjee, Chapter 29 – A comparative study on IoT-aided smart grids using blockchain platform, Editor(s): Rajiv Pandey, Sunil Kumar Khatri, Neeraj Kumar Singh, Parul Verma, Artificial Intelligence and Machine Learning for EDGE Computing, Academic Press, 2022, Pages 443–467, ISBN 9780128240540, https://doi.org/10.1016/B978-0-12-824054-0.00006-X

[22] Zhong Chen, C.B. Sivaparthipan, BalaAnand Muthu, IoT-based smart and intelligent smart city energy optimization, Sustainable Energy Technologies and Assessments, Volume 49, 2022, 101724, ISSN 2213-1388, https://doi.org/10.1016/j.seta.2021.101724

[23] Arman Behnam, Sasan Azad, Mohammadreza Daneshvar, Amjad Anvari-Moghaddam, Mousa Marzband, 5 – Artificial intelligence–enabled Internet of Things technologies in modern energy grids, Editor(s): Mohammadreza Daneshvar, Behnam Mohammadi-Ivatloo, Kazem Zare, Amjad Anvari-Moghaddam, IoT Enabled Multi-Energy Systems, Academic Press,2023, Pages 69–86, ISBN 9780323954211, https://doi.org/10.1016/B978-0-323-95421-1.00004-5

[24] Fursan Thabit, Ozgu Can, Asia Othman Aljahdali, Ghaleb H. Al-Gaphari, Hoda A. Alkhzaimi, Cryptography algorithms for enhancing IoT security, Internet of Things, Volume 22, 2023, 100759, ISSN 2542-6605, https://doi.org/10.1016/j.iot.2023.100759

[25] Tarun Kumar, Ravi Srinivasan, Monto Mani, An emergy-based approach to evaluate the effectiveness of integrating IoT-based sensing systems into smart buildings, Sustainable Energy Technologies and Assessments, Volume 52, Part C, 2022, 102225, ISSN 2213-1388, https://doi.org/10.1016/j.seta.2022.102225

[26] Botazhan Satuyeva, Chingis Sauranbayev, Ikechi A. Ukaegbu, H.S.V.S. Kumar Nunna, Energy 4.0: Towards IoT applications in Kazakhstan, Procedia Computer Science, Volume 151, 2019, 909–915, ISSN 1877-0509, https://doi.org/10.1016/j.procs.2019.04.126

[27] Suvarna Patil, Mandeep Kaur, Katarina Rogulj, Fairness-driven link scheduling approach for heterogeneous gateways for digital twin enabled industry 4.0, International Journal of Intelligent Networks, Volume 4, 2023, Pages 162–170, ISSN 2666-6030, https://doi.org/10.1016/j.ijin.2023.06.001

[28] Xiangdong Zhang, Gunasekaran Manogaran, BalaAnand Muthu, IoT enabled integrated system for green energy into smart cities, Sustainable Energy Technologies and Assessments, Volume 46, 2021, 101208, ISSN 2213-1388, https://doi.org/10.1016/j.seta.2021.101208

[29] Sana Benhamaid, Abdelmadjid Bouabdallah, Hicham Lakhlef, Recent advances in energy management for Green-IoT: An up-to-date and comprehensive survey, Journal of Network and Computer Applications, Volume 198, 2022, 103257, ISSN 1084-8045, https://doi.org/10.1016/j.jnca.2021.103257

[30] Sakshi Popli, Rakesh Kumar Jha, Sanjeev Jain, A comprehensive survey on Green ICT with 5G-NB-IoT: Towards sustainable planet, Computer Networks, Volume 199, 2021, 108433, ISSN 1389-1286, https://doi.org/10.1016/j.comnet.2021.108433

[31] Qianzhi Zhang, Jinyue Yan, H. Oliver Gao, Fengqi You, A systematic review on power systems planning and operations management with grid integration of transportation electrification at scale, Advances in Applied Energy, Volume 11, 2023, 100147, ISSN 2666-7924, https://doi.org/10.1016/j.adapen.2023.100147

[32] Wenming Zhao, Jin Chen, Tao Hai, Muamer N. Mohammed, Zaher Mundher Yaseen, Xuelan Yang, Jasni Mohamad Zain, Ruihua Zhang, Qiang Xu, Design of low-energy buildings in densely populated urban areas based on IoT, Energy Reports, Volume 8, 2022, 4822–4833, ISSN 2352-4847, https://doi.org/10.1016/j.egyr.2022.03.139

[33] Sri Harsha Mekala, Zubair Baig, Adnan Anwar, Sherali Zeadally, Cybersecurity for Industrial IoT (IIoT): Threats, countermeasures, challenges and future directions, Computer Communications, Volume 208, 2023, 294–320, ISSN 0140-3664, https://doi.org/10.1016/j.comcom.2023.06.020

[34] Swati Sharda, Mukhtiar Singh, Kapil Sharma, Demand side management through load shifting in IoT based HEMS: Overview, challenges and opportunities, Sustainable Cities and Society, Volume 65, 2021, 102517, ISSN 2210-6707, https://doi.org/10.1016/j.scs.2020.102517

[35] S.C. Dharmadhikari, Veerraju Gampala, Ch. Mallikarjuna Rao, Syed Khasim, Shefali Jain, R. Bhaskaran, A smart grid incorporated with ML and IoT for a secure management system, Microprocessors and Microsystems, Volume 83, 2021, 103954, ISSN 0141-9331, https://doi.org/10.1016/j.micpro.2021.103954

[36] Mohammed A. M. Sadeeq, Subhi R.M. Zeebaree, Design and implementation of an energy management system based on distributed IoT, Computers and Electrical Engineering, Volume 109, Part A, 2023, 108775, ISSN 0045-7906, https://doi.org/10.1016/j.compeleceng.2023.108775

Chapter 11

Coordinated Response Strategies: Swarm Robotics for Crisis Management

Renu R. Dandge, Tushar R. Mahore, Ankush Vasant Dahat, Pallavi H. Dhole, Nikhilesh P. Mankar, and Sagar Dhanraj Pande

11.1 Introduction

Swarm perception is a novel idea. In this type of knowledge, simple powers that trail local rules agree to successfully perform difficult tasks in an arranged way. This idea, following the way in which ants, bees, fowls, and fish work together in nature, has surpassed to move something forward in machine intelligence and androids [1]. Researchers have created algorithms that can solve questions accompanying adeptness, plan routes, and confirm that tasks are accomplished well by copying the habit these open swarms present image of a group. This attitude has had a grown impact on the progress of swarm machine intelligence, at which point groups of natural androids agree to do belongings like surveying, probing for distracted objects, and keeping an eye on the environment.

 DOI: 10.1201/9781003479062-11

In fault-finding sketches, it's very important to intelligently act fast by accompanying others. Swarm machine intelligence is a powerful capacity to handle deadlocks because it uses the elasticity of swarm demeanor [2]. In contrast to schemes that focus on a single point of deterioration, swarms can maintain functionality even if some bots get inactive. They can likewise change how they act as a group to conform to changing surroundings and new impediments.

This portion explains how swarm machine intelligence may be used in different districts, "in the way that," "toward," "at sea," and "on the way" [3]. It examines in what way or manner these wholes can improve functional abilities, resilience, and effectiveness to form defense, freedom, and crisis administration to work better. By comparing various approaches, this stage further examines, in a better way, how swarm machine intelligence methods have transformed over time and got divergent. This work sets the entertainment industry for new algorithms that will improve the field by critically checking the expert and cons of former approaches [4].

Response plans that work and are arranged are very crucial for handling accidents, workplace incidents, and various crises. The established methods are not very compatible to resolve the complicatedness and instability of these crises [5]. Swarm android is a progressive order that uses the plans of swarm knowledge to boost the strength to handle crises.

The study of swarm robotics uses groups of plain, self-forceful robots that agree to receive belongings approved without a main boss. These, done or made by machine plans, can respond fast, flexibly, and dependably to troubles by understanding how organic swarms symbolize a group [6]. This arrangement works well for everything, especially when it is mainly expected to smartly change fast by adapting to new situations and to be powerful even when individual part forsakes [7]. Easy-to-use and active deployment, one of significant features in the study of swarm computing for trouble management, is that it starts fast on a variety of backgrounds. Swarm machines may be deployed or transported to various locations as needed [8]. Since they are limited and light, they can disturb places that folk would find hard or hazardous to get to. Different sensors may be added to swarm machines so they can draw useful facts in absolute-occasion and help community take smart decisions. Swarm electronics schemes are typically so strong that they do not need any governance [9]. They have auxiliary plans and are flexible. Each robot in a swarm has allure to set rules of his own by the virtue of what it endures for guiding the other androids in allure swarm. If individual robot forsakes, efforts are made to deploy new ones in operation.

This backup makes sure that the swarm can keep working even if some units get lost or broken [10]. For example, during a wildfire, a group of aerial drones can keep an eye on how the fire is spreading, help firefighters work together, and send updated reports. If some drones fail because of bad weather, others can easily take over their jobs.

Swarm robots are able to adapt to changing environments. Crisis events are often marked by conditions that are hard to predict and change quickly. Because they can change as a group, swarm robotics systems work well in these kinds of settings [11]. Swarm robots can change how they act in real time because they always sense and respond to their surroundings. This ability to change is very important in situations like chemical spills or nuclear accidents, where things can go badly very quickly, and prompt, flexible reactions are needed to reduce damage [12].

Swarm robots are used in various ways in a range of media. They are useful for handling emergencies because they can be used in air, on land, and at sea. Drone swarms can be used in the air to do things like map making, monitoring, and delivery. In the ground, robots can move through trash, help with search and rescue, and fix things after a disaster. Marine protection, search and rescue, and keeping an eye on the environment can all be helped by robots that can work on or under water [13].

Swarm robotics has the ability to do operations in an improved manner. When used in crisis management, swarm robotics greatly improves operational skills. These tools can give people constant, real-time information about what's going on, which can help them work together better and respond more effectively [14]. In the event of an urban disaster, for example, swarm robots can make moving maps of the affected areas, point out dangers, and direct first rescuers to safe routes and important sites. At this level of practical improvement, lives can be saved and damage is kept to a minimum level [15].

Swarm robotics is a new field that uses the way simple robots act together to resolve complicated jobs. Several publications are available on this growing topic, which are shown exclusively in Figure 11.1.

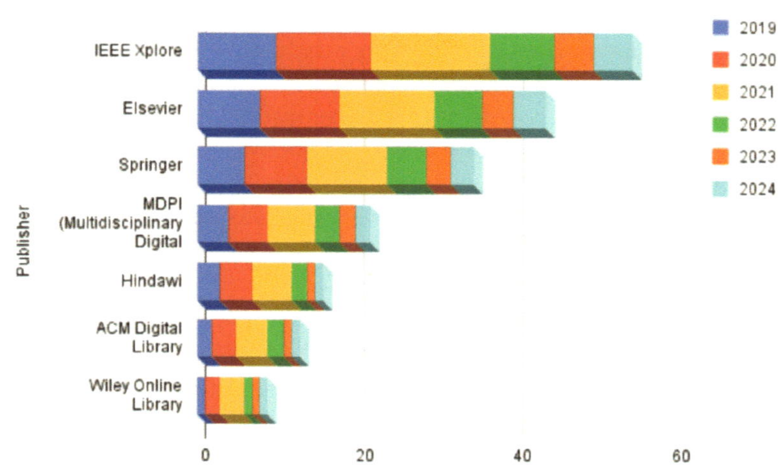

Figure 11.1 Graphical representation of publications in recent years on swarm intelligence for crisis management.

11.2 Literature Survey

Swarm robotics can be applied, in the air, to tasks, including environmental monitoring, agricultural surveys, and search and rescue using drones or UAVs (unmanned aerial vehicles). These drones communicate with one another to cover large regions fast and adjust to changing circumstances [16]. Several published papers on the subject have been compared. Swarm robotics finds application on land in logistics, monitoring, and exploration. Ground-based robots can navigate difficult terrain, cooperate on construction projects, and even locate survivors and transport supplies to aid in disaster relief. Swarm robots in maritime settings employ autonomous underwater vehicles (AUVs) and surface vessels [17]. These can map undersea, study oceanography, and monitor the environment. Even under difficult conditions like deep sea, these robot groups can cooperate to collect and transmit helpful data exploration. Various works that were completed in the field were contrasted against one another, and Table 11.1 was created out of the most important notices.

11.3 Important Things about Swarm Robotics for Crisis Management

There are several uses of swarm intelligence in this fast-moving and growing world. It has many applications throughout the systems. It's quite useful in several critical situations due to its unique features and abilities [18]. Some of the major reasons are mentioned below:

- One of the benefits of swarm systems is that they can easily be scaled up or down based on the size and scope of the crisis.
- Flexibility: Swarms can change with their surroundings and meet new needs, which is very important in emergencies where the outcome is hard to predict.
- Redundancy and Robustness: More than one robot can do the same job, so if one or more of them break, the goal doesn't get harmed.
- Robustness: The system can handle failures in individual robots without a big drop in total performance.
- Autonomy and Decentralization: "Autonomy" means that robots can make their own choices and work without being watched all the time by a person.
- Decentralization: There isn't a single point of failure, and the swarm makes decisions as a whole.
- Collective Intelligence: **Swarms** have collective intelligence, which means that simple behaviors by individuals lead to complex behaviors by the group as a whole. This makes it easier to solve problems and distribute resources during a crisis.

Table 11.1 Overview

Title	*Authors*	*Year*	*Publication Type*	*Methodology*	*Overview*
Knowledge-Guided Evolutionary Optimization for Large-Scale Air Defense Resource Allocation	Wenhua Li et al.	2024	Journal Article (IEEE Transactions on Artificial Intelligence)	Evolutionary Optimization	Efficient allocation of air defense resources
Adaptive Trust Model for Multi-Agent Teaming Based on Reinforcement-Learning-Based Fusion	Chin-Teng Lin et al	2024	Journal Article (IEEE Transactions on Emerging Topics in Computational Intelligence)	Reinforcement Learning	Improved trust evaluation in multi-agent teams
Research on Exploratory Simulation Technology of Command Decision Making Oriented to Game Confrontation	W. Qingqing et al.	2023	Conference Paper (Jiangsu Annual Conference on Automation (JACA))	Simulation Technology	Develops training for command decision making
An Integrated Monitoring System for Aerial Drones and Underwater ROVS	Niccolò Cecchinato et al.	2023	Conference Paper (IEEE International Workshop on Technologies for Defense and Security (TechDefense))	Integrated Monitoring System	Centralized monitoring for drones and ROVS
An Improved Actor-Based Programming Framework for Simultaneous Management of Multiple Heterogeneous Robot Swarms	Wei Yi et al.	2023	Conference Paper (IEEE International Conference on Robotics and Biomimetics (ROBIO))	Actor-Based Programming Framework	Manages multiple robot swarms
Based on Intelligent Attack and Defense Command Technology System for Multi-Rotor Tilting	Zhi-Hao Chen	2023	Conference Paper (IEEE International Conference on e-Business Engineering (ICEBE))	Intelligent Attack and Defense Command Technology System	Coordinated attack and defense for UAV swarms

11.3.1 Core Capabilities for Crisis Management

There are several in which swarm intelligence can make novel moves in this field. A few of them are as follows:

1. Search and Rescue: Swarm machines can quickly cover substantial extents to find survivors by utilizing sensors and ruling their shifts together [19]. They can go places where the public can't harmlessly go, like into hazardous places and like destroyed constructions.
2. Surveillance and Listening: Swarm drones can take pictures from above in evidence and attend regions that are affected, bestowing the public the news they need to resolve what's continuing and create selections [20]. Drones that bother the ground can find hazardous matters and check how powerful houses are.
3. Communication and Socializing for Professional or Personal Gain: Swarm machines can form limited ways for nations to warn each one in places where foundation isn't skilled or is crushed [21]. They can maintain communication lines open between command centers and rescue groups by mobile from individual place to another.
4. Logistics and Delivering Supplies: Swarms can take feed, water, healing provisions, and added belongings that society needs to move to places that are hard to access by land or air. Based on the present needs and aims, they can agree to confirm that money is joint highest-rank habit attainable [22].
5. Response to Natural Disasters: After earthquakes, storms, and other natural disasters, many drones or robots work together to find survivors and figure out how much damage was done.
6. Epidemic Control: When there is a health emergency like a disease outbreak, swarm intelligence can help model how the disease spreads and figure out the best way to use medical resources and vaccination programs [23].
7. Resource Allocation and Deployment: Swarm intelligence algorithms can help automatically assign resources like food, medical supplies, and rescue teams based on data that is collected in real time. For instance, if more medical help is needed in one place, more resources can be sent there right away.
8. Routing and Logistics: Swarm-based algorithms can find the best way for goods and people to get from one place to another, even when roads are closed or new dangers are found by getting information and sharing it.
9. Collecting Data: Drones or self-driving cars can collect data from places that have been affected. These bots can quickly cover large areas, find hotspots, and send data back to a central system by using swarm intelligence [24].
10. Real-Time Reporting: People can act as agents by sharing their state, needs, and observations through apps or social media. This data can be put together by swarm intelligence to find trends and decide which answers are most important.

11. Problem-Solving in Multiple Locations: Swarm-based platforms help volunteers and affected people work together to solve problems in multiple locations, such as organizing shelters or distributing resources [25].

11.4 Dataset

Several datasets were considered to perform these various applications [25–30]. All are listed in detail in Table 11.2 and the major used datasets and their distribution are shown in Figure 11.2.

11.5 Methodology

There are several applications that use several models to serve the need and purpose. For one task many algorithms can be used and one out of them can be more generalized, optimal, and reliable. Table 11.3 is exclusively made to show the purpose and the major algorithms used based on the study of various journals and published works.

11.5.1 Challenges and Limitations

Besides Ant Colony Optimization (ACO), there are several other swarm intelligence models and methods that can be used to help with search and rescue. These models use the way robots act as a group to cover large areas, get through dangerous areas, and find survivors [31]. We'll talk about a few important types in the following sections

11.5.1.1 Particle Swarm Optimization (PSO)

Particle Swarm Optimization (PSO) simulates particles moving in a search area. It was inspired by the way birds flock. Each particle remembers where it was best and where the swarm was best. Based on this memory, particles change how they move, and as a group, they move toward the best answer [32]. PSO is used to find the best way to do something, to build machines, and to optimize functions.

11.5.1.2 Bee Colony Algorithm

Inspired by how honeybees find food and work with other bees to take advantage of food sources that are easy to find. BA uses scout bees to find possible food sources (solutions) and worker bees to take advantage of the best ones. This is based on how

Table 11.2 Various Datasets Used to Perform the Purpose Using Swarm Intelligence

Task	*Dataset Type*	*Example Datasets*	*Description*
Search and rescue (finding survivors)	Satellite imagery: high-resolution post-disaster imagery	WorldView, DigitalGlobe	Identifies potential survivor locations based on building collapse, debris patterns, or changes in vegetation.
Search and rescue (entering dangerous areas)	3D building scans: detailed models of building interiors	Stanford 3D Indoor Dataset (S3DIS)	Aids navigation and risk assessment for robots entering damaged buildings.
Surveillance and monitoring (aerial monitoring)	Real-time weather data: temperature, wind speed, precipitation	National Oceanic and Atmospheric Administration (NOAA)	Optimizes drone flight paths and data collection considering weather conditions.
Surveillance and monitoring (ground monitoring)	Structural health monitoring data: sensor readings on bridges, buildings	National Center for Earthquake Engineering Research (NCEER)	Helps robots assess structural damage and identify potential hazards.
Communication and networking	Social media data: tweets, location check-ins	Twitter, Facebook	Provides insights into affected areas, resource needs, and communication disruptions.
Logistics and supply delivery	Digital elevation models (DEMs): terrain data	USGS National Elevation Dataset (NED)	Facilitates path planning for robots delivering supplies, considering elevation and potential obstacles.
Epidemic control (resource allocation)	Epidemic spread models: Disease transmission rates and demographics	SEIR models	Informs resource allocation for medical teams, vaccinations, and quarantine zones.

(*continued*)

Table 11.2 (Continued)

Task	*Dataset Type*	*Example Datasets*	*Description*
Routing and logistics (finding optimal paths)	Traffic data:real-time traffic congestion information	Waze, Google Maps	Optimizes delivery routes for robots or aid vehicles, considering blocked roads and congestion.
Data collection (gathering information)	Sensor data: temperature, air quality, radiation levels	Environmental Protection Agency (EPA) AirData	Provides insights into environmental hazards and areas needing immediate attention.
Real-time reporting (alternative)	Crowdsourced data: reports from affected people through apps	Ushahidi, Swiftpoint	Provides real-time information on damage, needs, and locations of people requiring assistance.

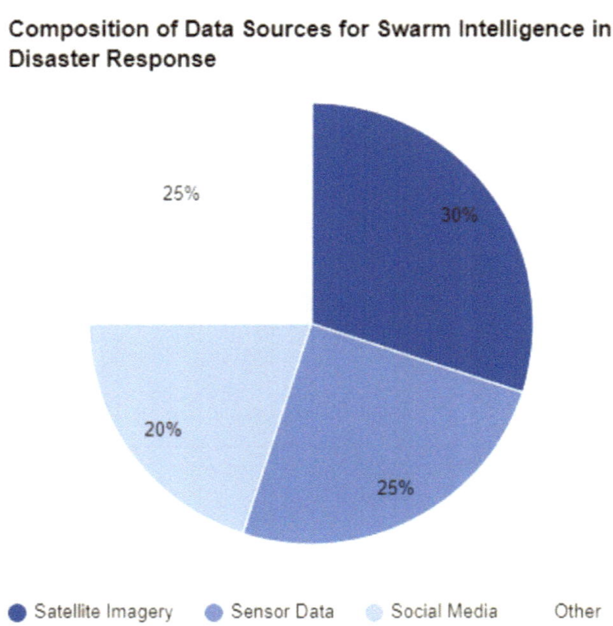

Figure 11.2 Pie chart showing the composition of data sources.

Table 11.3 Various Algorithms Used for the Particular Task

Task	*Major Algorithm*	*Possible Alternatives*
Search and rescue (finding survivors)	Ant Colony Optimization (ACO)	Bacterial Foraging Optimization (BFO), Artificial Bee Colony (ABC) Optimization
Search and rescue (entering dangerous areas)	Bacterial Foraging Optimization (BFO)	Artificial Bee Colony (ABC) Optimization, Glowworm Swarm Optimization (GSO)
Surveillance and monitoring (aerial monitoring)	Particle Swarm Optimization (PSO)	Artificial Bee Colony (ABC) Optimization, Firefly Algorithm (FA)
Surveillance and monitoring (ground monitoring)	Artificial Bee Colony (ABC) Optimization	Ant Colony Optimization (ACO), Termite Colony Optimization (TCO)
Communication and networking	Termite Colony Optimization (TCO)	Ant Colony Optimization (ACO), Artificial Fish Swarm Algorithm (AFSA)
Logistics and supply delivery	Glowworm Swarm Optimization (GSO)	Artificial Bee Colony (ABC) Optimization, Ant Colony Optimization (ACO)
Epidemic control (resource allocation)	Firefly Algorithm (FA)	Bacterial Foraging Optimization (BFO), Artificial Fish Swarm Algorithm (AFSA)
Routing and logistics (finding optimal paths)	Artificial Fish Swarm Algorithm (AFSA)	Ant Colony Optimization (ACO),Particle Swarm Optimization (PSO)
Data collection (gathering information)	Cuckoo Search (CS)	Firefly Algorithm (FA), Particle Swarm Optimization (PSO)
Communication networks (alternative)	Greedy algorithm: prioritizes establishing connections with the most critical nodes first.	Genetic algorithm: evolves network configurations based on fitness for communication.

(*continued*)

Table 11.3 (Continued)

Task	*Major Algorithm*	*Possible Alternatives*
Problem-solving (alternative 1)	Consensus-based decision making: swarm intelligence helps aggregate information and reach collective decisions on resource allocation or shelter organization.	Multi-agent systems: independent agents collaborate and share information to solve problems.
Problem-solving (alternative 2)	Market-based approaches: swarm intelligence simulates a virtual market where resources are allocated based on "supply" and "demand" from different locations.	Social learning models: swarm helps model how people learn from each other and adapt solutions across different areas.

honey bees find food [33]. The information that scout bees share with worker bees helps the workers come up with better ideas. Assigning resources, making plans, and organizing data are all things that BA is used for.

Artificial Bee Colony (ABC): Like BA, ABC uses worker bees to look for food, spectator bees to make decisions based on the information from worker bees, and scout bees to find new places to live. Exploration and use of the search area can be done efficiently with this combination [34]. To improve power systems, handle images, and solve scheduling issues, ABC is used.

Bacterial Foraging Optimization (BFO): BFO uses fake bacteria to look for nutrients by copying the way bacteria move, which is called chemotaxis. Bacteria move away from bad solutions (those with less nutrients) and toward ones with more nutrients. To keep from getting stuck in local optima, mutation operators add variety. BFO is used to create control systems, mine data, and find the best way to do things.

Glowworm Swarm Optimization (GSO): GSO uses virtual glowworms to attract each other, based on the way glowworms flash. Brighter glowworms draw in other glowworms, which then group together. The goal is to find the best answer by balancing global discovery (random movement) with local exploitation (attraction). GSO is used to plan robot paths, optimize functions, and separate parts of images [35].

The Firefly Algorithm: The Firefly Algorithm moves like a firefly, which is very beautiful. Bright bugs are considered possible answers to a problem. The brightness of the fireflies shows how good they are, with brighter ones showing better answers [36]. Fireflies are drawn to each other by how bright they are, but the pull gets weaker as they move away. They move toward the brightest firefly thanks to this interaction, which leads them to the best option.

Artificial Fish Swarm Algorithm: Fish groups move together, which is where the Artificial Fish Swarm Algorithm gets its idea. In this case, fake fish stand for possible answers. These fish act in certain ways, like swarming closer to healthier fish that they can see as better choices, following the center of the swarm to move together as a group, and even going off on a whim (like looking for food) if they can't find any better options. By doing these things, the fish "school" around the fittest individuals, coming to the best answer in the end.

Cuckoo Search: Cuckoo Search is sneakier, based on how some cuckoo birds feed their young on other birds' eggs. In the algorithm, cuckoos stand for possible answers. Like real-life cuckoos, some cuckoos lay their eggs in the nests of other birds [37]. Symbolically, these eggs stand for new ideas that might replace old, less-effective ones. The quality of these "eggs" is checked, and if they're better, they replace the ones that are already there, like a good egg replacing a bad one. This "parasitic" behavior forces the algorithm to keep looking for better answers until it finds the best one.

11.6 Challenges

The first thing is coordination and control. It is very hard to come up with algorithms that can effectively coordinate and direct large swarms. More research is being done to make these programs better for use in real world.

1. Communication: It is very important that robots and human controllers can communicate effectively. To ensure strong connections, wireless communication systems need to be improved.
2. Energy Management: For long-term activities in crises, it's important to have good energy management and charging plans.
3. Ethical and Legal Considerations: Using swarm robots brings up ethical and legal issues, such as privacy issues and who is responsible for what the robots do on their own.

11.7 Conclusion

In conclusion, swarm robots could change the way crises are handled by making search and rescue, surveillance, communication, and transportation more efficient. Their ability to be expanded, adaptable, and make decisions together with others makes it possible to quickly and thoroughly explore disaster areas, accurately identify hazards, and quickly find survivors. Swarm robots keep an eye on things all the time, set up important communication networks, and improve logistics by getting goods to places that are hard to get to.

Even though there are problems like how to coordinate them and how to make them adapt to different environments, swarm robotic systems are getting stronger and more reliable thanks to ongoing study and technological progress. Swarm intelligence models, which are based on how natural groups act together and are driven by complex algorithms, provide effective ways to coordinate many robots in changing environments. Robotic search and rescue missions are now more reliable and successful thanks to these improvements.

Swarm robots and swarm intelligence will continue to improve over time, making them even more important in disaster response and crisis management. This will make it much easier to save lives and lessen the effects of disasters.

11.8 Future Scope

In the future, swarm robots could be used in a lot of different ways to handle crises. As machine learning and artificial intelligence keep getting better, swarm behaviors will get even better. This will let them make smarter decisions on their own. It will be easier for swarm robots and human rescues to talk to each other and share data in real time if they can connect to new technologies like Wi-Fi and 5G networks. Building robotic platforms that are bigger and more adaptable will also let them be used in more places, even harsh ones like space and the deep sea.

References

1. I. S. Rahat, et al. "A Step towards Automated Haematology: DL Models for Blood Cell Detection and Classification." *EAI Endorsed Transactions on Pervasive Health and Technology* 10 (2024), 1–16.
2. K. S. R. Kumar, et al. "IoT and Data Mining Techniques to Detect and Regulating of Solar Power System." 2023 International Conference on Inventive Computation Technologies (ICICT). IEEE, 2023.
3. M. Rubenstein, A. Cornejo and R. Nagpal, Programmable self-assembly in a thousand-robot swarm. *Science*, 345(6198) (2014), 795–799.
4. T. P. Okoumassoun, I. A. Ridhawi, A. Abbas and I. Al-Oqily, "Blockchain-Enabled SAGIN Communication for Disaster Prediction and Management," 2023 IEEE Intl Conf on Dependable, Autonomic and Secure Computing, Intl Conf on Pervasive Intelligence and Computing, Intl Conf on Cloud and Big Data Computing, Intl Conf on Cyber Science and Technology Congress (DASC/PiCom/CBDCom/CyberSciTech), Abu Dhabi, United Arab Emirates, 2023, pp. 0678–0683, doi: 10.1109/DASC/PiCom/CBDCom/Cy59711.2023.10361502
5. S. Huang, Z. Zhang, T. Wang, F. Yao, D. Shen and Y. Du, "Optimizing Nurse Scheduling in Organizations a Heuristic Algorithm Assisted by Adaptive Integer Programming Reconstruction," 2023 IEEE 3rd International Conference on

Digital Twins and Parallel Intelligence (DTPI), Orlando, FL, USA, 2023, pp. 1–5, doi: 10.1109/DTPI59677.2023.10365473

6. H. Zhu, B. Akbari, L. Wan and Y.-J. Pan, "Hybrid Human/Robot Team Establishment Using E-CARGO and Role-Based Collaboration," 2023 IEEE Symposium Series on Computational Intelligence (SSCI), Mexico City, Mexico, 2023, pp. 908–913, doi: 10.1109/SSCI52147.2023.10371915
7. Q. T. Do, et al. "UAV-Satellite Integration for Communication System: Potential Applications and Key Challenges," 2023 14th International Conference on Information and Communication Technology Convergence (ICTC), Jeju Island, Korea, Republic of, 2023, pp. 663–665, doi: 10.1109/ICTC58733.2023.10392976
8. U. B. Gohatre, V. P. Patil and S. Y. Wadekar, "Effective Information Gathering in a Three-Dimensional Setting Through the Unmanned Aerial Vehicles and Wireless Sensor Networks," 2023 1st DMIHER International Conference on Artificial Intelligence in Education and Industry 4.0 (IDICAIEI), Wardha, India, 2023, pp. 1–6, doi: 10.1109/IDICAIEI58380.2023.10406473
9. Y.-H. Chou, Y.-C. Jiang, Y.-R. Hsu, S.-Y. Kuo and S.-Y. Kuo, "A Weighted Portfolio Optimization Model Based on the Trend Ratio, Emotion Index, and ANGQTS," *IEEE Transactions on Emerging Topics in Computational Intelligence*, vol. 6, no. 4, pp. 867–882, Aug. 2022, doi: 10.1109/TETCI.2021.3118041
10. M. S. Hosen and Y. Peng, "Dynamic Channel Allocation Technique for Cognitive Radio based UAV Networks," 2021 International Conference on Artificial Intelligence and Big Data Analytics, Bandung, Indonesia, 2021, pp. 152–155, doi: 10.1109/ICAIBDA53487.2021.9689719
11. C.-T. Lin, H. Zhang, L. Ou, Y.-C. Chang and Y.-K. Wang, "Adaptive Trust Model for Multi-Agent Teaming Based on Reinforcement-Learning-Based Fusion," *IEEE Transactions on Emerging Topics in Computational Intelligence*, vol. 8, no. 1, pp. 229–239, Feb. 2024, doi: 10.1109/TETCI.2023.3319253
12. W. Li, R. Wang, Y. Heng, T. Zhang and L. Wang, "Knowledge-Guided Evolutionary Optimization for Large-Scale Air Defense Resource Allocation," *IEEE Transactions on Artificial Intelligence*, doi: 10.1109/TAI.2024.3375263
13. A. D. Brown, "Radar Challenges, Current Solutions, and Future Advancements for the Counter Unmanned Aerial Systems Mission," *IEEE Aerospace and Electronic Systems Magazine*, vol. 38, no. 9, pp. 34–50, 1 Sept. 2023, doi: 10.1109/MAES.2023.3289928
14. M. Doğrul and H. Yalçm, "Future Trends and Challenges of Autonomous and Robotic Systems in Air and Space Technologies," 2023 10th International Conference on Recent Advances in Air and Space Technologies (RAST), Istanbul, Turkiye, 2023, pp. 1–4, doi: 10.1109/RAST57548.2023.10197880
15. F. Kong, H. Chen, H. Li, J. Yan, X. Wang and J. Fang, "Flocking with Obstacle Avoidance for Fixed-Wing Unmanned Aerial Vehicles via Nonlinear Model Predictive Control," 2023 42nd Chinese Control Conference (CCC), Tianjin, China, 2023, pp. 5957–5962, doi: 10.23919/CCC58697.2023.10240689
16. H. Lei, Y. Guoxing, N. Yi and W. Hao, "Combat Situation Suppression of Multiple UAVs Based on Spatiotemporal Cooperative Path Planning," *Journal of Systems Engineering and Electronics*, vol. 34, no. 5, pp. 1191–1210, Oct. 2023, doi: 10.23919/JSEE.2023.000119

17. Z.-H. Chen, "Based on Intelligent Attack and Defense Command Technology System for Multi-Rotor TiltingVertical Takeoff and Landing Swarm UAVs," 2023 IEEE International Conference on e-Business Engineering (ICEBE), Sydney, Australia, 2023, pp. 276–280, doi: 10.1109/ICEBE59045.2023.00012
18. W. Yi et al., "An Improved Actor-Based Programming Framework for Simultaneous Management of Multiple Heterogeneous Robot Swarms," 2023 IEEE International Conference on Robotics and Biomimetics (ROBIO), Koh Samui, Thailand, 2023, pp. 1–8, doi: 10.1109/ROBIO58561.2023.10354588
19. N. Cecchinato, A. Toma, I. Scagnetto, C. Drioli and G. L. Foresti, "An Integrated Monitoring System for Aerial Drones and Underwater ROVs," 2023 IEEE International Workshop on Technologies for Defense and Security (TechDefense), Rome, Italy, 2023, pp. 187–191, doi: 10.1109/TechDefense59795.2023.10380896
20. W. Qingqing, W. Zhengdong and C. Zhongwei, "Research on Exploratory Simulation Technology of Command Decision Making Oriented to Game Confrontation," Jiangsu Annual Conference on Automation (JACA 2023), Changzhou, China, 2023, pp. 31–35, doi: 10.1049/icp.2024.0053
21. J. Chen, X. Li, J. Xu and Y. Wang, "Deployment for NOMA-UAV Base Stations Based on Hybrid Sparrow Search Algorithm," *IEEE Transactions on Aerospace and Electronic Systems*, vol. 59, no. 5, pp. 6138–6149, Oct. 2023, doi: 10.1109/TAES.2023.3272530
22. S. N. Makhadmeh, H. N. Fakhouri and M. A. Al-Betar, "A Modified White Shark Optimizer for Gene Selection Optimization Problem," 2024 2nd International Conference on Cyber Resilience (ICCR), Dubai, United Arab Emirates, 2024, pp. 1–6, doi: 10.1109/ICCR61006.2024.10533161
23. M. K. Pandey, R. N. Mahia and O. P. Mahela, "Optimal Placement and Sizing of RE Generators to Minimize Loss and Improve Voltage Profile in Radial Distribution Network Using PSO," 2024 2nd International Conference on Device Intelligence, Computing and Communication Technologies (DICCT), Dehradun, India, 2024, pp. 623–628, doi: 10.1109/DICCT61038.2024.10532810
24. S. Vimalkumar and R. Latha, "Enhancing Maize Leaf Disease Detection Using Cat Swarm Optimization with Deep Learning Model," 2024 International Conference on Cognitive Robotics and Intelligent Systems (ICC – ROBINS), Coimbatore, India, 2024, pp. 182–188, doi: 10.1109/ICC-ROBINS60238.2024.10533930
25. M. K. Raja, M. Shermatova, M. S. Al Ansari, S. Abdufattokhov, V. S. Rao and I. I. Raj, "Optimization of Neural Networks using Swarm Intelligence Techniques for Achieving Energy Efficiency in Smart Building Architecture," 2024 International Conference on Cognitive Robotics and Intelligent Systems (ICC – ROBINS), Coimbatore, India, 2024, pp. 14–21, doi: 10.1109/ICC-ROBINS60238.2024.10533932
26. Y. FengYing, A. Din, L. HuiChao, M. Babar and S. Ahmad, "Decentralized Consensus in robotic swarm for Collective Collision and Avoidance," in IEEE Access, doi: 10.1109/ACCESS.2024.3402564
27. B. A. S. Emambocus et al. "A Clustering Algorithm Employing Salp Swarm Algorithm and K-Means," 2024 20th IEEE International Colloquium on Signal Processing & Its Applications (CSPA), Langkawi, Malaysia, 2024, pp. 108–113, doi: 10.1109/CSPA60979.2024.10525648

28. F. Pangerang, F. A. Samman, Z. Zainuddin and R. S. Sadjad, "The Performance of a Brushless DC Motor with Six Step Commutation-Based Speed Controller Optimized by PSO Algorithm," 2024 IEEE International Conference on Artificial Intelligence and Mechatronics Systems (AIMS), Bandung, Indonesia, 2024, pp. 1–5, doi: 10.1109/AIMS61812.2024.10513026
29. M. A. Dief, W. Alomoush, H. Atta, O. A. Khashan, A. Solyman and R. ELGohary, "PID Controller Tuning Using Multi-Objective Ant Colony Optimization for Blood Glucose Level of a Diabetic Patient," 2024 2nd International Conference on Cyber Resilience (ICCR), Dubai, United Arab Emirates, 2024, pp. 1–10, doi: 10.1109/ICCR61006.2024.10532878
30. Z. Aung, A. K. Myo, N. N. War and G. O. Mikhailovna, "Experimental Study of Algorithms for Planning the Trajectory of a Warehouse Mobile Robot Based on Reinforcement Learning," 2024 IEEE Conference on Computer Applications (ICCA), Yangon, Myanmar, 2024, pp. 1–6, doi: 10.1109/ICCA62361.2024.10533116
31. T. Tablada, C. Davis and N. Palumbo, "An Ant Colony Algorithm Approach to the Vehicle Routing Problem and Variants," 2024 Systems and Information Engineering Design Symposium (SIEDS), Charlottesville, VA, USA, 2024, pp. 51–56, doi: 10.1109/SIEDS61124.2024.10534666
32. S. A. Khan, S. S. Pande, M. S. Al Ansari, A. H. Siddiqui, B. Abdurasul and P. V. M, "Ant Colony Optimization for Network Customization in Cognitive Human-Computer Communication," 2024 International Conference on Cognitive Robotics and Intelligent Systems (ICC – ROBINS), Coimbatore, India, 2024, pp. 22–28, doi: 10.1109/ICC-ROBINS60238.2024.10534027
33. X. Yang and H. Jiang, "Optimization Research on the MDVRPTW Problem in Urban Cold Chain Logistics under Time-Varying Road Networks," in IEEE Access, doi: 10.1109/ACCESS.2024.3402833
34. I. Bahlaouane, A. Belhabib, S. Ibnyaich and A. Zeroual, "Design and Optimization of Microstrip Patch Antenna Using Ant Colony Optimization," 2024 International Conference on Global Aeronautical Engineering and Satellite Technology (GAST), Marrakesh, Morocco, 2024, pp. 1–5, doi: 10.1109/GAST60528.2024.10520747
35. K. H. Shedbalkar and D. S. More, "Efficient Solar Panel Maintenance: A Data-Driven Strategy using Ant Colony Optimization and Innovative Cleaning Index," 2024 11th International Conference on Reliability, Infocom Technologies and Optimization (Trends and Future Directions) (ICRITO), Noida, India, 2024, pp. 1–6, doi: 10.1109/ICRITO61523.2024.10522416
36. Z. Li, Y. Wei and J. H. Park, "An Improved Bilevel Algorithm Based on Ant Colony Optimization and Adaptive Large Neighborhood Search for Routing and Charging Scheduling of Electric Vehicles," *IEEE Transactions on Transportation Electrification*, doi: 10.1109/TTE.2024.3398113
37. Y. Zhang, "Research on Spark-Based Cultural Ant Colony Algorithm for Capacity Vehicle Routing Problem," 2024 IEEE 7th Eurasian Conference on Educational Innovation (ECEI), Bangkok, Thailand, 2024, pp. 24–27, doi: 10.1109/ECEI60433.2024.10510838

Chapter 12

Smart Farming and Human Bioinformatics Systems Based on IoT and Sensor Devices

G. Bhavya, N. Ashwini, K. Sumana, Mukesh Soni, and Sagar Dhanraj Pande

12.1 Introduction

Internet of Things (IoT) is one of the specific functions of the healthcare domain. IoT is the occurrence in a complement of medical fields, such as incurable diseases, remote fitness detection, fitness records, and aged care [1–5]. The total number of medical accessories used along with those observed as disorder and imaging therapeutic tools and devices [6–10]. The IoT-based care domain could be expected to reduce the quality of life embellished in the trials [11–15]. The collaborator editor can be comparable and verify that the analysis of the publishing paper was done by Muhammad Aleem—the aspect of translation of medical care accessories [16]. IoT can decrease the interlude of medical devices over remote supply. IoT represents the patient's efficiency with limited reserved services [17–20]. Healthcare systems and hospitals are logically connected and interact with patients concerned with the cost of insurance. The latest cellular technologies in healthcare systems help with initial diagnostics, medical accidents, prolonged diseases, and real-time examinations [9,

 DOI: 10.1201/9781003479062-12

21–24]. The current report on IoT healthcare domains should be familiar with the valuable contributions of many associators for further analysis. Besides initiating IoT in areas such as healthcare relevance, machine research offers device techniques that can support therapeutic conjecturing and interpretation issues in many types of domains of health concern [25]. The apparatus studying can be used to examine the scientific parameters and sequence to indicate the abstraction of learning medical analysis and issues through the report, patient general administration, outlining medication, and abutment of diving into the sickness evolution [26]. The synthesis of gadget analysis in the healthcare division to the pharmaceutical experience and enlarged aspect of expertise in the medical care domain.

The development of technology can be used for satisfaction and convenience to speed up the evolution process [27–30]. To examine the many urban hospitals' reductions in medical equipment for patients, the other city moved towards another region to integrate with demand relevant to medical apparatus [31–34]. IoT has immense benefits in healthcare; it can develop the current orientation of themes and be able to develop the latest machinery at a lower cost for all conditions of health concern [35]. Further IoT in life science studies as essential for the pharmaceutical treatment of humans can be analysed as recognizing, regulating, penetrating, and immediately taking away the structure and later gathering the data through reducing the medical errors and damages.

IoT is the auxiliary of systems securing sensors, micro-organizers, and detection devices, a particular hook to transfer the data to assist the specific report. In applying the different types of health concerns to implicate the disease in the process of caution, perception of disease at the time of self-remedial trial, and curing disease in the act of medicinal solving, a carried body sensor is a division of IoT that can be defined as a proper treatment for suffering people. IoT established the service of smart sensors that can drain foreign embellishments, rap or hold the skin, or bury or impact a body, usually in clothing and threads. This smart sensor should be combined with data gathering, data mailing, and collecting reports to create a smart declaration. These sensors carry the body, which develops into a crucial fragment, away from the IoT machinery. Then they develop to be effective against many functional and usage services taken from plain medical accessories. This acute wearable could be combined with other sorts of sensors related to smart mobiles for computing functions and transmission.

Biodevices are useful for examining diseases, low-priced proteins, and other investigations early. These devices are assembled and recover the regular specialized trails that are restricted to the environment and demand for expertise in science, a conclusion of primary charges to this various region. Biodevices are the growing range of multi-ordered applications that have been presented for the probable transformation of users, fitness care, and mechanical trails. The pandemic biodevice function currently has a valuation of around 10 billion dollars each year. The assessment of biodevices is the main hazard to an expansive service. The current prices can be higher for the functional devices of medical supplies with the assurance of greater materials and penetrating indicators. The variety of biosensors can reduce the cost. The specific preference of the part of the structure tested to change the cost of each test is very low. Biodevice transformation is the automation of the latest technology to reduce costs and advance the commercialization system. When it comes to the use of biosensors for suffering people, the cost of devaluation is more essential. The other demand for biosensor machinery can be updated in nature to healthcare domains and symptomatic tools for ailments to affect the high death and morbidity obligation.

The reward of brilliant wearables has been determined by the machinery wearables at demand, the length of computerized devices and sensors, and the evolution of low-speed functions of networks. The investigation of the high development of wearable sensors transformed many regulations throughout a short period. Illustration of wearable commodities to update the many usages as alert jewellery, hand and consumed sensors, body belts, earbuds, headsets, eye frictions, and wrist straps. Despite critical metamorphic functions that can be organized in pharmaceutical and health industries, detecting essential facts of wearable machinery.

The wearable sensors are highly fortified in day-to-day life; the regular transmission networks are major contributors to human-balanced function and face curious health issues related to inflexible postponement, particular system capacity, and low battery life. The next generation of 5G cellular connections will therefore follow the later trails of Wi-Fi communications to protect the huge comparability, short inactivity, and uncommonly higher quantity. 5G can modify efficient and individual lives, particularly with innovative functions, such as remote health concerns, Wi-Fi robots, combined homes, and riderless cars. These sensors provide high-quality video technology, therapeutic services, and telemedical treatment at efficient prices.

12.2 Revolutionizing Agriculture with IoT and Sensor Technology: The Future of Farming

12.2.1 Internet of Things—Definition and Architecture

IoT is one of the fast-growing technologies like artificial intelligence (AI). It offers different paradigms of technical applications through integration of the real world and the cyber world. Large-scale innovations and value-added services are easily enabled by IoT via a series of network communications with other sensor devices. IoT serves traffic, supply chain logistics, healthcare, agriculture, retail industries, smart cities, remote sensing, and process automation in manufacturing. The network of physical items or "things" that are implanted with sensors, software, and other technologies to gather and exchange data via the internet is referred to as the IoT. The IoT architecture describes how these gadgets are linked together, how

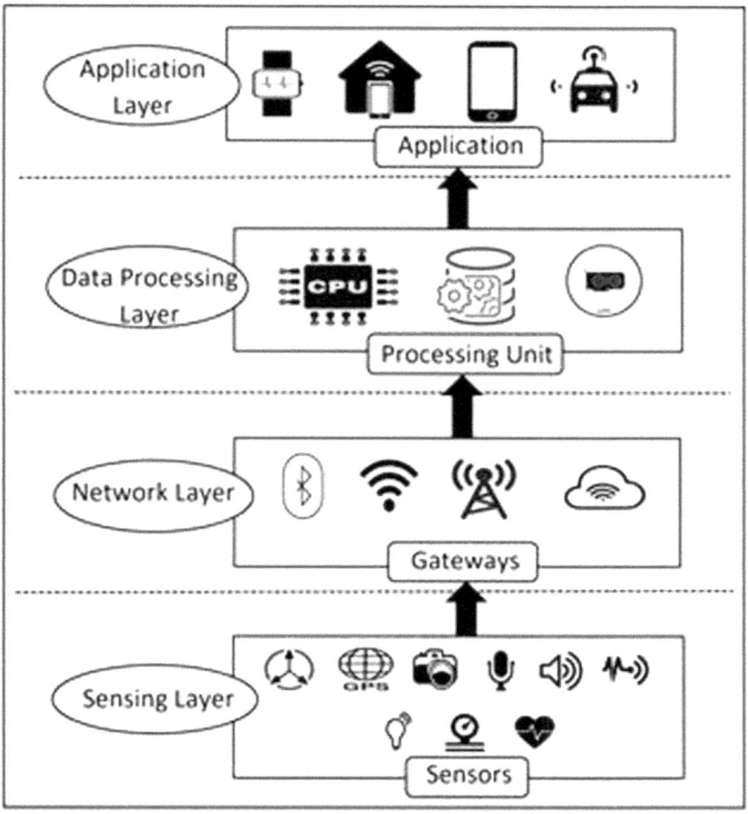

Figure 12.1 IoT system architecture.

information is collected, processed, and used, and how various system parts communicate with one another. The functioning of IoT systems and devices is enabled by several levels and components. The basic architecture of IoT systems has four layers (sensing layer, network layer, processing layer, and application layer), as shown in Figure 12.1.

12.2.2 Sensing Layer

The sensing layer, which is the initial layer in the IoT architecture, is in charge of gathering data from various sources. This layer consists of actuators and sensors that are positioned in the surrounding space to record data on environmental factors like moisture, temperature, sound, light, and other physical characteristics. These devices use wired or wireless communication methods to connect to the network layer.

12.2.3 Network Layer

It is the responsibility of the network layer in an IoT architecture to provide communication and connectivity among IoT system components. It consists of protocols and technologies that provide connections and communication within and among networks of electronic devices. IoT makes use of several network technologies, including Wi-Fi, Bluetooth, Bluetooth Low Energy, Zigbee, and various cellular networks, including 4G and 5G. The network layer may also include security measures like authentication to prevent unwanted access, routers, and gateways that serve as middlemen between devices and the internet, and other components (Uckelmann et al., 2011).

12.2.4 Processing Layer

The hardware and software elements in charge of gathering, analysing, and interpreting data from IoT devices are referred to as the data analysis layer in an IoT architecture. This layer is responsible for gathering raw data from the devices, processing it, and then providing it available for additional study or action. According to Uckelmann et al. (2011), such technologies as data lakes and machine learning algorithms are employed to extract useful information from the data and utilize it as the basis for conclusions.

12.2.5 Application Layer

The application layer is the topmost layer that interacts with end users in an IoT setup. It offers a user interface (UI) and a wide range of control IoT devices. This layer comprises many applications and platforms that are intended to connect with the underlying IoT infrastructure. Moreover, the computation and analytics tools in this layer enable data analysis and the creation of informative information. This might incorporate complex analytics technologies like machine learning algorithms and data presentation methods.

12.3 Components of IoT Architecture

IoT has five major components: device, gateway, cloud, analytics, and UI (Uckelmann et al., 2011).

12.3.1 Device

Things or devices are the main tangible objects being tracked. Smart sensors are affixed to objects or devices, continuing to gather data from the latter and sending it to the portal or gateway at the next layer. New developments in microelectronics have made it feasible to create small smart sensors for several uses. Some frequently used sensors include the following:

- thermometers and temperature sensors
- pressure gauges
- moisture/humidity level
- detectors of light intensity
- humidity detectors
- proximity sensing
- RFID labels

12.3.2 User Interface

A user-facing application that enables the user to monitor and alter data is known as a user interface, or UI. The visible, physical part of an IoT device that users can interact with is called the UI. To encourage more interactions, developers must offer a well-designed UI that needs little effort from consumers.

12.3.3 Cloud

Data that has been gathered from various objects or devices is stored in the cloud. Simply described, cloud computing is the continual (24/7) operation of a network of linked servers. IoT users, apps, and devices produce enormous volumes of data that need to be effectively handled. IoT clouds are responsible for a variety of tasks, including data gathering, processing, administration, and archiving.

Industries and services may remotely access the data, enabling them to make important decisions at any time. In its most basic form, an IoT cloud is a network of computers designed to handle data at fast rates for a variety of devices, manage traffic, and do highly accurate data analysis (Figure 12.2). A decentralized management database system is necessary for an IoT cloud to function properly (Uckelmann et al., 2011).

12.3.4 Analytics

The cloud-based data is handled after it has been received. Here, data is processed using a variety of methods, including machine learning and others. Analogue information is transformed into useful insights that may be processed, evaluated, and thoroughly examined through the use of linked sensors and devices. Monitoring and improving the IoT requires the evaluation of raw information or data for further processing. Real-time smart analysis, allowing designers to discover abnormalities in information collection and respond promptly to avert an unfavourable situation, is one of the most important advantages of a well-constructed IoT system. Network

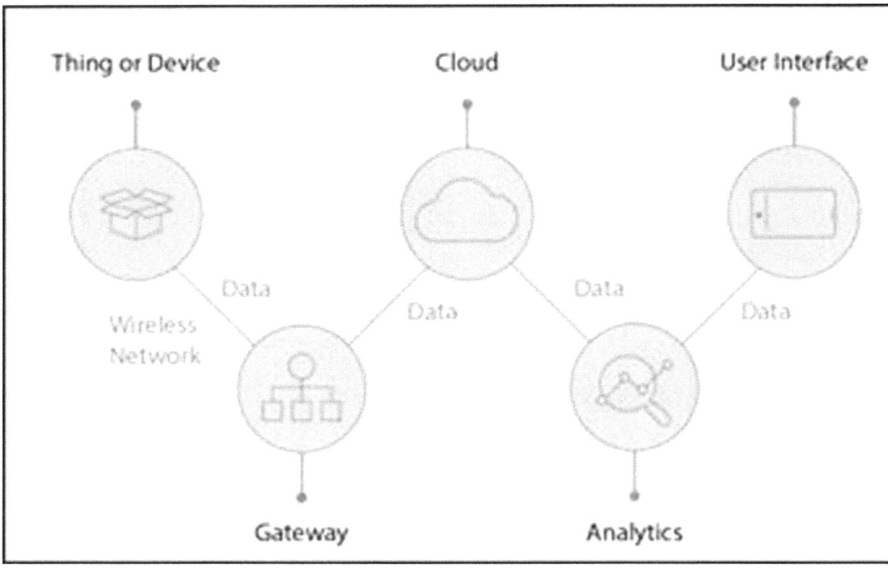

Figure 12.2 Components of IoT architecture.

operators can prepare for the following steps if the information is gathered accurately and at the appropriate time.

12.4 Interconnection of Networks

The number of linked and managed devices by the internet has significantly increased in the IoT during the last several years. Even though IoT devices include a wide range of purposes, they also have significant similarities and distinctions. A multitude of technologies allow IoT. Many wireless or wired technologies can serve the crucial role of the network needed to connect with additional devices in an IoT deployment (Uckelmann et al., 2011).

12.4.1 Network Security

Although security is an essential part of IoT deployment, this security perspective is much too frequently ignored throughout the design phase. IoT's vulnerabilities are being exploited daily with malicious intent, although the majority of those may be simply and reasonably resolved. IoT device flaws must be fixed, and tools must be provided to withstand, identify, and recover from damaging assaults to create a secure network.

12.4.2 Hardware Central Control

A control panel controls two or more data flows among various channels and interfaces. A control panel's additional responsibilities include converting different wireless interfaces and ensuring that connected sensors and devices are available (Uckelmann et al., 2011).

12.5 Benefits of Smart Farming

Precision agriculture, commonly referred to as smart farming, is transforming the agricultural sector by utilizing cutting-edge technologies to improve farming operations, boost production, and reduce resource use and environmental effects. To empower farmers to make defensible decisions based on real-time data, this method makes use of several technologies, like IoT devices, sensors, analytics of data, and automation. Smart farming has a wide range of uses and advantages that apply to several facets of agricultural management [4]. Smart farming applications include the following:

12.5.1 Precision Crop Management

Precision crop management is made possible by smart farming. It allows for exact crop monitoring and control. Sensors positioned throughout the fields gather information on the temperature, nutrient content, and soil moisture. With the use of this information, farmers may modify their irrigation and fertilization schedules, resulting in ideal growth conditions and increased yields.

12.5.2 Livestock Monitoring

Livestock energy and behaviour are monitored by way of IoT-outfitted gadgets and wearable sensors. Farmers can visualize manifestations of sickness and insulate the energy of their animals on account of the evident-time data on their neighbourhood, temperature, and endeavour.

12.5.3 Irrigation Automation

Weather forecasting and news on soil moisture are used by IoT-located watering structures to automate dampening processes. This avoids overwatering, cuts water waste, and keeps the soil active.

12.5.4 Detect Crop Diseases

Early manifestations of land illnesses and epidemic infestations are found through sensors and image electronics. This reduces agricultural misfortunes by permissive prompt mediation and stopping the spread of ailments [3].

12.5.5 Smart Devices

Operations for setting, spraying, and gathering are advanced by automated and GPS-led supplies. In addition to hateful labour expenses and growing efficiency, this likewise uses less fuel.

12.5.6 Data-Driven Decision-Making

Data from diversified beginnings is used in smart ranching to produce observations that support being in charge. This composes previous facts, weather guesses, and advertising patterns. Crop yields increase on account of savvy farming methods, and animal administration is embellished. Plants and mammals are healthier and more fruitful when capability exercise is optimized and prompt attacks are fashioned. Minimizing waste and lessening the effect on the environment are attained by exact listening and administration of resources containing water, fertilizers, and pesticides. Labour payments and functional expenses are reduced through computerization and data-compelled decision-making. Utilizing possessions efficiently too results in cost harvests. Smart farming approaches support tenable farming and safeguard environments by decreasing the use of chemicals and system waste [3]. Farmers can humble risks connected with bad weather and crop ailments by utilizing early affliction diagnosis and correct weather prophecies. Farmers can clarify operations and reduce their strategies for looming seasons on account of data science or logical analysis, which offers perceptive news about previous actions.

12.6 Different Kinds of Agricultural Sensors

The inclusion of abundant sensor electronics has substantially upgraded up-to-date farming. These sensors are essential for managing and monitoring key variables in land positions. These sensors allow peasants to make well-conversant conclusions, exaggerate ability use, and boost crop yields by bestowing real-opportunity data on soil environments, weather patterns, and vitamin levels. Here, we look at common people types of sensors repeatedly used in farming. They are soil sensors, weather sensors, nutrient sensors, and crop well-being sensors [5].

12.6.1 Soil Sensors

Moisture sensors gauge the amount of dampness in the soil. Farmers may counter either overwatering or underwatering by listening to dampness levels and regulating watering schedules, promising high-quality, likely increasing environments for their crops. Nutrient sensors measure the aggregation of essential fibres in the soil, to a degree nitrogen, planet seen at dawn, and potassium. Monitoring mineral levels allows ranchers to regulate propagation methods for maximum crop tumour. Soil temperature sensors gauge the hotness of the soil. Temperature data helps growers decide on high-quality occasions for setting and gathering, in addition to labelling potential frost risks. pH sensors measure the acidity or alkalinity of the soil. Understanding soil pH is essential for correctly answering vitamin rudeness by plants, as it influences the chance of essential details [5].

12.6.2 Weather Sensors

Temperature sensors allow farmers to predict frost, heat waves, and the best times to plant by using weather stations with temperature sensors, which offer real-time temperature data. Humidity sensors monitor the amount of moisture in the air. For disease prediction and irrigation control, this information is essential. Sensors for measuring wind speed and direction: Sensors for measuring wind speed and direction offer information on wind patterns that may impact crop pollination, evaporation rates, and possible wind-related damage. Rainfall sensors measure the amount of precipitation and rainfall. Planning irrigation and water management are made easier by using precipitation data.

12.6.3 Nutrient Sensors

Nitrate sensors assess the amount of nitrate in the soil. Tracking nitrate levels helps to optimize fertilizer use, as nitrate is a crucial component for plant development. Ammonium sensors measure the levels of ammonium in the soil. Monitoring ammonium, a different kind of nitrogen utilized by plants, helps with nitrogen control. Potassium sensors assess the amount of potassium in the soil. For crops to be healthy, potassium is essential since it affects things like water control and disease resistance.

12.6.4 Plant Health Sensors

Leaf area index sensors measure the area of the crop's leaves, which helps farmers determine the biomass and general state of the plant. Chlorophyll sensors monitor the amount of chlorophyll, which reveals the plant's capacity for photosynthesizing and the state of its general health. A few sensors employ spectroscopy to look for symptoms of stress or illness in plants, enabling primary intervention and prevention [5].

12.7 Collection, Transmission, and Analysis of Sensor Data

Precision agriculture and the IoT are two examples of modern technology-driven industries that rely heavily on the gathering, transmission, and interpretation of sensor data [8]. Real-time data from the environment is collected by sensors, which are subsequently transferred for processing to produce valuable insights and guide decision-making. Several steps in this process improve the effectiveness, precision, and worth of data-driven systems.

12.7.1 Sensor Data Gathering

Utilizing a variety of sensors, sensor data collection entails gathering information from the physical world. These sensors may be used to track variables like temperature, humidity, soil moisture, and more by embedding them in objects or setting them out in the environment. Sensors turn physical events into processed and evaluated electrical impulses [9].

12.7.2 Transmission of Sensor Data

Once sensor data has been gathered, it must be shipped to a main location for the supplementary handling. Different ideas and methods, containing Wi-Fi networks (Wi-Fi, Bluetooth, and Zigbee), natural networks, and low-capacity general technologies like LoRaWAN, may be used to send this data. The selection of a broadcast arrangement is influenced by variables containing arrangement atmosphere, data rate, range, and capacity use [10].

12.7.3 Analysis of Sensor Data

After the broadcast, reasoning is performed on the sensor data to prolong the main information. The procedures used for data study might be as plain as mathematical computations or as refined as machine learning algorithms [10]. Finding the data's styles, patterns, deviations, and equations that offer valuable news for decision-making is the aim.

12.8 Human Bioinformatics and Healthcare

12.8.1 Understanding Bioinformatics

To analyse and learn organic data, the multidisciplinary region of bioinformatics integrates the study of animals, robotics, arithmetic, and statistics. It requires the concoction and use of computational finishes, databases, and algorithms to address disputing organic and biomedical issues [14]. Understanding biological processes, defining historical data, and expanding medical research and situations all believe in bioinformatics.

12.8.2 Part of Bioinformatics in Healthcare, the Study of Animals, and Embodied Medicine Genomic Analysis

Decryption and analysing genomic sequences are made likely by bioinformatics. For a better understanding of genetic variants, gene functions, and evolutionary links, it incorporates genome assembly, annotation, and comparative genomics [13].

12.8.2.1 Structural Biology and Proteomics

To forecast protein structures, activities, and interactions, bioinformatics methods are applied. This helps with the development of proteins, drug discovery, and the comprehension of molecular mechanisms [15].

12.8.2.2 Applied Genomics

Bioinformatics aids in the discovery of gene functions, regulatory networks, and biological pathways by combining gene expression data. It helps to find the genes linked to certain illnesses [12].

12.8.2.3 Biology's Theory of Evolution and Phylogenetics

The creation of phylogenetic trees, which reveal evolutionary links between species, is made possible by bioinformatics. It aids in understanding biodiversity and tracking historical genetic changes [12].

12.8.2.4 Personalized Medicine and Medical Genomics

Finding disease-related genes, variations, and biomarkers is made easier by bioinformatics. It contributes to customized medicine by adjusting prescriptions to patients' unique genetic profiles [11].

12.8.2.5 Bioinformatics for Structures

With the use of bioinformatics, biomolecules' three-dimensional structures may be predicted, revealing information about their relationships and activities [12].

12.8.2.6 Drug Development and Discovery

Potential drug candidates are identified, and their interactions with target proteins are predicted via virtual screening, molecular docking, and simulations [11].

12.8.2.7 Biological Systems

By facilitating the modelling and simulation of biological systems, bioinformatics contributes to our knowledge of the intricate connections that occur inside living things [11].

12.8.2.8 Analysis of the Microbiome and Metagenomics

The analysis of metagenomic data by bioinformatics techniques to comprehend the functions of microbial communities in health and ecosystems advances the study of microbiomes [11].

12.9 Sensor Devices in Healthcare

Healthcare sensor technology has changed the way medical care is provided and encouraged patients to take charge of their health management. These technologies improve patient care by offering ongoing surveillance and real-time data, which improves diagnostic abilities, treatment plans, and all aspects of patient care [19]. By providing real-time monitoring, data collecting, and remote patient management, sensor devices have revolutionized healthcare. These gadgets collect environmental and physiological information, giving medical practitioners knowledge of prognosis, therapy, and preventative care (). Personalized healthcare interventions and better patient outcomes have resulted from the integration of sensor devices.

12.9.1 Wearable Sensors for Health Monitoring (Heart Rate, Glucose, Temperature)

12.9.1.1 Wearable Health Sensors

These include the heart rate, activity level, sleep patterns, and even electrocardiograms (ECGs) that are monitored via fitness trackers, smartwatches, and patches. These gadgets enable ongoing monitoring and give people information about their health [16].

12.9.1.2 Blood Sugar Monitors

Devices that measure glucose continuously (CGM) help diabetic individuals control their blood sugar levels [17]. Since these sensors give real-time glucose readings, less frequent fingerstick measurements are required.

12.9.1.3 Blood Pressure Monitors

Users may check their blood pressure throughout the day with wearable blood pressure monitors and smart blood pressure cuffs. These tools support the management of hypertension.

12.9.1.4 Temperature Sensors

Smart thermometers and wearable temperature sensors can assist in identifying fever, illnesses, or changes in health state by monitoring body temperature [19].

12.9.1.5 Blood Oxygen Meters

Pulse oximeters assess the blood's oxygen saturation. These tools are essential for detecting low oxygen levels and monitoring respiratory disorders [19].

12.9.1.6 Breathing Sensors

The care of respiratory problems is aided by respiratory rate and pattern sensors, which track breathing rates and spot anomalies [19].

12.10 Isolated Patient Monitoring and Telemedicine

Telemedicine and detached patient listening have emerged as transformative uses of bioinformatics, exploiting cutting-edge science to help the habit of healthcare being provided. The study and understanding of the enormous books of data created by remote following and telemedicine massively rely on bioinformatics, the association of computational methods accompanying biological data. These methods present healing specialists the ability to, by chance, monitor the insides of their patients, continuous aches, and comprehensive health state, reconstructing patient care and hateful down on in-body visits. These bioinformatics-located patterns offer several benefits. By giving victims current information on their ailments, they allow victims to take an active function in their situation. Remote listening of patients with incessant illnesses allows healing experts to see early warning signals and take appropriate operations. This results in itemized treatment regimes, fewer hospital stays, and better patient effects [20]. Additionally, bioinformatics electronics manage the possibility of combining data from telemedicine and remote patient listening into electronic medical records (EMRs), advancing complete and data-driven healthcare.

12.11 Integration of Smart Farming and Human Bioinformatics

12.11.1 The Synergy between Smart Farming and Human Bioinformatics

An important cross-corrective partnership that has the potential to change both farming and healthcare is the junction of smart farming and human bioinformatics. Human bioinformatics focuses on analysing hereditary and strength-connected data, whereas smart breeding engages IoT science to improve agricultural forms. These regimens work well together because they both rely on data-compelled acumen and share the aim of enhancing human health [22]. Data sharing and interoperability between agricultural and healthcare systems

The plan of data giving and interoperability is one of the main forces advancing the cooperation between smart breeding and human bioinformatics. Large volumes of data about soil traits, weather patterns, and crop fitness are created by land systems. Similar news is assembled by healthcare orders on sickness profiles, healing histories, and a human person's family tree. New intuitions that are favourable to both sectors may become known when these miscellaneous data sources are replaced and absorbed. By detecting links between cultivation orders and human health consequences, models, and environmental facts from smart farms can help healthcare research [21]. Research has connected respiratory sicknesses, irritant uncovering, and air character. Shared data can be managed smoothly for epidemiological studies to completely understand these links. Additionally, combining data from smart agriculture with accompanying human bioinformatics can help in predicting outbreaks of ailments. Researchers can identify well-being concerns and be cautious by trying material elements and styles of cruel fitness. The union of human bioinformatics and smart cultivation offers a chance to get all-encompassing observations into the relations among the environment, farming, and human energy by advancing cooperation and giving information money. This cross-corrective policy has the potential to bring about breakthroughs that improve tangible sustainability, human happiness, and cooking results. Predictive modelling for crop yields and disease outbreaks is another promising application of AI in agriculture.

A powerful plan for bridging the break among land practices, environmental determinants, and human strength belongings is predictive shaping. Researchers can design models that anticipate crop yields and affliction outbreaks by handling the extensive amounts of data assembled in the land and healing areas. In addition to reconstructing land productivity, this multidisciplinary approach likewise climaxes the complex friendships among farming, the surroundings, and human energy.

A study conducted by Hassel et al. [23] revealed the correlation between the natural environment and human society that is stimulating the emergence of novel microbes. The role of forecasting in public health is largely debated. Genomics is

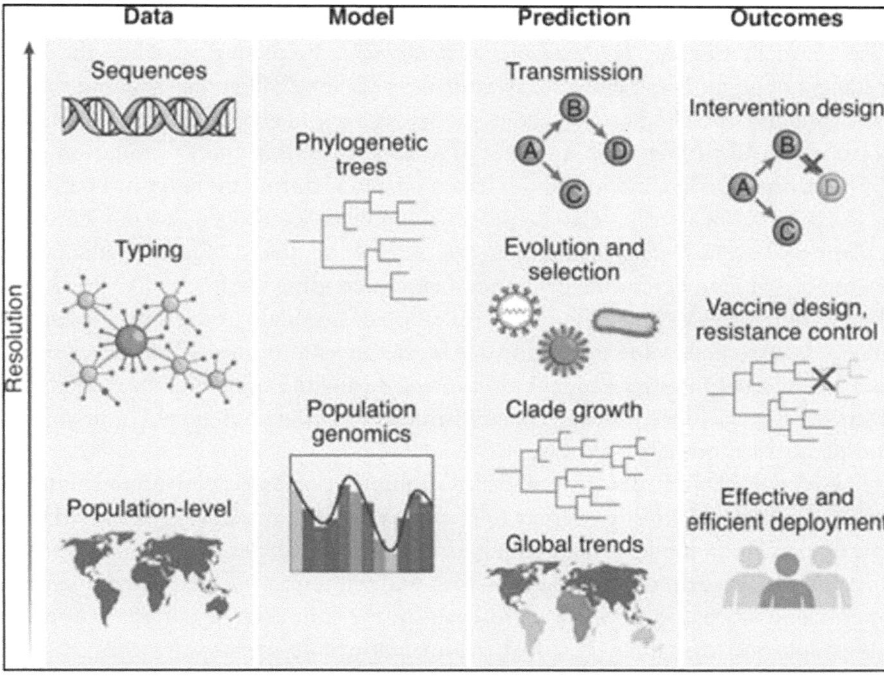

Figure 12.3 Forecasting models for crop yield and human health.

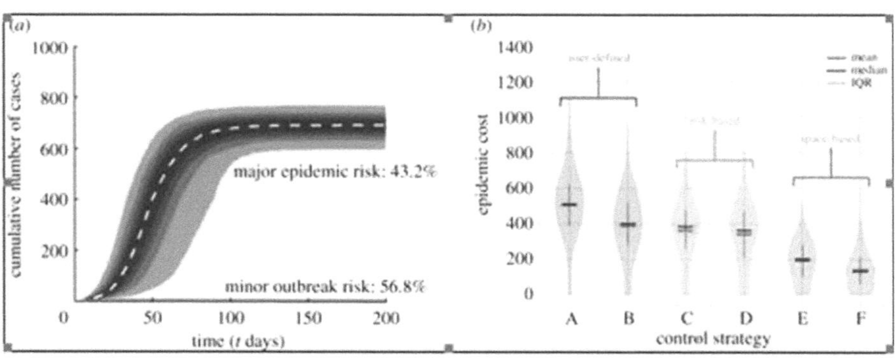

Figure 12.4 Forecasting and control strategy for disease.

an excellent tool for epidemiological analysis of the infestation of crops affecting humans (Figure 12.3).

Mathematical modelling is widely used for epidemic forecasting and outbreak analytics. Such modelling often begins when a new pathogen evolves and affects humans, which is spread through plants and animals (Figure 12.4).

Plant-based epidemics threaten food security. The primary pathogen of plant-based human diseases is *Pseudomonas aeruginosa*. Predicting whether the early instances of an outbreak will fade away or whether they will go on to start a significant epidemic is difficult when the microbe has been identified among the human population. All pathogens in a variety of systems share this query. Thompson et al. [35] described the likelihood of a significant epidemic during the entry of a pathogen to the population after a previous epidemic of a related pathogen like influenza and a plant pathogen, *Podosphaera plantaginis*, infects the host. Once an epidemic has begun, it is vital to determine its current magnitude using the frequently sparse data that is available. Bourhis et al.'s [34] explanation of how data from recurrent samples may be used to determine the magnitude of the current outbreak. Though such data may be collected through frequent monitoring during the epidemic, the data may be fetched by frequently surveilling the transmission of infectious agents from animals and plants by thorough screening.

Gaydos et al. [33] demonstrated the application of participative modelling for forecasting and monitoring a variety of model types, from abstract models to 3-D stochastic simulation models, highlighting its capability. Lessler et al. addressed the issue of examining epidemiologic subtleties with and without a planned control strategy when using stochastic simulations to evaluate the influence of current actions on epidemiological dynamics. Potential treatments include preventative measures like immunization, which Baker et al. [31] and Getz et al. [30] investigated in the epidemic models of wildlife rabies and Ebola in people and its preventive transmission, while Chapters et al. evaluated it in the context of cattle illnesses. Probert et al. [28] and others utilizing reinforcement learning and optimal control theory, respectively, and Bussell et al. [27] also concentrate on utilizing models to lead therapies; similar strategies could also apply to pathogens regardless of the kind of host.

Ansarifar et al. [36] have developed a machine learning predictive model to forecast crop yield through weather, soil, and agronomic data. This model effectively predicted the soya bean yield with less than an 8% root mean square error value. The data was gathered from the gridded soil survey and environmental Mesonet database. The developed regression model uses weather, soil, and statistical explanatory variables like precipitation, temperature, and clay percentage. The relative root mean square error of the crop yield, such as soya beans and corn, during 2015–2018 is illustrated in Figure 12.5.

The yield increases from April to July due to ambient weather conditions. The predictions were till the end of December to depict the variations in yield based on the environmental variables.

A study conducted by Martin et al. predicted the yield of Irish potatoes and maize using machine learning models, such as support vector machines, random forests, and polynomial regression models. Results suggested that the random forest model has high R squared values of 0.87 and 0.81 for potatoes and maize, respectively. The above graph illustrates the relationship between rainfall and the yield of potatoes and maize. Maize yield increases with the rainfall, whereas the yield of potatoes does not

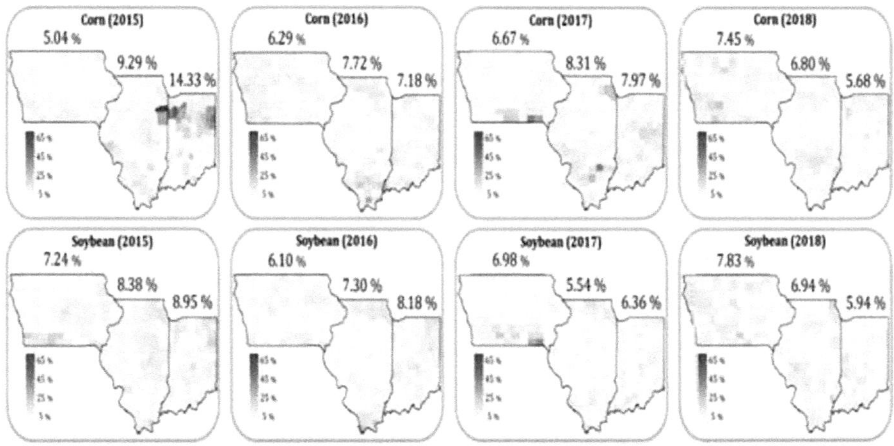

Figure 12.5 The root mean square error.

require rainfall after the first month. (Refer to Figure 12.6.) Rainfall ranging from 50 to 200 mm aids in the optimal yield of potatoes.

12.12 Challenges, Future Directions, and Conclusion

12.12.1 Challenges

Smart agricultural and human bioinformatics integration offers transformational prospects, but it also raises several difficulties and ethical issues that require careful thought. Addressing these issues is essential as these professions develop to make sure that society benefits and appropriate outcomes are achieved.

12.12.1.1 Challenges

12.12.1.1.1 Data Privacy and Security

The gathering and exchange of private data is a requirement for both smart farming and human bioinformatics. To avoid unauthorized access, security breaches, and abuse, it is crucial to ensure the privacy and security of genetic information, agricultural data, and personal health information [26].

12.12.1.1.2 Interoperability

Standardized formats and protocols are needed to integrate various data sources, sensors, and platforms. Data exchange, analysis, and teamwork can all be hampered by a lack of interoperability [25].

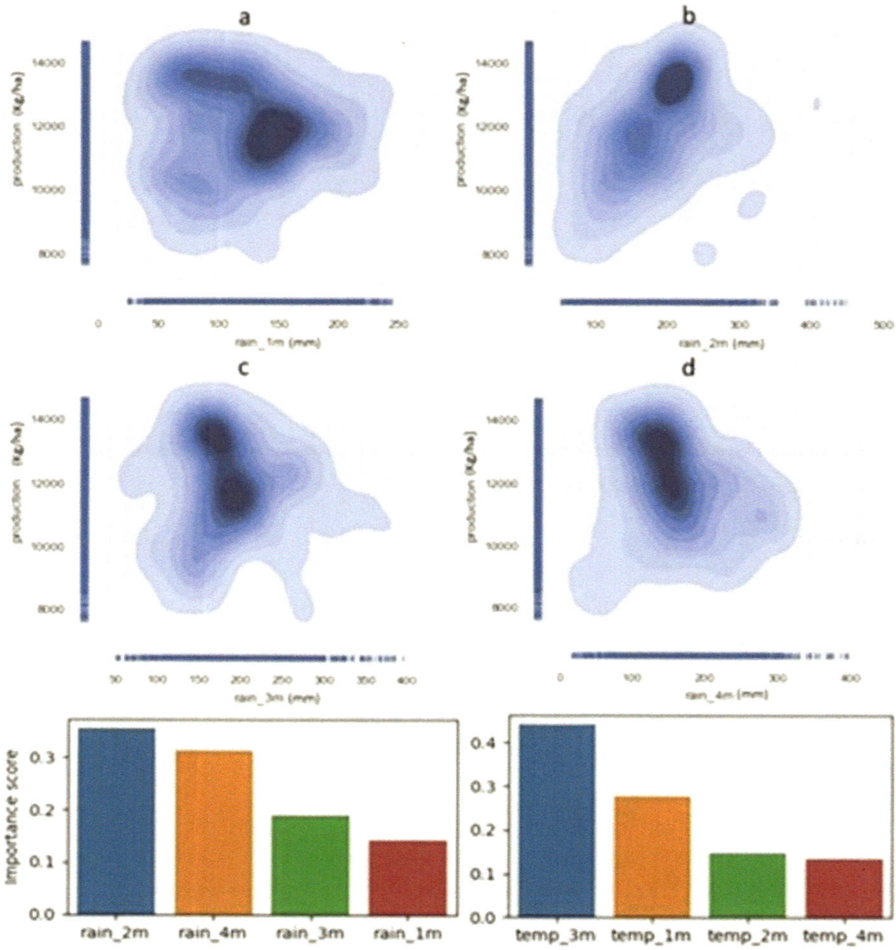

Figure 12.6 The rainfall.

12.12.1.1.3 Data Accuracy and Quality

The dependability of data gathered by sensors and gadgets is crucial. Inaccurate/ Incomplete or inaccurate data might result in erroneous conclusions and judgements [25].

12.12.1.1.4 Regulatory Frameworks

The dynamic nature of these industries puts pressure on the effectiveness of current laws and policies. To avoid legal issues, explicit rules must be established for data ownership, consent, and sharing.

12.12.1.2 Ethical Concerns

12.12.1.2.1 Informed Consent

Whether for agricultural or medical purposes, collecting data from people requires their informed permission. People should be able to choose how their data is used and should be aware of how it will be used.

12.12.1.2.2 Data Ownership and Control

Determining who owns the data created by sensor devices presents ethical concerns, particularly in the context of health and genetics. How data is shared and utilized should be at the authority of the individual [25].

12.12.1.2.3 Equity and Access

There is a chance that improved technology may favour certain groups more than others while denying access to or technological competence to others.

12.12.1.2.4 Fairness and Bias

Data analysis algorithms may unintentionally reinforce prejudices, producing uneven results. Fairness assurance and bias reduction are moral requirements [26].

12.12.1.2.5 Accountability and Transparency

Stakeholders should be responsible for their activities and transparent about data practices. Public trust is damaged by a lack of openness. Large-scale sensor network deployment may hurt the environment due to energy use and electronic waste [25].

12.12.2 Future Opportunities and Trends

Future directions in this field involve the creation of sophisticated prediction models that use socioeconomic and demographic information in addition to environmental and genetic information. Governments, healthcare experts, and producers may form determinations based on best choice-free data with the help of these models, which can offer inclusive insights into the elaborate connections between farming, the surroundings, and human strength [24]. Additionally, as technology advances, the use of more leading sensors, analytics powered by AI, and edge calculating will increase the accuracy and efficiency of data assemblage and interpretation [9]. This will likely manage to monitor conditions in real-time, label diseases early, and form exact situations in both land and healing settings.

12.12.3 Conclusion

A trouble cross-corrective strategy that has excellent potential for mutating healthcare, farming, and the relationship between the surroundings and human comfort is the consolidation of smart farming and human bioinformatics. A better understanding of the friendships among agricultural practices, incidental issues, and human strength has made attainable a piece of cooperation across these subdivisions, which is compelled by data-compelled insights. Several issues need to be ironed out in addition to the marvellous potential. Due to the interconnectedness of these commerces, data protection and privacy are critical. Sensitive data must be shipped and stored fixedly. For smooth data exchange and reasoning, it is also owned by determining unity across differing sensor systems, databases, and manifestos.

References

1. Rahat, I. S., et al. (2024). A step towards automated haematology: DL models for blood cell detection and classification. *EAI Endorsed Transactions on Pervasive Health and Technology* 10, 1–24.
2. Kumar, K. S. R., et al. (2023). IoT and Data Mining Techniques to Detect and Regulate of Solar Power System. In *2023 International Conference on Inventive Computation Technologies (ICICT)*. IEEE.
3. Stojanovic, Z., Milinkovic, D., & Logar, J. (2021). Internet of things in agriculture: A comprehensive overview. *IEEE Access*, 9, 19895–19913.
4. Gubbi, J., Buyya, R., Marusic, S., & Palaniswami, M. (2013). Internet of Things (IoT): A vision, architectural elements, and future directions. *Future Generation Computer Systems*, 29(7), 1645–1660.
5. Wani, S. P., Rockström, J., & Oweis, T. (2009). Rainfed agriculture: unlocking the potential. *Comprehensive Assessment of Water Management in Agriculture Discussion Paper 3*. International Water Management Institute.
6. Evett, S. R., & Schwartz, R. C. (2016). Precision agriculture and water management. *Precision Agriculture*, 17(6), 581–584.
7. Srinivasan, A., & Velmurugan, A. (2016). Precision Agriculture: An IoT-Based Approach. In *Proceedings of the International Conference on Informatics and Analytics* (pp. 211–217). Springer.
8. Wang, Y., Kung, L., & Byrd, T. A. (2018). Big data analytics: Understanding its capabilities and potential benefits for healthcare organizations. *Technological Forecasting and Social Change*, 126, 3–13.
9. Zanella, A., Bui, N., Castellani, A., Vangelista, L., & Zorzi, M. (2014). Internet of Things for Smart Cities. *IEEE Internet of Things Journal*, 1(1), 22–32.
10. Sun, L., Luo, J., & Chen, L. (2018). A review of data-driven smart farming: Past, present and future. *Journal of Cleaner Production*, 187, 603–613.
11. Baxevanis, A. D. (2005). *Bioinformatics: A Practical Guide to the Analysis of Genes and Proteins*. Wiley-Interscience.

12. Attwood, T. K., & Parry-Smith, D. J. (2001). *Introduction to Bioinformatics*. Pearson Education.
13. Pevsner, J. (2015). *Bioinformatics and Functional Genomics*. Wiley.
14. Altman, R. B., & Dunker, A. K. (Eds.). (2018). *Practical Bioinformatics*. CRC Press.
15. Ranganathan, S. (2010). *Bioinformatics: Methods and Applications*. Alpha Science International Ltd.
16. Gao, Y., & Zhang, H. (2019). A survey of wearable EEG and ECG monitoring systems. *Journal of Neuroscience Methods*, 325, 108353.
17. Nguyen, D., & Kilic, E. S. (2019). Wearable medical devices for tele-home healthcare. *Sensors*, 19(17), 3746.
18. McManus, D. D., Lee, J., Maitas, O., & Esa, N. (2020). A novel application for the detection of an irregular pulse using an iPhone 4S in patients with atrial fibrillation. *Heart Rhythm*, 9(1), 23–27.
19. Klonoff, D. C., & Kerr, D. (2017). Smartwatches for monitoring heart rate. *Journal of Diabetes Science and Technology*, 11(1), 144–147.
20. Goldsack, J. C., Coravos, A., Bakker, J. P., Bent, B., Dowling, A. V., Fitzer-Attas, C., ..., & Thwaites, A. (2020). Verification, analytical validation, and clinical validation (V3): the foundation of determining fit-for-purpose for Biometric Monitoring Technologies (BioMeTs). *Digital Medicine*, 3(1), 1–13.
21. Li, J., & Li, J. (2019). Smart farming: Technologies, policies, and socioeconomic impacts. *Sustainability*, 11(17), 4700.
22. Alencar, A. B., Dantas, M. A., & Lima, M. A. S. (2019). Precision agriculture and public health: An epidemiological approach using satellite images to describe the spatiotemporal dynamics of dengue. *PLoS ONE*, 14(6), e0217852.
23. Hassell, J. M., Newbold, T., Dobson, A. P., et al. (2021). Towards an ecosystem model of infectious disease. *Nature Ecology and Evolution* 5, 907–918. https://doi.org/10.1038/s41559-021-01454-8
24. Hossain, M. A., & Muhammad, G. (2020). Applications and challenges of IoT in agriculture: A comprehensive review. *IEEE Access*, 8, 107593–107614.
25. Kocaballi, A. B., Berkovsky, S., Quiroz, J. C., & Rezazadegan, D. (2019). Ethical considerations in AI-based healthcare: A case study of smart hospitals. *IEEE Engineering in Medicine and Biology Magazine*, 38(1), 83–87.
26. Weckert, J. (2020). Ethical issues in agriculture. In Reference Module in Food Science.
27. Bussell, E. H., Dangerfield, C. E., Gilligan, C. A., & Cunniffe, N. J. (2019). Applying optimal control theory to complex epidemiological models to inform real-world disease management. *Philosophical Transactions of the Royal Society B*, 374, 20180284. https://doi.org/10.1098/rstb.2018.0284
28. Probert, J., Lakkur S., Fonnesbeck, J., Shea, K., Runge, C., Tildesley, J., & Ferrari, J. (2019) Context matters: Using reinforcement learning to develop human-readable, state-dependent outbreak response policies. *Philosophical Transactions of the Royal Society B*, 374, 20180277. https://doi.org/10.1098/rstb.2018.0277
29. Chapters, L. (2019). Analysing livestock network data for infectious disease control: an argument for routine data collection in emerging economies. *Philosophical Transactions of the Royal Society B*, 374, 20180264.
30. Getz, M., Salter, R., & Mgbara, W. (2019). Adequacy of SEIR models when epidemics have spatial structure: Ebola in Sierra Leone. *Philosophical Transactions of the Royal Society B*, 374, 20180282.

31. Baker, L., Matthiopoulos, J., Müller, T., Freuling, C., & Hampson, K. (2019). Optimizing spatial and seasonal deployment of vaccination campaigns to eliminate wildlife rabies. *Philosophical Transactions of the Royal Society B*, 374, 20180280.
32. Kaminsky, J., Keegan, L. T., Metcalf, C. J. E., & Lessler, J. (2019). Perfect counterfactuals for epidemic simulations. *Philosophical Transactions of the Royal Society B*, 374, 20180279.
33. Gaydos, A., Petrasova, A., Cobb, C., & Meentemeyer, K. (2019). Forecasting and control of emerging infectious forest disease through participatory modelling. *Philosophical Transactions of the Royal Society B*, 374, 20180283.
34. Bourhis, Y., Gottwald, T., & van den Bosch F. (2019). Translating surveillance data into incidence estimates. *Philosophical Transactions of the Royal Society B*, 374, 20180262.
35. Thompson, N., Thompson, P., Pelerman, O., Gupta, S., & Obolski, U. (2019). Increased frequency of travel in the presence of cross-immunity may act to decrease the chance of a global pandemic. *Philosophical Transactions of the Royal Society B*, 374, 20180274.
36. Ansarifar, J., Wang, L. &Archontoulis, S. V. (2021). An interaction regression model for crop yield prediction. *Scientific Reports* 11, 17754. https://doi.org/10.1038/s41598-021-97221-7

Chapter 13

Machine Learning Techniques Applied in Predictive Maintenance: A Review

P. Chenga Reddy, Karamala Naveen, Naveen Kilari, and Nagendra Panini Challa

Abbreviations

A2-LSTM	Attribute-attentioned long short-term memory network
AI	Artificial intelligence
ANN	Artificial Neural Network
CBM	Condition-Based Maintenance
CMS	Condition monitoring systems
CNN-BLSTM	Convolution Neural Network Bi-Directional Long Short-Term Memory
DHL	Deep hybrid learning
DL	Deep learning
DSS	Decision Support System
DT	Decision Tree
FCFS	First-Come, First-Served
IIoT	Industrial Internet of Things

DOI: 10.1201/9781003479062-13

KNN	K-nearest neighbor
KPIs	Key performance indicators
LDA	Linear discrimination analysis
ML	Machine learning
MTTF	Mean Time to Failure
OEE	Overall equipment effectiveness
PdM	Predictive maintenance
RF	Random forest
RNN	Recurrent Neural Network
ROCs	Receiver Operating Characteristics
SVM	Support vector machine

13.1 Introduction

There was little to no scientific foundation for the 1.0 Industrial Revolution and the majority of advancements technology that came before it. It produced power machinery without thermodynamics, a ferrous business devoid of metallurgy, and a chemical enterprise devoid of chemistry [1]. Until 1850, engineering, medicine, and agriculture were practical domains of practical knowledge where objects were recognized by function but their causes were rarely understood. Thus, a considerable deal of energy and inventiveness was wasted on things like fountains of youth, perpetuum mobiles, stones of the wise, and alchemy. People also frequently did not realize which strategies didn't work. Research didn't go in a different path until science proved that such pipedreams were unreal [2]. Furthermore, even when something was known to function, it was usually sluggish and rigid. The second Industrial Revolution brought about a significantly larger range of products and activities than the relatively limited and localized successes of the first Industrial Revolution [3]. As new technology became more ubiquitous in the daily lives of middle-class and working-class people, both their spending power and standard of life increased rapidly. The expansion of industrial organization is the second notable feature of the second Industrial Revolution. Because of economies of scale, certain industries saw tremendous expansion during the second Industrial Revolution [4].

The manufacturing sector of an economy is in charge of creating highly automated and mechanized goods. Since the beginning of industrialization, paradigm shifts brought about by technological advancements have given rise to what are now known as Industrial Revolutions: the first was typified by mechanization [5], the second by the extensive utilization of electrical energy [6], and the third by the widespread adoption of digital technology [7]. The convergence of contemporary Internet technologies with intelligent objects, such as machinery and products, signals a significant shift in industrial production paradigms. This transformation is underpinned by advanced digitalization within factory settings, as illustrated in

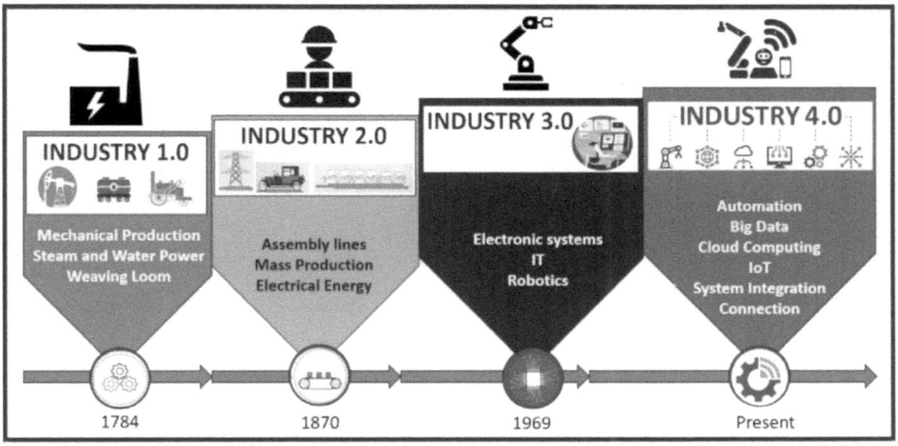

Figure 13.1 Architecture of industrial revolution.

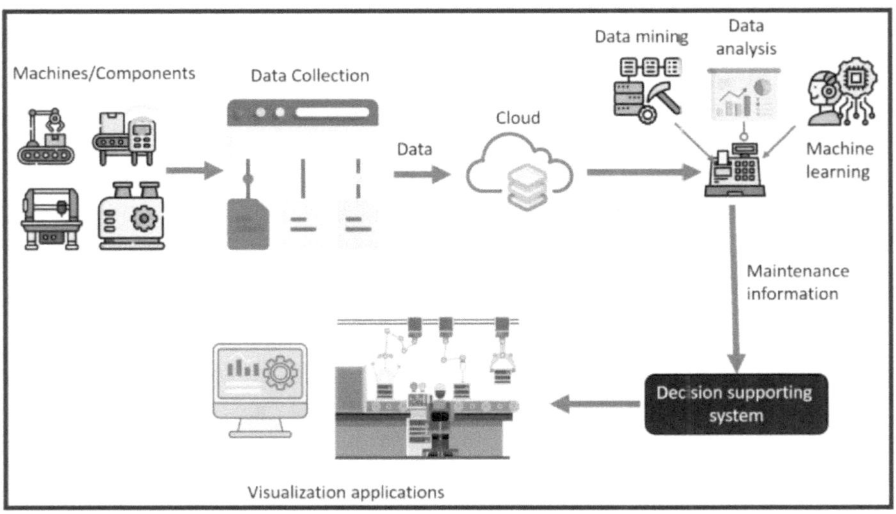

Figure 13.2 Architecture of predictive maintenance.

Figure 13.1. Future production scenarios are characterized by modular and efficient manufacturing processes, and they feature goods that are in control [8].

From Figure 13.2, it can be understood how machine learning (ML) approaches have changed in every engineering process as well as in our personal life. Domains including industrial design, manufacturing, inspection, and monitoring management control, industrial asset upkeep and repairs, as well as product testing and evaluation, have all had their fair share of attention [9]. Equipment maintenance

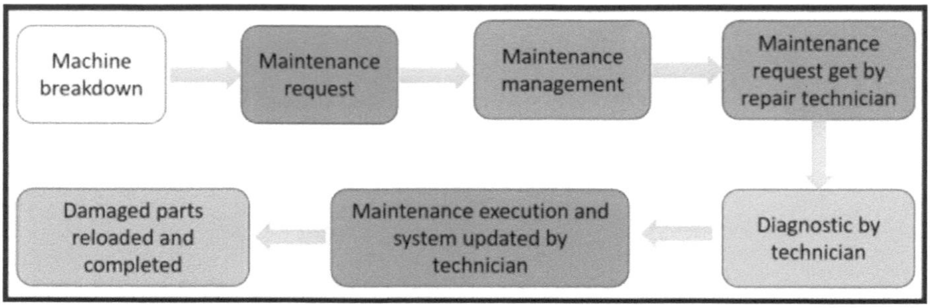

Figure 13.3 Architecture of traditional maintenance.

plays a critical role in industries, impacting both the equipment's operational duration and efficiency as shown in Figure 13.3. Therefore, in order to prevent manufacturing processes from being shut down, equipment problems must be located and fixed. Techniques for ML have shown promise as an instrument for predictive maintenance (PdM) applications to stop equipment breakdowns in the manufacturing lines on the factory floor. However, the right ML method selection affects how well PdM applications perform [10].

A multitude of applications have been made possible by the latest advancements in maintenance modeling, which are driven by data-driven strategies like ML. Ensuring operational safety throughout the product lifecycle and reducing maintenance costs have emerged as key priorities in the automotive industry. PdM stands out as a critical strategy to achieve this goal [11]. The equipment under study is the coiler drums at Steckel mills, which are crucial components of the hot rolling process used to manufacture stainless steel sheets. Since these parts need to be replaced often, maintenance procedures must be carefully scheduled to maintain high plant efficiency. An ML prediction model can be fitted using the vast amounts of data generated by the plant, once it has been condensed and made more readable. This choice was taken after considering the difficulties that Industry 4.0's PdM systems—which the Discrete Bayes Filter (DBF) attempts to address—presented [12].

Joerg Leukel et al. carried out a thorough search from 2012 to 2020 for experimental studies published in peer-reviewed journals. Total thirty four of the 1,024 articles that we evaluated satisfied the requirements for inclusion. Our main goals were to comprehend the datasets under analysis, produce features through preprocessing, and train and assess prediction models. The findings show that a wide range of systems and domains were covered, modern preprocessing and training techniques were used, performance evaluation was lacking in some cases, which helped to mitigate the overfitting issue and there was a significant degree of variation in the reporting of experimental designs and outcomes [13]. ML algorithms have significantly improved in terms of performance and usefulness in the industrial sector, particularly in maintenance. By enabling PdM, its implementation delivers notable increases in efficiency. Nevertheless, obtaining the necessary information through rigorous data preparation, interdisciplinary teamwork, and effective employee communication is still necessary for the successful application of such solutions [14].

In the Industry 4.0 ecosystem, Convolution Neural Network Bi-Directional Long Short-Term Memory (CNN-BLSTM) based defect prognosis evaluation of equipment is an efficient and potent ML model that was suggested by Vivek Justus et al. Compared to current tactics, the evidence found throughout this inquiry indicates that ML technique has a significant capability to handle adverse outcomes resulting from manufacturing defects [15]. A real-time PdM system, called AdaBoost, is introduced that uses ML algorithms to categorize various machine stop types. It is intended to be used with knitting machines. The ML model was trained with pre-processed machine speed and step data from the machines, different types of machine stops are idle, lycra, completed roll, gate, feeder, needle, and needle stop. The results show that the recommended system has the ability to precisely categorize machine pauses and facilitate prompt maintenance procedures, which will raise the textile industry's general productivity and efficiency [16]. A PdM system driven by data was created for manufacturing lines. Through the use of actual data given by IoT sensors, the system seeks to identify probable failure signals using ML techniques before they happen. The evaluation's conclusions indicate that the PdM system can help prevent some production halts and was successful in identifying potential failures' warning indications. The models of the boosting technique XGBoost and the bagging ensemble algorithm random forest (RF) seemed to outperform the individual methods in the comparative analyses of ML algorithms [17].

In order to solve a PdM task—such as monitoring vibration, current, temperatures, etc. and running to malfunction data to forecast the Lasting Usable Life of Components—establish and evaluate a Decision Support System (DSS) with an eye toward an actual case study for industry that involves cutting-edge processing and measuring equipment. The primary elements of the proposed DSS are feature extraction, data collecting, cloud storage, predictive modeling, and data analysis. Unlike other relevant studies, DSS innovative technique is built on an extraction of features strategy and ML prediction model that are driven by particular themes that are gathered from both the higher and lower levels of the production chain. By directly assisting the maintainer/operator, the suggested cloud-based architecture with an ML methodology enables the optimization of the machining excellence procedures. These benefits could lower service expenses for manufacturers by boosting productivity and uptime, streamlining maintenance schedules, and sending out real-time alerts about operational issues [18].

Figure 13.4 shows the schematic view of the ML techniques is employed in various manufacturing processes. The Industrial Internet of Things (IIoT), which is based on condition monitoring systems (CMS), has supplanted traditional maintenance with the revolution in Industry 4.0, which is becoming more and more prevalent in the manufacturing sector. Reliable and simpler maintenance are

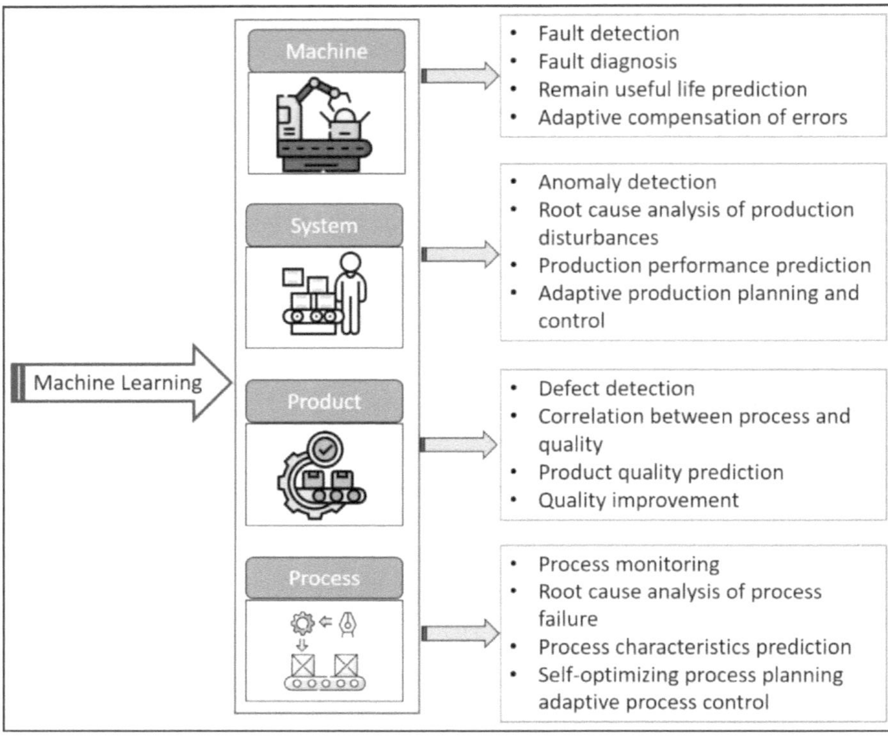

Figure 13.4 Diagram illustrating the application of machine learning across diverse manufacturing processes.

offered by the IIoT concept. In contrast to old style maintenance, IIoT systems that monitor in real time can benefit the organization greatly by alerting the relevant workshop maintenance squad members before a major breakdown happens. Finding malfunctioning bearings before they become problematic during rotation is crucial. This study resulted in the creation of a low-cost, IIoT-based, Industry 4.0-compatible CMS that has three main components: prioritizing the experimental setup, ML models, and IIoT-based condition monitoring applications (CMAs). On an Android-powered mobile smartphone, with wireless data acquisition from the setup, the planned CMA offers actual monitoring and recording. In the event that critical criteria are surpassed, the CMA can also notify the supervisors of the maintenance team via SMS and email on their mobile devices. Finally, well-known ML methods such as support vector machine (SVM), linear discrimination analysis (LDA), RF, decision tree, and k-nearest neighbor (KNN) were used to model the data gathered from the experimental setup for classification. Success was achieved by ML models, while LDA and DT models produced results considerably more quickly than other models. However, the LDA model's classification success rate is not very high. However, the DT model's ability to provide quick and respectable classification rates, along with its ease of threshold value determination, make it the best option for CMS [19].

Keleko et al. centered on bibliometric analysis to offer helpful advice that might aid in understanding the most important scientific questions and major obstacles that come with an effective execution of artificial intelligence (AI) in PdM4.0 for researchers and practitioners. The PdM system's structure methods are not covered in the majority of published publications, despite the fact that they concentrate on AI approaches used in PdM. PdM4.0 is rapidly expanding, especially in the areas of hybrid models driven by data and digital twin frameworks used for anomaly or prognostic diagnosis. Emerging fields like deep learning (DL) and ML also had a big impact on PdM4.0 development [20]. Using the suggested processing approach and data from a population of 29 comparable devices spanning more than four years, failure and Condition-Based Maintenance (CBM) statistical models are built. These models are applied in simulation to determine how various tactics will affect particular key performance indicators (KPIs) over the long run. In comparison to an optimal time-based strategy, the studied example can achieve a 40% cost reduction with six times fewer breakdowns with the use of the simple CBM approach [21].

ML-based PdM is among the most well-known data-driven analytical techniques for industrial system monitoring that aims to maximize efficiency and dependability. In actuality, PdM maximizes equipment life in order to save operating expenses in addition to lowering equipment failure rates. Decision trees enable the examination of many situations in order to determine the circumstances in which a PdM policy—based on the ML algorithm—is financially advantageous in contrast to helpful maintenance is believed to be the present condition. The findings demonstrate that the suggested procedure, which evaluates the various degrees of the frequency and seriousness of errors, is a straightforward and uncomplicated instrument to enhance the method of making decisions. In order to determine the best maintenance approach to adopt at each level, savings and possible expenses have been assessed at the tree's leaf nodes [22]. The application of ML to sophisticated monitoring and diagnostic technologies has grown in popularity. Implementing ML-based PdM can be challenging and costly, particularly for companies facing constraints in manpower, funding, and skill sets. Therefore, a cost-focused analysis is essential to determine the optimal timing for adopting ML-based PdM as the most efficient maintenance strategy. Depending on how well the ML model classifier performs, the implementation of this method will require investments in IT technology in addition to typical maintenance expenditures. To estimate expenses associated with maintenance procedures, an error matrix is employed. Furthermore, the mathematical model delivers a quantitative method based on expenses and is based on the Receiver Operating Characteristics (ROC) curve. In contrast to conventional decision threshold optimization techniques, this optimizes the ML model classifier's decision threshold, allowing for a reduction in maintenance expenses.

PdM may be the best maintenance approach since it aims to reduce maintenance costs and boost production system availability. The quantity of data accessible for the condition of machines and their components, as well as the industry and research interest in ML, have grown significantly in the last few years [23].

Mohsen Nikfar uses a two-phase ML technique (as shown in Figure 13.5) for PdM after measuring vibrations with sensors mounted to low voltage motors. First stage involved analyzing the data to look for any odd activity, and the second stage involved attempting to identify the specific types of faults that might happen. The goal of the suggested PdM system is to assist in identifying the type of issue that is occurring and reduce the time it takes to notice a failure. To identify aberrant motor behavior, we employed and evaluated three ML algorithms: RF, back propagation neural network, and SVM. An SVM was employed to forecast the potential occurrence of particular motor defects. Promising results were shown in detecting anomalous patterns in small-voltage motors using this two-phase ML approach. In order to improve operator and machine safety, lower maintenance costs, and offer

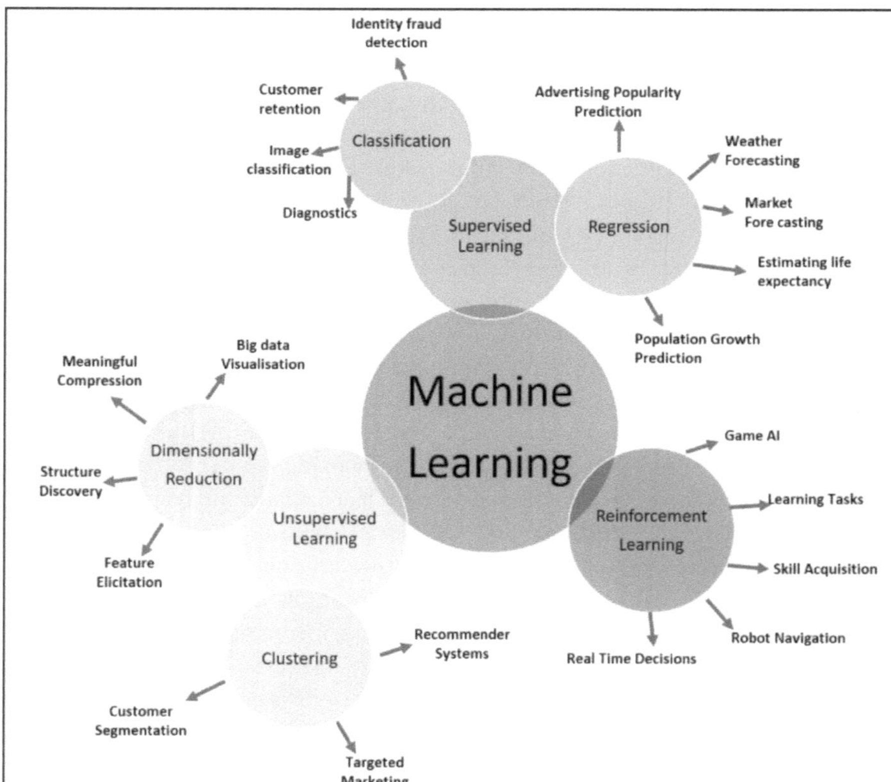

Figure 13.5 Machine learning algorithms.

high confidence regarding the motor status, this ML component can be integrated into a PdM system [24].

By effectively predicting likely machine failure, manufacturers may prevent idle time caused by uneven tool wear or inferior workpiece quality and identify possible difficulties during operations. By forecasting probable system failures based on particular traits or system settings (input variables), it could be able to reduce downtime and prevent future failures. The comparison of different DL, ML, and deep hybrid learning (DHL) models for fault detection was exercised. The algorithms that were used in the end, deep forest and gradient boosting, produced results that were comparatively accurate—more than 90% [25]. Thus, ML is shown to be an essential part of IIoT, with applications in quality control and management, maintenance cost reduction, and industrial process optimization [26]. Due to the increasing need to lower downtime and associated expenses, PdM is a popular approach to maintenance concerns. Creating the "Health factors," often known as quantitative indicators, of a system's condition related to a particular maintenance problem and figuring out among the difficulties with PdM are its relationships to operating costs and failure risk. The suggested PdM method makes it possible to apply dynamic decision rules to maintenance management—even in situations where there are problems with filtered and high-dimensional data. In order to achieve this, several classification modules with different time horizons for prediction are taught. This results in different performance tradeoffs regarding the regularity of unplanned pauses and unused lifetime. The operating cost-based maintenance decision system utilizes this data to minimize anticipated expenses [27].

The fundamental ideas, tools, and techniques used to build an Industry 4.0 architecture focused on PdM—with a particular focus on cost for small manufacturing enterprises—were presented by Erim Sezer. This research led to the development of an affordable cyber-physical system architecture. This architecture monitors temperature and vibration factors during machining processes in a Haas CNC turning center, storing the data in the cloud. Subsequently, this data is utilized to predict machined part rejections based on predefined quality criteria. Using a technique called recursive partitioning and regression tree (RPART) modelling, promising predictive accuracy is demonstrated when machining quality is forecasted based on temperature and/or vibration data, and then compared to each machined product's average surface roughness [28]. Using ML techniques, Shashidhar Kaparthi developed a comprehensive framework for the design of a structure for creating PdM programs. Furthermore, he provided a case study that uses several ML algorithms to analyze a particular instance scenario in order to demonstrate the suggested approach and assess its effectiveness. Finally, he discovered steady gains in the precision of failure predictions using such ML approaches, indicating the effectiveness of the suggested approach [29]. These days, PdM and the timings may be integrated into automated monitoring through ML techniques, eliminating the need for sights or different senses. The integration of sensor observation data with ML methods offers a viable solution applicable to a toothpaste manufacturing plant equipped with tube filling equipment. The goal of this research project is to forecast which elements will sustain damage in order to raise up to 10% in terms of overall equipment effectiveness (OEE). In this investigation, two ML methodologies, namely RF regression and linear regression, were scrutinized. The findings indicate that employing RF regression for Prognostics and Health Management (PHM) prediction accurately forecasts 88% of the genuine information, whereas linear regression achieves a prediction accuracy of only 59% of the actual data [30].

Earlier algorithms' execution is frequently constrained by their failure to take temporal dependence and attribute contribution to final results into account. Yuchen Jiang devised a comprehensive PdM framework utilizing cloud computing, Internet of Things (IoT) technology, and total productive maintenance to tackle the issue. In this framework, an A2-LSTM (attribute-attentioned long short-term memory network) is introduced. The A2-LSTM is tasked with extracting attributes from a series of electrical recordings provided as input. Subsequently, distinct attributes are included in the attribute attention network, enabling automatic adjustment of each attribute's significance. The health prediction component then receives the reweighted attributes, which establishes the manufacturing system's temporal dependence. Lastly, the A2-LSTM's output—that is, its remaining usable life—can assist personnel in maintaining equipment. Real-world examples confirm the method's efficacy, and the comparison results indicate the A2-LSTM's promise [31]. Salama Mohamed Almazrouei et al. emphasize the advantages of PdM based on AI for work-in-progress at the Oil and Gas Industry (OGI). It provides precise RUL estimation, early fault identification, and enhanced maintenance scheduling. Case studies from real-world applications demonstrate increased precision, reduced downtime and expenses, improving equipment dependability and environmental effectiveness. For further advancement, however, data issues and ethical implementation must be addressed. Decision-making and operations can be optimized by adopting these suggestions. In this chapter, we looked at the latest developments in AI and ML in the oil and gas sector, with an emphasis on how they might be used to foresee and diagnose work-in-progress faults. Our research revealed a significant increase in the adoption of AI for defect diagnostics in work-in-progress (WIPs), particularly within the oil and gas sector, over the past decade. Industry 4.0-driven AI is improving data processing, computational capacity, and storage while revolutionizing automation and mechanical production. It enhances productivity, supply chain intelligence, quality assurance, safety, and problem identification. The 4.0 Revolution is growing rapidly due to the synergy between AI and mechanical production [32].

Applications of ML in the manufacturing sector are shown in Figure 13.6. AI is revolutionizing the manufacturing sector in Industry 4.0 era: engineering systems can now monitor mechanical operations and make intelligent decisions by collaborating in actual time with humans, machines, sensors, and other entities, thanks to the development of the IoT and ML techniques. With AI, firms can use MI techniques

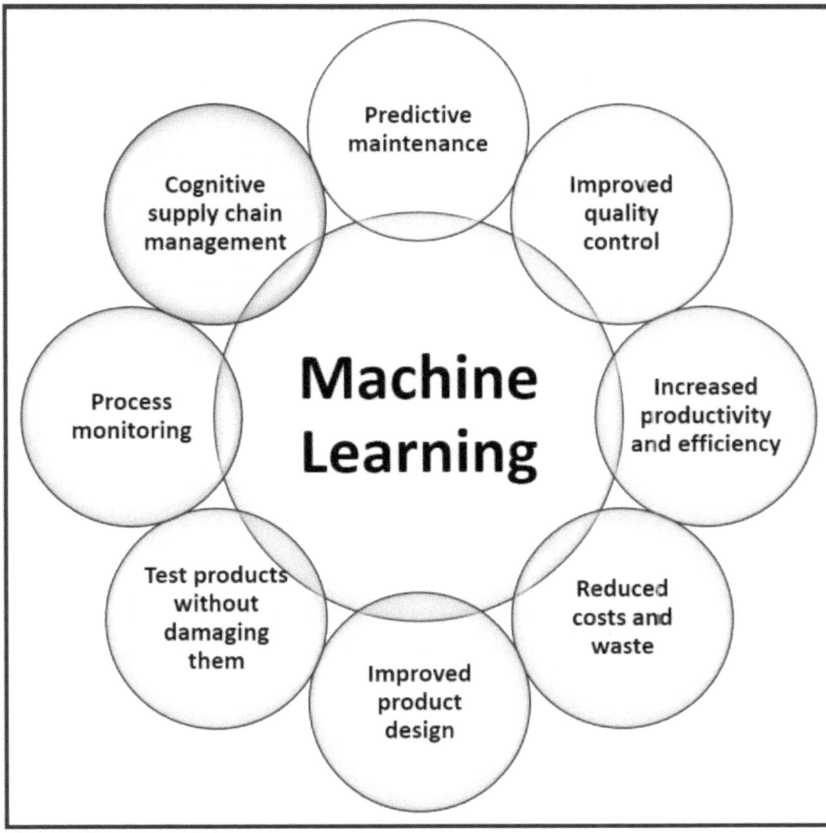

Figure 13.6 Applications of machine learning in the manufacturing sector.

that learn from previous experience to cut design cycles, enhance the supply chain, detect production errors, and minimize equipment maintenance. The creation of PdM systems is one of the last uses for these technologies. Through the use of MI and industrial IoT technologies, PdM forecasts when manufacturing equipment will require repair, enabling problem-solving and flexible decision-making [33]. A factorial experiment was carried out using two years' worth of high-dimensional production data from Joerg Leukel in order to develop an industrial application for the manufacturing of exceptionally accurate milled and turned parts. The impacts of different reading and prediction windows were investigated in connection to four variables (accuracy, precision, recall, and F-score) and three ML techniques (RF, SVMs, and logistic regression). The experimental work's findings demonstrated the predicted window plays a crucial role in connection with the program field, the non-monotonic interaction between performance and the reading window, and the methodical application of sliding window choosing to enhance various performance aspects [34]. Engineering is one of the many intricate fields that make up the IIoT, statistics, and information and operation technologies. The industrial data management system builds a system intended for use in industry using five core layers: the things layer, edge layer, fog layer, communication layer, and cloud services. In order to construct an industrial scenario-based smart enterprise, the cloud helps recover and acquire huge amounts of industrial data produced on the shop floor by multiple equipment in the industry. This particular is then retrieved using a context-aware approach. Umesh W. Hore introduced a novel IoT-based solution for industry that use ML to increase productivity in the workplace and forecast and remotely maintain a variety of industrial environmental characteristics. The solution also helps to reduce the amount of labor that must be done by hand. More precisely, based on controllers, this IIoT solution gathers data on using data from one sensor node; the fog layer assesses air quality, humidity, outside temperature, and boiling temperature to provide a rapid assessment of the intelligence required to operate the system and boost production line productivity. Additional data sources include light intensity, smoke, humidity, indoor temperature, and object detection [35].

In order to forecast the performance enhancement of shop floor management and preventive maintenance operations, the greatest computational modeling (AI/ML and quantum computing) approaches were offered by Biswaranjan Senapati at the industrial production facility's manufacturing sites. Shop floor activities have a major impact on the operational efficiency and productivity of any industrial production site or facility. Factory supervisors and managers in manufacturing facilities have a limited amount of time to forecast production pauses and preventative maintenance. Occasionally, a little milling bit breaks on a production line, causing the line to stop. In certain instances, unscheduled downtime not only results in expensive transactions but also hinders supervisors' ability to manage unscheduled delays and downtime at manufacturing facilities. Biswaranjan Senapati presented a failure detection method in this study that is intended to rank the most likely failure states at maximum capacity. Incoming jobs to the backup unit can also be handled by the system, which allows the overloaded unit to recuperate and start afresh in a new state. Our suggested plan increases overall system durability in the event of a component failure by adding an extra portion of the parallel system using split protocol [36]. PdM has advanced significantly thanks to the latest developments in the digital twins (DT). Reliability is increased through proactive fault prediction and accurate equipment status recognition made possible by DT for PdM. The transition from reactive to proactive services boosts company profitability and competitiveness, streamlines maintenance schedules, and reduces downtime. However, the field of DT for PdM study and application is still in its infancy, most likely due to a lack of thorough industry and academic investigation into the part that ML plays in DT for PdM. The application of DT for PdM was supplied by Chong Chen, and it has the potential to significantly improve the industry's upkeep and operation [37].

PdM seeks to lower expenses so that businesses can compete more fiercely. It optimizes the maintenance intervention plan by combining sensor data with analytics approaches. The implementation of this maintenance strategy necessitates the collaboration of multiple agents and calls for expertise in various domains, as it covers everything from the shop floor's relevant signal averaging incorporating processing, transmission, storage, and analysis to extract valuable insights. Three distinct viewpoints were connected to Nunes et al.: architectures, prognostic methods, and anomaly detection. Prognostics techniques, which deal with models to forecast the state of industrial machinery; anomaly detection, which enables the removal of noisy or inaccurate data and the identification of pertinent events that can be used to enhance the prognostics methods; and architectures, which may permit the deployment of the anomaly detection and prognostics methods in real time and in various industrial scenarios. One of the biggest problems is that the abnormalities could be brought on by a variety of things, including transmission mistakes, equipment malfunctions, stops for curative care, modifications to the working environment, and issues of sensors brought on by extreme climatic circumstances [38]. For automated decision-making, ML systems are being used extensively. In particular, crucial for safety, autonomous vehicles in the automobile sector need testing in order to guarantee their quality. ML systems pose additional difficulties that could impact test maintenance, which is the procedure for updating. The test cases reflect the changing architecture of the system. Lukas Berglund et al. conducted a preliminary case analysis in the automotive industry to identify the variables influencing ML system test and maintenance. Their study aimed to provide recommendations for improving the maintenance procedure. The test's enhanced maintenance, particularly in the area of safety, was evident from the results [39].

Jon Bokrantz et al. suggested an empirically based research agenda that took into account the variety of ways that smart maintenance used and implemented in industry. In focus groups as well as interviews, over 110 specialists from more than 20 distinct organizations participated to find out the responses, implications for performance, and backup plans for smart maintenance. A study agenda comprising five primary topics was developed based on the results, which became a model for contingencies. These included environmental contingencies; institutional isomorphism; change, investment, and interface implementation issues; the four aspects of preventive maintenance; and consequences for plant and company performance. Maintenance has effectively advanced the area of research in industrial maintenance management in terms of its social function in facilitating manufacturing processes that are more ecologically, financially, and socially viable [40]. Innovative solutions leveraging the IoT are paving the way for intelligent and efficient plant management. In this regard, manufacturing organizations need to use those technologies in an inventive way in order to endure in an environment of fierce competition in business. PdM, as it is known, which is driven by these specifications, is crucial and presents a large opportunity for innovation to get around the constraints of conventional maintenance practices. Nevertheless, actual shop floors frequently face challenges in offering information that help smart companies manage their assets more effectively. Despite the potential of sophisticated ML algorithms to help the diagnosis of failure indicators to minimize maintenance scheduling, firms have limited options when it comes to the analysis of manufacturing data. Furthermore, the data generated by individual machines is highly varied, which poses a challenge in terms of integration for the purpose of providing PdM with data-driven decision support. The hybrid ML approach offers a combination of semi-supervised and unsupervised learning, spurred by these issues [41].

An adaptive ARIMA model-based ML strategy is offered to enable adaptive error prediction model by altering the windowing method, hence anticipating future breakdowns before they occur. This is achieved by anticipating the key signature characteristics in the machinery. A high-pressure hydraulic sand molding machine in an automotive gray casting manufacturing foundry uses the suggested maintenance procedure. In the foundry's high pressure sand molding process, the parameter designated for analysis is the degree of oil contamination. After applying the suggested strategy, an analysis is conducted on the monthly breakdown occurrences, measured in minutes, and the overall count of breakdowns attributed to various phenomena. Promising outcomes are guaranteed by the suggested solution, which achieves zero downtime in industries by increasing mean time between failure by 800% [42]. PdM has gained importance in industries as a means of achieving sustainable operational management and reducing maintenance costs. PdM involves predicting the next failure and scheduling the necessary maintenance ahead of time. Facility managers commonly employ reactive or preventative maintenance methods in the development of maintenance management strategies. Reactive and preventive maintenance, on the other hand, are unable to stop failure or forecast how plumbing, electrical, or mechanical parts would behave in the future [43].

Thus, in order to extend the life of the facilities, these parts are fixed ahead of time with the use of PdM planning models that are created with clever techniques. An intelligent PdM planning model—which consists of five consecutive stages: data cleansing, data normalization, optimal feature selection, precision network decision-making, and prediction—was introduced by Mustufa Haider Abidi et al. PdM helps manufacturers become more sustainable by reducing malfunctions, breakdowns, and material waste. Time and material waste can be minimized with an effective PdM. The research endeavor made use of two databases: one for Li-ion batteries and the other for aviation engines. Following their cleaning, PdM data was standardized. To reduce redundant data, the recommended J-SLnO approach was used for optimal feature selection. The anticipated values varied, making it difficult for the ML system to generate the optimum results. An SVM was used to identify a network that could handle prediction values with different ranges. Therefore, a Recurrent Neural Network (RNN) was employed to provide predictions throughout a variety of ranges, and the J-SLnO algorithm that was proposed was used to optimize the weight of the RNN. The J-SLnO-RNN algorithm that was constructed demonstrated an Root Mean Square Error (RMSE) that was 95.3% higher than that of NN, 95.1% higher than KNN, 96.9% higher than RNN, and 61.8% higher than SVM-RNN, according to analysis by proving the J-SLnO-RNN's efficacy for PdM planning that has been suggested [44].

With PdM, the owner of the company can make decisions like replacing or fixing a component before it actually breaks and impacts the entire production line. For Industry 4.0 to optimize work allocations and PdM models, an efficient asset management system is therefore necessary. The resource management system based on Genetic Algorithms (GA) and utilizing ML with fog computing for preventative maintenance is presented in this research. The FogWorkflowsim simulates the energy performance, cost, and time of GA in addition to MinMin, MaxMin, FCFS (First-Come, First-Served) , and RoundRobin. Employing real-time datasets, a two-class logistic regression approach is employed to construct the PdM model. The findings exhibit superior performance of the proposed method over FCFS, RoundRobin, MinMin, MaxMin, and alternative approaches in execution time, cost-effectiveness, and energy efficiency [45]. An ML algorithm can track how the properties of the monitoring sensors deviate from a state of health. By automating the quality control process, ML may be utilized to track the production's quality by tracking the additive manufacturing process to identify printing flaws and enable mitigation by reproducing the affected region and machining it in real time [46].

The recurring failures of the operation must be carefully considered in production systems. Both the brief interventions and the prolonged dysfunctional states come at a disproportionate time and financial expense. Real-time Big Data obtained from machinery is a major component of Industry 4.0 technology,

Table 13.1 Papers Published in Various Fields

Field	*Publications*
Engineering	419,146
Biomedical and Clinical Sciences	191,802
Chemical Sciences	171,551
Materials Engineering	111,019
Biological Sciences	96,713
Agricultural, Veterinary	88,985
Information and Computing	79,576
Chemical Engineering	64,881
Food Sciences	52,333
Environmental Sciences	50,331
Physical Chemistry	48,973

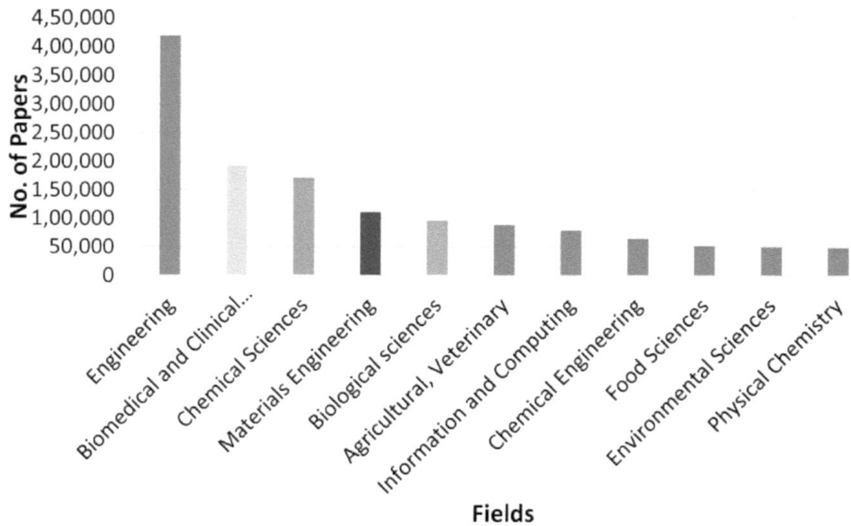

Figure 13.7 No. of papers published in various fields.

allowing prospective issues to be addressed and fixed before they cause an avalanche for the business. It is possible to create lasting solutions, which enables the establishment of production efficiency. The chapter leverages failure records and Mean Time to Failure (MTTF) data from the manufacturing line's robot system to construct an Artificial Neural Network (ANN) model for predicting system failures [47]. In the past decade many research papers were published in various fields such as Engineering, Biomedical and Clinical Sciences, Chemical Sciences, Materials Engineering, and Biological Sciences according to Table 13.1 and Figure 13.7.

13.2 Conclusion

Original research and analysis were conducted to comprehensively examine the pertinent challenges and advancements in utilizing ML and reasoning for PdM within the framework of Industry 4.0. This chapter discussed about the theories, tools used in this field and also difficulties encountered when implementing it in the actual world. Finding architectures or frameworks that employ ontology adoption or reasoning based on ML models was the main goal of this review. This research was limited to PdM of cyber-physical systems and omitted related works that apply PdM in other situations, such as software failure prediction.

References

1. Crafts, Nicholas. "The industrial revolution." In *The economic history of Britain since 1700*, vol. 1, pp. 1–16. Oxford University Press, 1994.
2. Stearns, Peter N. *The industrial revolution in world history*. Routledge, 2020.
3. Hague, Richard, Philip Dickens, and Neil Hopkinson, eds. *Rapid manufacturing: An industrial revolution for the digital age*. John Wiley & Sons, 2006. Schwab, Klaus. *The fourth industrial revolution*. Currency, 2017.
4. Clark, Gregory. "The industrial revolution." In *Handbook of economic growth*, vol. 2, pp. 217–262. Elsevier, 2014.
5. Lasi, Heiner, Peter Fettke, Hans-Georg Kemper, Thomas Feld, and Michael Hoffmann. "Industry 4.0," *Business & Information Systems Engineering* 6, pp. 239–242, 2014.
6. Freeman, Christopher, and Francisco Louçã. *As time goes by: From the industrial revolutions to the information revolution*. Oxford University Press, 2001.
7. McDonough, William, and Michael Braungart. "The next industrial revolution." In *Sustainable solutions*, pp. 139–150, Routledge, 2017.
8. Popkova, Elena G., Yulia V. Ragulina, and Aleksei V. Bogoviz. "Fundamental differences of transition to industry 4.0 from previous industrial revolutions." In *Industry 4.0: Industrial revolution of the 21st century*, pp. 21–29, Springer, 2019.
9. Dalzochio, Jovani, Rafael Kunst, Edison Pignaton, Alecio Binotto, Srijnan Sanyal, Jose Favilla, and Jorge Barbosa. "Machine learning and reasoning for predictive maintenance in Industry 4.0: Current status and challenges," *Computers in Industry* 123, p. 103298, 2020.
10. Çınar, Zeki Murat, Abubakar Abdussalam Nuhu, Qasim Zeeshan, Orhan Korhan, Mohammed Asmael, and Babak Safaei. "Machine learning in predictive maintenance towards sustainable smart manufacturing in industry 4.0," *Sustainability* 12, no. 19, p. 8211, 2020.
11. Theissler, Andreas, Judith Pérez-Velázquez, Marcel Kettelgerdes, and Gordon Elger. "Predictive maintenance enabled by machine learning: Use cases and challenges in the automotive industry," *Reliability Engineering & System Safety* 215, p. 107864, 2021.
12. Ruiz-Sarmiento, Jose-Raul, Javier Monroy, Francisco-Angel Moreno, Cipriano Galindo, Jose-Maria Bonelo, and Javier Gonzalez-Jimenez. "A predictive model for the maintenance of industrial machinery in the context of industry 4.0," *Engineering Applications of Artificial Intelligence* 87, p. 103289, 2020.
13. Leukel, Joerg, Julian González, and Martin Riekert. "Adoption of machine learning technology for failure prediction in industrial maintenance: A systematic review," *Journal of Manufacturing Systems* 61, pp. 87–96, 2021.
14. Welte, Rebecca, Manfred Estler, and Dominik Lucke. "A method for implementation of machine learning solutions for predictive maintenance in small and medium sized enterprises," *Procedia CIRP* 93, pp. 909–914, 2020.
15. Justus, Vivek, and G. R. Kanagachidambaresan. "Machine learning based fault-oriented predictive maintenance in industry 4.0," *International Journal of System Assurance Engineering and Management* 15, no. 1, pp. 462–474, 2024.

16. Elkateb, Sherien, Ahmed Métwalli, Abdelrahman Shendy, and Ahmed EB Abu-Elanien. “Machine learning and IoT – Based predictive maintenance approach for industrial applications,” *Alexandria Engineering Journal* 88, pp. 298–309, 2024.
17. Ayvaz, Serkan, and Koray Alpay. “Predictive maintenance system for production lines in manufacturing: A machine learning approach using IoT data in real-time,” *Expert Systems with Applications* 173, p. 114598, 2021.
18. Rosati, Riccardo, Luca Romeo, Gianalberto Cecchini, Flavio Tonetto, Paolo Viti, Adriano Mancini, and Emanuele Frontoni. “From knowledge-based to big data analytic model: A novel IoT and machine learning based decision support system for predictive maintenance in Industry 4.0,” *Journal of Intelligent Manufacturing* 34, no. 1, pp. 107–121, 2023.
19. Cakir, Mustafa, Mehmet Ali Guvenc, and Selcuk Mistikoglu. “The experimental application of popular machine learning algorithms on predictive maintenance and the design of IIoT based condition monitoring system,” *Computers & Industrial Engineering* 151, p. 106948, 2021.
20. Keleko, Aurelien Teguede, Bernard Kamsu-Foguem, Raymond Houe Ngouna, and Amèvi Tongne. “Artificial intelligence and real-time predictive maintenance in industry 4.0: A bibliometric analysis,” *AI and Ethics* 2, no. 4, pp. 553–577, 2022.
21. Schmidt, Bernard, and Lihui Wang. “Predictive maintenance of machine tool linear axes: A case from manufacturing industry,” *Procedia Manufacturing* 17, pp. 118–125, 2018.
22. Arena, Simone, Eleonora Florian, Ilenia Zennaro, Pier Francesco Orrù, and Fabio Sgarbossa. “A novel decision support system for managing predictive maintenance strategies based on machine learning approaches,” *Safety Science* 146, p. 105529, 2022.
23. Florian, Eleonora, Fabio Sgarbossa, and Ilenia Zennaro. “Machine learning-based predictive maintenance: A cost-oriented model for implementation,” *International Journal of Production Economics* 236, p. 108114, 2021.
24. Nikfar, Mohsen, Julia Bitencourt, and Konstantinos Mykoniatis. “A two-phase machine learning approach for predictive maintenance of low voltage industrial motors,” *Procedia Computer Science* 200, pp. 111–120, 2022.
25. Hosseinzadeh, Ali, F. Frank Chen, Mohammad Shahin, and Hamed Bouzary. “A predictive maintenance approach in manufacturing systems via AI-based early failure detection,” *Manufacturing Letters* 35, pp. 1179–1186, 2023.
26. Samatas, Gerasimos G., Seraphim S. Moumgiakmas, and George A. Papakostas. “Predictive maintenance-bridging artificial intelligence and IoT.” In *2021 IEEE world AI IoT congress (AIIoT)*, pp. 0413–0419. IEEE, 2021.
27. Kanawaday, Ameeth, and Aditya Sane. “Machine learning for predictive maintenance of industrial machines using IoT sensor data.” In *2017 8th IEEE international conference on software engineering and service science (ICSESS)*, pp. 87–90. IEEE, 2017.
28. Sezer, Erim, David Romero, Federico Guedea, Marco Macchi, and Christos Emmanouilidis. “An industry 4.0-enabled low cost predictive maintenance approach for smes.” In *2018 IEEE international conference on engineering, technology and innovation (ICE/ITMC)*, pp. 1–8. IEEE, 2018.

29. Kaparthi, Shashidhar, and Daniel Bumblauskas. "Designing predictive maintenance systems using decision tree-based machine learning techniques," *International Journal of Quality & Reliability Management* 37, no. 4, pp. 659–686, 2020.
30. Natanael, David, and Hadi Sutanto. "Machine learning application using cost-effective components for predictive maintenance in industry: A tube filling machine case study," *Journal of Manufacturing and Materials Processing* 6, no. 5, p. 108, 2022.
31. Jiang, Yuchen, Pengwen Dai, Pengcheng Fang, Ray Y. Zhong, Xiaoli Zhao, and Xiaochun Cao. "A2-LSTM for predictive maintenance of industrial equipment based on machine learning," *Computers & Industrial Engineering* 172, p. 108560, 2022.
32. Mohamed Almazrouei, Salama, Fikri Dweiri, Ridvan Aydin, and Abdalla Alnaqbi. "A review on the advancements and challenges of artificial intelligence based models for predictive maintenance of water injection pumps in the oil and gas industry," *SN Applied Sciences* 5, no. 12, p. 391, 2023.
33. Traini, Emiliano, Giulia Bruno, Gianluca D'antonio, and Franco Lombardi. "Machine learning framework for predictive maintenance in milling," *IFAC-PapersOnLine* 52, no. 13, pp. 177–182, 2019.
34. Leukel, Joerg, Julian González, and Martin Riekert. "Machine learning-based failure prediction in industrial maintenance: Improving performance by sliding window selection," *International Journal of Quality & Reliability Management* 40, no. 6, pp. 1449–1462, 2023.
35. Hore, Umesh W., and D. G. Wakde. "Intelligent predictive maintenance for Industrial Internet of Things (IIoT) using machine learning approach," In International Conference on Intelligent Cyber Physical Systems and Internet of Things, pp. 897–913. Cham: Springer International Publishing, 2022.
36. Senapati, Biswaranjan, and Bharat S. Rawal. "Adopting a deep learning split-protocol based predictive maintenance management system for industrial manufacturing operations," In International Conference on Big Data Intelligence and Computing, pp. 22–39. Singapore: Springer Nature Singapore, 2022.
37. Chen, Chong, Huibin Fu, Yu Zheng, Fei Tao, and Ying Liu. "The advance of digital twin for predictive maintenance: The role and function of machine learning," *Journal of Manufacturing Systems* 71, pp. 581–594, 2023.
38. Nunes, P., J. Santos, and E. Rocha. "Challenges in predictive maintenance – A review," *CIRP Journal of Manufacturing Science and Technology* 40, pp. 53–67, 2023.
39. Berglund, Lukas, Tim Grube, Gregory Gay, Francisco Gomes de Oliveira Neto, and Dimitrios Platis. "Test maintenance for machine learning systems: A case study in the automotive industry," In 2023 IEEE Conference on Software Testing, Verification and Validation (ICST), pp. 410–421, IEEE, 2023.
40. Bokrantz, Jon, Anders Skoogh, Cecilia Berlin, Thorsten Wuest, and Johan Stahre. "Smart maintenance: A research agenda for industrial maintenance management," *International Journal of Production Economics* 224, p. 107547, 2020.
41. Cho, Sangje, Gökan May, Ioannis Tourkogiorgis, Roberto Perez, Oscar Lazaro, Borja de La Maza, and Dimitris Kiritsis. "A hybrid machine learning approach for predictive maintenance in smart factories of the future," In Advances in Production Management Systems. Smart Manufacturing for Industry 4.0: IFIP WG 5.7

International Conference, APMS 2018, Seoul, Korea, August 26–30, 2018, Proceedings, Part II, pp. 311–317. Springer International Publishing, 2018.

42. Roosefert Mohan, T., J. Preetha Roselyn, R. Annie Uthra, D. Devaraj, and K. Umachandran. "Intelligent machine learning based total productive maintenance approach for achieving zero downtime in industrial machinery," *Computers & Industrial Engineering* 157, p. 107267, 2021.
43. Dogan, Alican, and Derya Birant. "Machine learning and data mining in manufacturing," *Expert Systems with Applications* 166, p. 114060, 2021.
44. Abidi, Mustufa Haider, Muneer Khan Mohammed, and Hisham Alkhalefah. "Predictive maintenance planning for industry 4.0 using machine learning for sustainable manufacturing," *Sustainability* 14, no. 6, p. 3387, 2022.
45. Teoh, Yyi Kai, Sukhpal Singh Gill, and Ajith Kumar Parlikad. "IoT and fog computing based predictive maintenance model for effective asset management in industry 4.0 using machine learning," *IEEE Internet of Things Journal* 10, p. 33, 2021.
46. Gan, T. H., J. Kanfoud, H. Nedunuri, A. Amini, and G. Feng. "Industry 4.0: Why machine learning matters?" In *Advances in condition monitoring and structural health monitoring: WCCM 2019*, pp. 397–404. Springer Singapore, 2021.
47. Onur, K. O. C. A., Ozgur Turay Kaymakci, and Muharrem Mercimek. "Advanced predictive maintenance with machine learning failure estimation in industrial packaging robots," In 2020 International Conference on Development and Application Systems (DAS), pp. 1–6. IEEE, 2020.

Chapter 14

Optimization of Parameters During Tribological Investigations on *Azadirachta indica*-Based Bio-Composites

G. Venkatakoteswararao and R. Vinayagamoorthy

14.1 Introduction

Natural composites are currently the most used type of polymer composites because of their specially designed characteristics that are tailored to the needs of each application. When first introduced, bio-based reinforcements were meant to replace crystal and carbon, and they functioned similarly to artificial ones. The main applications for bio-resins are in medicine and food processing. Several bio-resins, including drumstick sap, catechu, amygdalus, banana sap, and soybean oil, have been processed into matrix resin for use in composites intended for structural applications [1–9]. The mechanical results of these composites are primarily determined by the interfacial bond that exists between the reinforcement and matrix. Soy protein has been

DOI: 10.1201/9781003479062-14

used as the matrix material to achieve a high degree of bonding, which improves the composites' ability to support loads. *Azadirachta indica* is a natural product collected from the neem tree by persuasion or natural harm. *A. indica* gum is a transparent, vivid, amber-coloured substance that dissolves in cold water and has a mild flavour. It is used as a bulking agent and in the manufacturing of superior purpose foods (such as diabetic foods). *A. indica* gum (neem gum) is a flavourless, soluble natural exudate of brilliant yellow to amber-coloured substance. It is a compound of heteropolysaccharide and protein that belongs to the galactan gum family. Many researchers have conducted tribological analyses of natural fibre polymer composites. The tribological performance of materials plays a vital place when subjected to severe wear conditions, and this research mainly focusses on the development of bio-composites with the introduction of new natural resin followed by tribological characterization [10–13]. Neem gum has pharmaceutical applications such as drug binder, inspissation agent, slow-release agent, and film coating. It can also be used to improve solubility. The development of functional resins from *A. indica* gum is the subject of the current study. Bio-resins are generally formulated by suitable chemical techniques. After the formulation, composites have been fabricated with different combinations of bio-reinforcements and are then tested for tribological properties under various working conditions.

14.2 Experimental

A nearby dealer supplies the gum *A. indica*, which is milled into a fine powder with an element size of 0.5 mm. After that, distilled water is combined with the finely powdered gum, and it is left to soak for a full day. The resulting homogeneous solution is then continuously treated in a 1:2 ratio with a general-purpose polyester solution. Heating for approximately 6 hours at 80°C while stirring. Following the functional resin's preparation, four different woven jute compositions have been used to create various composites, such as reinforcements. Cobalt naphthenate acted as a catalyst to complete the polymerization, and Methyl Ethyl Ketone Peroxide (MEKP) acted as a hardener to expedite the chemical reaction [14]. Table 14.1 lists the components of the composite samples. Utilizing a compression moulding

Table 14.1 Sample Compositions

S. No.	*Sample Identity*	*NP Matrix (%)*	*Jute Reinforcement (%)*
1.	NP4	50	50
2.	NP3	60	40
3.	NP2	70	30
4.	NP1	80	20

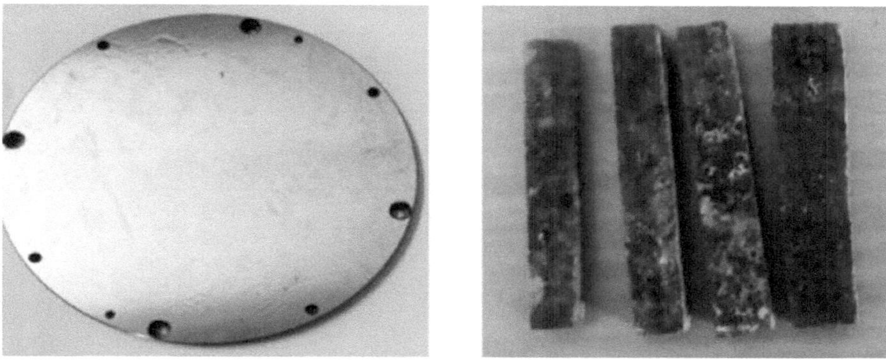

Figure 14.1 Pins and disc for the test of tribology.

machine, the composites were created by correctly arranging the fibre and resin layers and allowing them to settle for a full day. A pin-on-disc tribometer investigation system has been employed to analyse the tribological characters of composites and an ASTM G99 standard is adopted for testing of samples. The composites were cut into the shape of pins to create the samples, sized at 50 mm in height, 6 mm^2 in cross-section, and 165 mm-diameter for the stainless steel disc diameter and 8 mm in thickness, as represented in Figure 14.1.

14.3 Design of Experiments

A statistical method called RSM technique analyses the relationships between one or more response variable quantities and a number of explanatory variables. The primary goal of RSM is to find the finest response done in a sequence of planned experiments. To do this, Box and Wilson recommend utilizing a second-degree polynomial model. They use this model even though they are aware that it is only an approximation because it is simple to estimate and apply, even in situations where little is known about the process. One of the many iteration-based optimal designs that aim to less the variance of the estimated parameters is the D-optimal design. D-optimal method helps the researcher to eliminate unnecessary trials, thereby reducing the expenses and analysing the responses easier [15]. Furthermore, when there are restrictions on the design space and when both numerical and categorical factors, accurate predictions are produced by D-optimal design. With the help of these two- and three-dimensional surface plots, it is simple to examine how responses behave when given particular input parameters. Furthermore, it provides response equations that allow the responses for a specific set of input parameters and could be predicted with accuracy. The answer is that y might be approximated as a second-order mathematical model to input parameters as stated in Equation (14.1).

Table 14.2 Operating Factors

Input Factors	*Type*	*Levels*
Speed (m/s)	Numeric	1, 2, 3
Sliding distance SD (m)	Numeric	1000, 1500, 2000
Load (N)	Numeric	5, 10, 15
Sample	Categorical	I, II, III, IV

$$y = \beta + \beta 1\ x1 + \beta 2\ x2 + \beta 12\ x1\ x2 + \beta 11\ x12 + \beta 22\ x22 + \varepsilon \quad (14.1)$$

A list of the involvement constraints utilized in this design can be found in Table 14.2. Three numerical levels are assigned to load, speed, and sliding distance, while four categorical levels are assigned to the sample. A total of 32 trial runs have been conducted, and in each run, two parameters specific wear rate (SWR) and coefficient of friction (COF) were measured for analysis. Experimental conditions and corresponding measured responses were displayed in Table 14.3.

14.3.1 Specific Wear Rate

As shown in Table 14.4, ANOVA analysis is utilized to investigate how input components affect SWR. If input factors have a p-value is smaller than 0.05, they are considered significant. In this case, among the input factors except the sliding distance, all are affecting the SWR. Although the model does not seem to fit well, the mismatch is not significant. An adequate precision value higher than 4 gives the appropriate signal for the model and 5.87 is more than adequate. The model fits the information fine, as evidenced by the $R^2$and Adjusted R^2 values are very close to 1, conforming the fit of the model. Figures 14.2 and 14.3 show the 3D plots of responses and Figures 14.4 and 14.5 show the normal and residual plots for the responses.

As speed and load increase, SWR decreases. Transfer film that has formed on the steel disc is the cause of this. At this point, the frictional force enhances the temperature, resulting in softening of composite matrix. The transfer film forms more quickly and the counter face is adjusted easily when a maximum load is supplied. Wear rate is low at higher loads as the developed transfer film protects the soft matrix superficial from the tough metal surface. There is very little effect of sliding distance on the SWR. It lowers the fibre content to 40% and then raises it for an additional increase [16]. This indicates that to achieve the best wear resistance, a fibre content range of 40%–50% is required. Additionally, this suggests that the strongest bond between the matrix and fibre is present in this range. Since the point distribution of the normal plot is nearly straight, the developed model is applicable [17].

Table 14.3 Experimentation and Response Measurement

S.No	*Load*	*Speed*	*Sliding Distance*	*Sample*	*COF*	*SWR ($x 10^{-6}$) mm^3/Nm*
1	5	1	1000	IV	0.46	3.16
2	5	3	1500	II	0.56	3.42
3	5	2	1500	I	0.63	5.88
4	5	1	1000	II	0.55	4.92
5	5	1	1500	III	0.44	3.62
6	5	3	2000	III	0.53	3.14
7	5	1	2000	II	0.52	4.42
8	5	1	1000	IV	0.42	3.51
9	5	3	1000	I	0.59	4.84
10	5	2	2000	IV	0.53	3.16
11	5	2	1000	III	0.41	3.21
12	5	3	1500	II	0.52	3.67
13	10	3	2000	I	0.69	3.82
14	10	2	1500	III	0.44	2.92
15	10	2	2000	II	0.65	3.52
16	10	2	1000	II	0.52	3.39
17	10	1	2000	IV	0.48	2.96
18	10	1	1000	I	0.62	4.93
19	10	3	2000	I	0.59	3.63
20	10	3	1000	IV	0.51	3.18
21	15	1	1000	III	0.52	3.54
22	15	3	2000	IV	0.52	1.57
23	15	2	1500	I	0.59	2.44
24	15	3	1000	I	0.67	2.19
25	15	2	2000	III	0.49	4.52
26	15	1	1500	II	0.58	3.26
27	15	1	1000	III	0.43	4.62
28	15	3	1000	II	0.5	1.95
29	15	3	2000	IV	0.55	2.32
30	15	1	2000	I	0.62	3.36
31	15	3	1500	III	0.53	4.21
32	15	2	1000	IV	0.46	3.22

Table 14.4 Analysis of Variance for Specific Wear Rate

Source	*Sum of Squares*	*Degree of Freedom*	*Mean Squares*	*F Value*	*P Value*	*Source*	*Sum of Squares*
Model	24.25	21	1.15	4.83	0.0069	Model	24.25
Load	5.22	1	5.22	21.84	0.0009	Load	5.22
Speed	3.27	1	3.27	13.67	0.0041	Speed	3.27
SD	0.14	1	0.14	0.58	0.4649	SD	0.14
Sample	4.37	3	1.46	6.09	0.0126	Sample	4.37
Error	1.19	6	0.20	–	–	Error	1.19

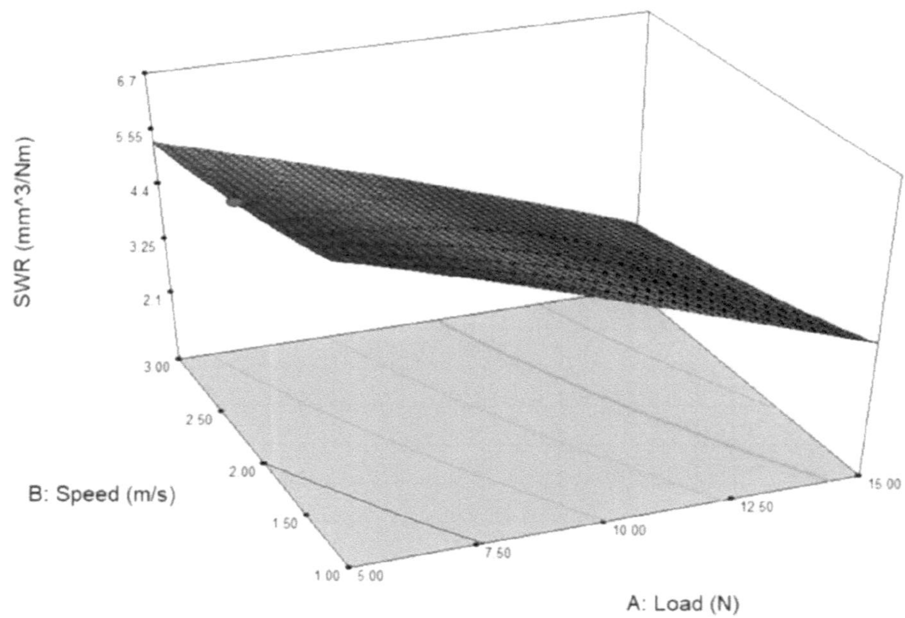

Figure 14.2 3D plots of load, speed, and SWR.

14.3.2 Coefficient of Friction

The findings of the ANOVA table show that among the input factors, only the sample significantly affects the friction coefficient in Table 14.5. Although the model does not seem to fit well, the mismatch is not significant. The model fits the information fine, as evidenced by the R^2 and adjusted R^2 values are very close to 1, suggesting that the model fits the data well. An acceptable signal for predicting responses and the model's sufficient precision of 10.44 is given by a value larger

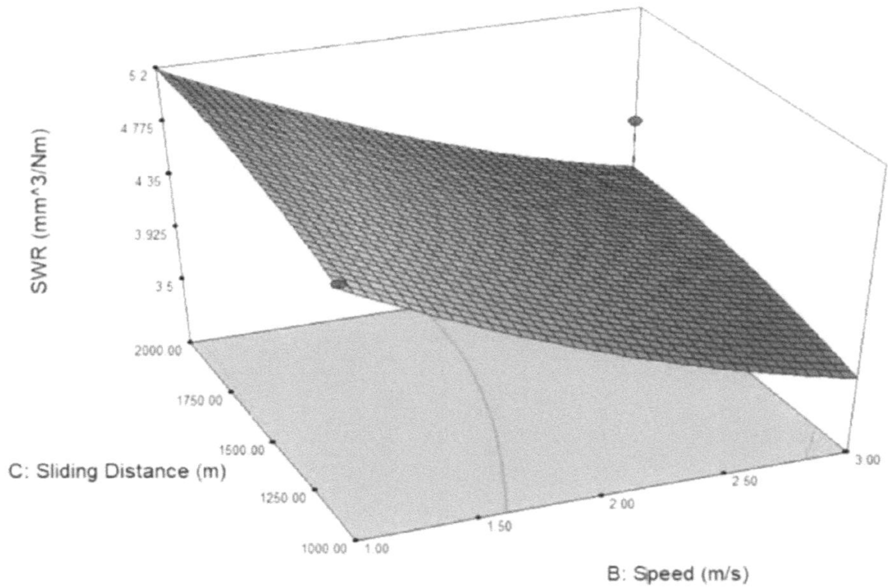

Figure 14.3 3D plots of sliding distance, speed, and SWR.

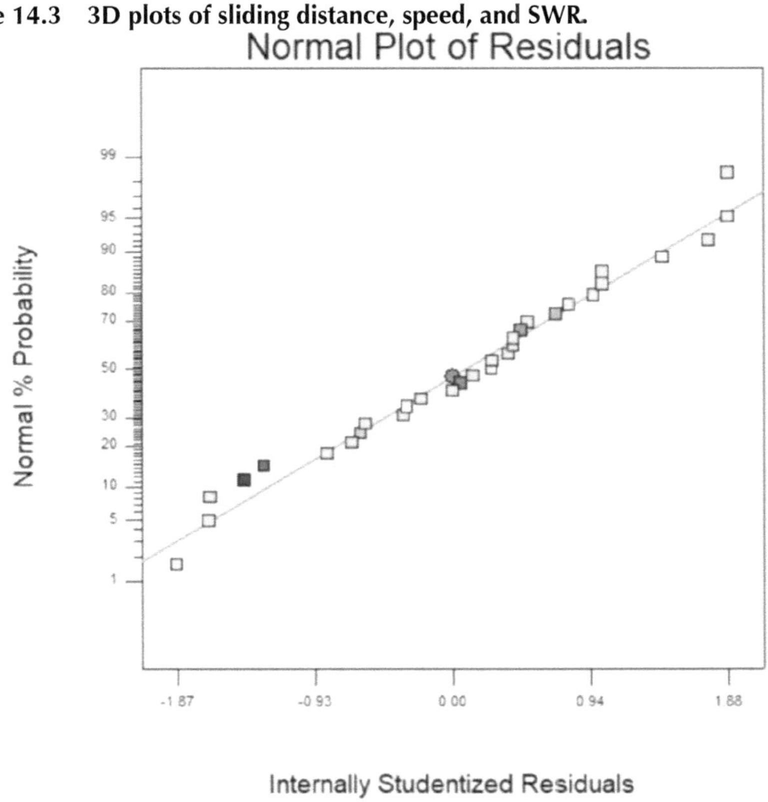

Figure 14.4 Normal plot residuals of SWR.

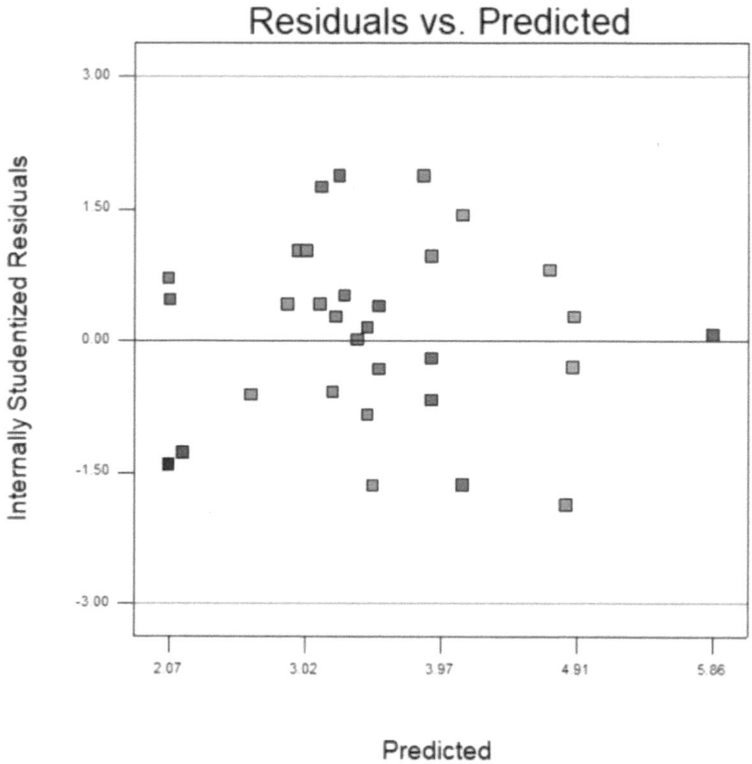

Figure 14.5 Residual vs predicted plot of SWR.

Table 14.5 Analysis of Variance for the Friction Coefficient

Source	*Sum ofSquares*	*Degrees of Freedom*		*Mean Squares*	*F-Value*	*p-Value*
Model	0.15	21		7.353×10^{-3}	3.02	0.0376
Load	3.376×10^{-3}	1		3.376×10^{-3}	1.38	0.2665
Speed	2.170×10^{-3}	1		2.170×10^{-3}	0.89	0.3677
SD	6.466×10^{-3}	1		6.466×10^{-3}	2.65	0.1345
Sample	0.093	3		0.031	12.68	0.0010
Error	0.011	6		1.861×10^{-3}	–	–

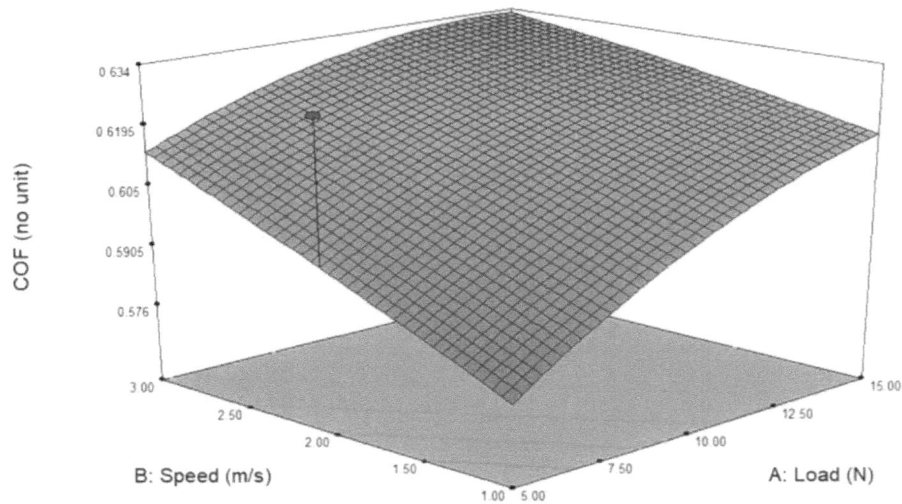

Figure 14.6 3D plots of load, speed, and COF.

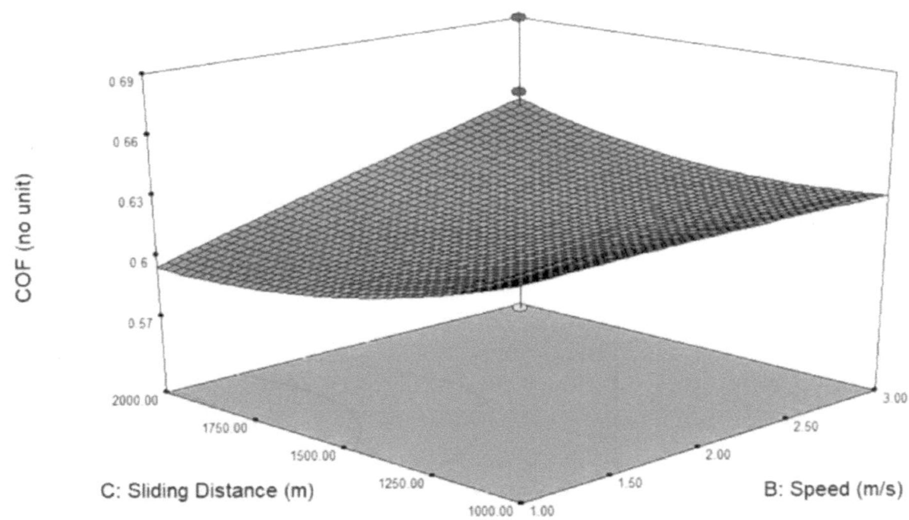

Figure 14.7 3D plots of load, speed, and COF.

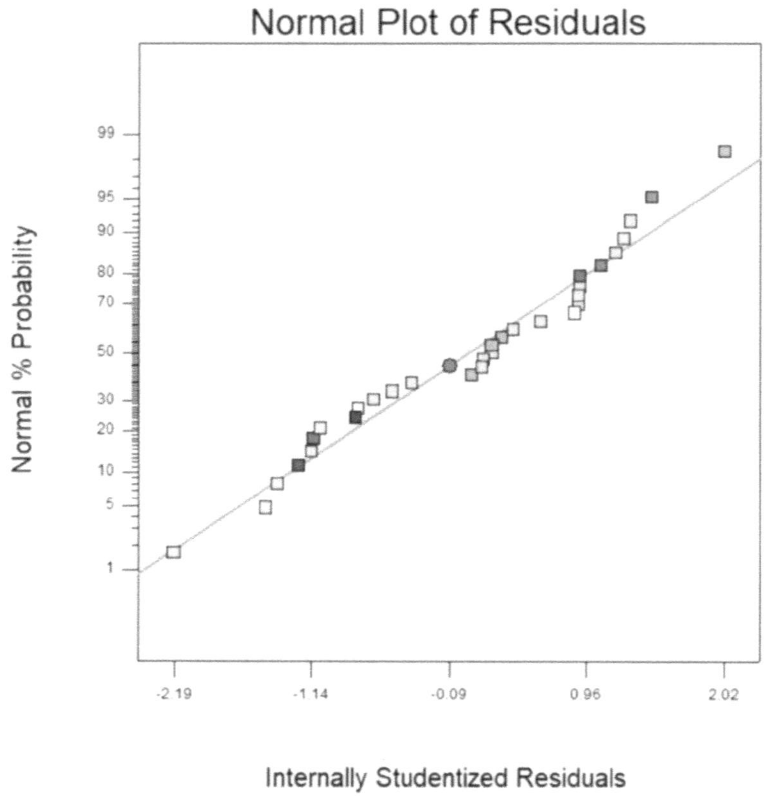

Figure 14.8 Normal plot residuals of SWR.

than 4. The friction coefficient results are displayed in three dimensions shown in Figures 14.6 and 14.7. Interaction plots and normal plots of residuals of COF are indicated in Figures 14.7 and 14.8. It has been discovered that there is a significant increase in the friction coefficient when the load and sliding distance are increased [18].

The area of contact would grow in response to the peripheral force if the normal load was raised. Thus, at that precise moment, the friction coefficient would also increase. As the sliding distance grows, the pin also travels a greater distance, which increases the force exerted by the pin against the disc surface. Speed changes do not seem to have much of an effect on the friction coefficient. The friction coefficient decreases with increasing fibre composition until it reaches 50%, at which point it increases for additional growth [19]. This indicates that 50% jute would be the optimal ratio for minimizing friction. The developed model is significant because the point distribution of the normal plot approaches a straight line.

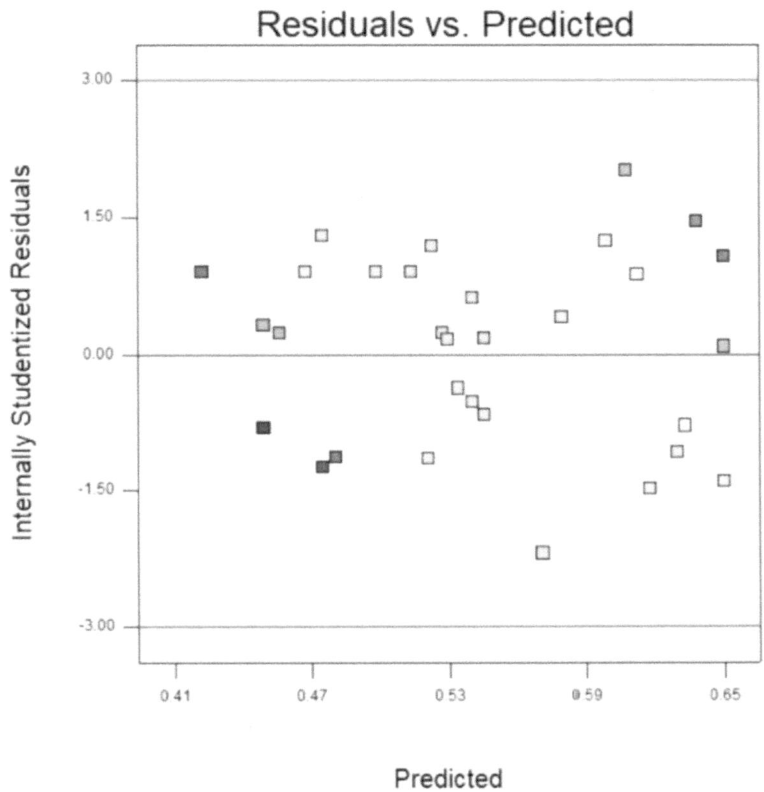

Figure 14.9 Residual vs predicted plot of SW.

Table 14.6 Adequate Variables

Parameters	*Goals*	*Lesser Limit*	*Higher Limit*
SD	Low-high	1000	2000
Speed	Low-high	1	3
Load	Low-high	5	15
FC	Minimalize	0.39	0.72
SWR	Minimalize	1.98	5.98

14.3.3 Confirmation and Optimization

Defining the goals indicated in Table 14.6. From the list of optimum conditions, the one with a desirability nearer to one was chosen once optimization was complete using the desirability method. After the confirmation, the findings indicate the

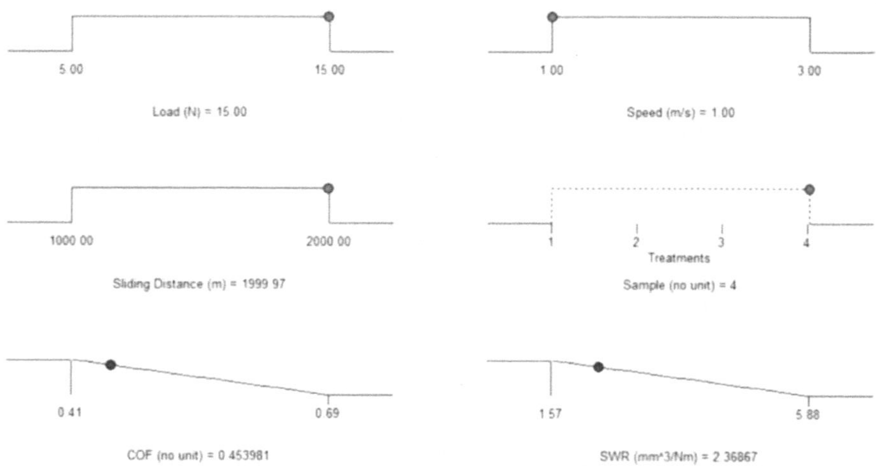

Figure 14.10 Categoric factor levels plot.

optimal parameters are 15 N load, 1 m/s speed, 2000 m sliding distance and sample 3 selection has a maximum desirability of 0.942.

It is discovered that the ideal output conditions for the friction coefficient and SWR are 0.4538 mm^3/Nm and 2.368 mm^3/Nm, respectively. Confirmatory trials have been conducted and performed three times for the optimal condition. Each time, the answers are retested, and the three values' average is calculated and the values are obtained in the categoric factor levels shown in the given Figure14.10.

14.4 Conclusion

Novel composites made of *A. indica* resin have been created and tested for wear and friction resistance. As the fibre content increases, the SWR first decreases by 30%–40% before rising even higher. This proves that for the best wear resistance, a fibre content range of 40%–50% is required. The precise wear rate is largely unaffected by the sliding distance. It lowers the fibre content to 40% and then raises it for an additional increase. The friction coefficient rises with elevation in load and sliding distance noticeably. Speed variations have little effect on the friction coefficient. Following optimization, sample 4 was chosen. The optimal conditions were determined to be a 2000 m siding distance, 1 m/s speed, and a 15 kN load. After conducting confirmatory runs, it is discovered that the results are identical to the anticipated standards.

References

1. Rahat, I.S., et al. "A step towards automated haematology: DL models for blood cell detection and classification," *EAI Endorsed Trans. Perv. Health Tech.*, vol. 10, 2024, p. 9.
2. Kumar, K.S.R., et al. "IoT and data mining techniques to detect and regulating of solar power system." *2023 International Conference on Inventive Computation Technologies (ICICT)*. IEEE, 2023. Manuf., vol. 114, 2018, pp. 40–48.
3. Vinayagamoorthy, R., K.S. Subrahmanyam, K. Murali Krishna Murthy, K. Arun Prajwar, P. Gopinath, M. Sai Lahari, and K. Pruthvi Rangan, "Influence of nanoparticles on the characters of polymeric composites." *IOP Conf. Series: Mater. Sci. Eng.*, vol. 954, 2020, p. 012026.
4. Silva-Guzman, J.A., R.R. Anda, and F.J. Fuentes-Talavera, "Properties of thermoplastic corn starch based green composites reinforced with barley (*Hordeum vulgare* L.) straw particles obtained by thermal compression," *Fibers. Polym.*, vol. 19, 2018, pp. 1970–1979.
5. Inamuddin, B. Rajender, and A.M. Asiri, *Green Sustainable Process for Chemical and Environmental Engineering and Science: Plant-Derived Green Solvents: Properties and Applications*, Elsevier, 2020, p. 64. ISBN 978-0-12-8219003.
6. Venkatakoteswararao, G. and R. Vinayagamoorthy, "Friction and wear behaviour of amygdalus-based green composites," in *Recent Advances in Mechanical Engineering*, Singapore: Springer, 2023, pp. 451–459. Doi: 10.1007/978-981-99-2349-6_41
7. Narwal, S.S., P. Tauro, and S.S. Bisla, *Neem in Sustainable Agriculture*. Scientific Publishers, 2003, p. 10. ISBN 978-93-87991-26-2.
8. Lee, J.H., G.H. Xu, and H. Liang. "Experimental and numerical analysis of friction and wear behavior of polycarbonate". *Wear*, vol. 251, 2001, pp. 1541–1556.
9. Aly, A.A., E.-S.B. Zeidan, A.A. Alshennawy, A.A. El-Masry, and W.A. Wasel, "Friction and wear of polymer composites filled by nano-particles: A review," *World J. Nano Sci. Eng.*, vol. 2, 2012, pp. 32–39.
10. Wang, Y. and F. Yan, "Tribological properties of transfer films of PTFE-based composites," *Wear*, vol. 261, 2006, pp. 1359–1366.
11. Xie, G.Y., G.S. Zhuang, G.X. Sui, et al. Tribological behavior of PEEK/PTFE composites reinforced with potassium titanate whiskers. *Wear*, vol. 268, 2010, pp. 424–430.
12. Pujan, S., N. Modak, and P. Sahoo. "Effect of aluminum filler on friction and wear characteristics of glass epoxy composites," *Silicon*, vol. 10, 2018, pp. 715–723.
13. Ibrahim, R.A. "Influence of natural fillers on tribological and mechanical performance of polyester composites," *Int. J. Adv. Mater. Res.*, vol. 2, 2016, pp. 27–32.
14. Vinayagamoorthy, R., "Bio-fibre reinforced polymeric composites for industrial, medicine and domestic applications," in *Bio-Fiber Reinforced Composite Materials, Composite Science and Technology*, Malaysia: Springer Nature, 2022, pp. 31–49.
15. Rios, N., P. Winker, and K.J. Dennis Lin, "TA algorithms for D-optimal Of a Mixture designs," *Comput. Stat. Data. Anal.*, vol. 168, 2022, pp. 1–10.
16.) Vinayagamoorthy, R. and G. Venkatakoteswararao, "Synthesis and property analysis of green resin-based composites," *J. Thermoplast. Compos. Mater.*, 2020, pp. 1–17.

17. Rajmohan, K.P., J.P. Davim, and A. Arun Premnath, "Modeling and optimization in tribological parameters of polyether ether ketone matrix composites using D-optimal design," *J. Thermoplast. Compos. Mater.*, 2014, pp. 161–188.
18. Sergey Panin, V., O. Vladislav Alexenk, and G. Dmitry Buslovich, "High performance polymer composites: A role of transfer films in ensuring tribological properties—A review," *Polymers*, vol. 14, 2022, pp. 1–43.
19. Vinayagamoorthy, R. "A review on the polymeric laminates reinforced with natural fibers," *J. Reinf. Plast. Compos.*, vol. 36, 2017, pp. 1577–1589.
20. Li, C.W. and G.J. Wang, "MEMS manufacturing techniques for tissue scaffolding devices," in *Mems for Biomedical Applications*, S. Bhansali and A. Vasudev, Eds. Cambridge: Woodhead, 2012, pp. 192–217.

Chapter 15

ANFIS Modelling Study on Surface Water Analysis

P. Meenakshi

15.1 Introduction

Water contamination is a problem which cannot be ignored and requires the attention of the government, not the community. Earth contains 70% of water, and more than 60% of the human body consists of water [1]. The World Health Organization stated that accessible, purified water is necessary for consumption, domestic usage, work, and exercise, as contaminated water creates illnesses in health like diarrhoea, hepatitis, skin diseases, and typhoid [2]. These pollutants can be produced by the movement of plants and people. Important factors affecting water quality include hydrology, weather, climate, topography, and lithology. The rapid growth of industrial development creates the constant development of contamination; the usage of fresh water for internal and commercial needs is increasing day by day [3]. Excessive use of water can lead to increased water demand. Many new toxins have emerged from human activities in recent years. Many industries, including dyeing, tanning, sugar, paper, and pulp, require a lot of water for contamination [4]. But these businesses need to be fixed. Our method of treating wastewater is to use physical, chemical, and organic methods. The purified polluted water must be recycled in industry before being discharged into the atmosphere. If these wastewaters are not

DOI: 10.1201/9781003479062-15

treated properly, they will contaminate soil [5]. Many factories in India are located near rivers. Water flow is the necessary characteristic of river water. Pollutants in the wastewater will be decomposed by the medicinal plants as the water has a self-cleaning effect. However, when too many pollutants enter the river, the river loses its ability to purify itself and all the wastewater flows out, causing serious problems for the environment [6]. Contaminated water is associated with some water problems. In the last decade water pollution has become necessary to have more quantitative methods to assess water quality. Scientists have developed many techniques to measure water quality by combining the electronic and organic properties of water [7]. Water quality is determined by the water quality index (WQI) value as a metric rather than considering many other factors [8]. Since many local and foreign licenced organizations regularly monitor and measure water quality, first-class water meters are used in various models. Finally, many WQIs with different software levels have been developed for specific regions around the world. The WQI is an important and accurate test that shows overall water quality over time and helps select best-in-class treatment equipment to solve problems. The importance is given below:

1. The method uses statistics from the best practices of various water bodies to generate numerical equations that measure water health based on a series of numbers.
2. The requirement for a particular service is less than the total primary function.
3. Good information to help provide drinking water to respective decision makers.
4. It is significant for the operative monitoring and administration of water properties as it shows that artificial neural networks (ANNs) are a powerful prediction formula that has started to be used in recent years. ANN is a mathematical method for predicting the results of input data. Many hidden neurons form a neural network used to express, try, and understand words. ANNs have become one of the most promising methods for estimating nonlinear and complex tasks. WQI is utilized to build neural community replicas using a different type of techniques, like neighbourhood networks, neural networks, and radial root ensembles.

Today's research introduced an excellent modelling procedure known as Adaptive Neuro-Fuzzy Inference System (ANFIS) that successfully predicts water content. ANFIS technology is a combined model of ANN and fuzzy logic structures. It has been accepted that in the years 2017–2020, the water superiority constraints total dissolved solids (TDS), pH, chloride, sulphate, and dissolved oxygen (DO); heavy metal iron and biological parameter total coliform count of the temple pool will be modelled for the WQI. The ANFIS model provides a good distribution of flow characteristics as well as training and management expertise. The results of the ANFIS method are relevant to research needs. The popular expectation value of the results is similar to the cost of research.

15.2 Study on Hybrid System ANN and ANFIS in Water Quality Analysis

Isaac Sajan R. and V. Bibin Christopher [9] described the determination of WQI using a fuzzy logic method. The statistical data used in our work include many constraints like pH, conductivity, TDS, calcium (Ca), magnesium (Mg), HCO_3, chloride (Cl), sulphate (SO_4), total hardness, Drinking Water Quality Index (DWQI), and different physicochemical parameters. This approach offers a general methodology for water class evaluation, allowing informed choices and more dependable and accurate measurements of groundwater [9].

Anugeetha Shine and G. Madhu [10] described the Periyar river in Kerala, India, using hybrid methods ANN and ANFIS. While the most important materials in DO estimation were determined to be the concentrations of phosphate and nitrate that determined from sensitivity test, chemical oxygen demand and total solids were found to be the relevant substances in biological oxygen demand estimation using sensitivity tests [10].

Banerjee, P. et al [11] assessed the feasibility of using ANN simulations instead of mathematical models to estimate water supply costs for managing salinity in island water bodies. The forecast accurateness of the ANN method was extended for another five year period to be similar to the limit value of the allowable level of groundwater purity [11].

Dogan, E. et al [12] developed a hybrid method for the Mellen stream(river) Basin for the period 2001–2002. To estimate biochemical oxygen demand (BOD), the actual data were divided into a direct study and an experiment with a garage. This way, the maximum number of hidden layers is obtained and all nodes two, three, five, and ten are tested. In this mode, the ANN procedure 8 entry and 1 secreted layer (3 nodes) provides the preferred method. An assessment of the outcomes shows that the ANN version provides an inexpensive estimate of BOD [12].

Miao, Q. et al [13] explained that due to the saturation of the development area of the old urban area, the northern of Jiaozhou Bay has a significant area for the improvement of Qingdao in Tuam, Ireland. The Black river became an important river in the region. Use neural networks to create predictive models for sewage quality and expect compensation for chemical oxygen demand, inorganic oil, and ammoniacal nitrogen. The derived results show that the harmful effects of emissions and the pollution retention capacity of the area can be determined, and all pollution can be taken under control after the accomplishment of these high-performance centres [13].

Singh, K.P. et al [14] Description of training, justification and reporting of ANN model for calculating DO and BOD stages in Gomti stream (river in India). The ANN detection model was used as a device to calculate the content of water flow [14].

Dogan, E. et al [15] concluded that in the water quality management and planning process, BOD is an important variable, but it is problematic to determine and takes long time (five days) to obtain outcomes. ANNs are increasingly utilized to estimate changes in water supply. Based on the concept of benchmarking, it has been shown that the hybrid method can be effectively utilized to expect the daily BOD in biochemical wastewater influent [15].

Elhatip. H, Komar, M.A [16] explained that the ANN model is used in water pollution caused by anaerobic conditions at the Mamasın Dam site due to the increase in nutrients and the decrease in O_2 levels in the water. It can promote the growth of algae in such waters, thus reducing the water level [16].

Palani, S [17] explained that this study was conducted off the coast of Singapore. ANN models can also provide simulated results for areas of interest where statistical tests are not available but the desired water quality is poor [17].

Yingyi Chen et al [18] explained that many innovations proposed in this ANFIS work, including loop models, hybrid structures, and compound models, can improve the playful development model [18].

Yeon et al [19] showed a modular neutral community prototype that was implemented to the qualitative records of period corresponding to the quantitative message, with the least error [19].

Ozgur Kisi et al [20] described the overall presentation of the hybrid model in accepting chemical oxygen demand equated to neuro-fuzzy and smallest squares auxiliary vector techniques. The results show that Chemical Oxygen Demand (COD) estimation is more accurate by means of total nitrogen, ammonia, and hot water as involvements. [20].

15.3 Water Quality Monitoring

Water monitoring can be described wireless as repeated analysis of water high-quality at permanent points; the processing of facts and prognosis of trends to aid movements centred on the interception and remediation of damaging anthropogenic impacts in the aquatic surroundings. This machine is composed of a particular sensor that can measure the water excellent parameter consisting of pH, turbidity, conductivity, DO value, and temperature. The readings taken from the sensors are handled by means of microcontroller and these processed numerals are conveyed.

15.4 Developments in Water Quality Monitoring

Monitoring water in conventional method consists of collection of samples on site, and performing bodily, chemical, and microbiological evaluations in a laboratory. The effects of checking out take some days [21]. The traditional strategies are carried out by the vital water commission. The traditional techniques want greater time, high price, usage of chemical and may not supply actual-time measurement values.

Therefore, the research delays the constant watching of systems. The investigator has done an investigation on water high-class investigation amongst conventional and contemporary procedures [22]. Contemporary strategies have greater blessings than conventional techniques on the grounds that cutting-edge approaches can give output values and investigate the water superiority constraints in actual period. Later a brief finding of lower water high rate of quality, quicker movement will be occupied to address undesired constituents inside the water. The conventional strategies are the reason to postpone physical errors, which probably take place through the methods. The technique evolved through the use of wi-fi sensor nodes to display water best. The structure has ten constraints which have been supervised by interior node bins, and it turned into related via with the wireless radar node. There has been a get entry to factor to ship statistics to the farmers. If some hassle took place, the disturbing model became used [22]. In comparison to the contemporary technique, where sensors can screen water exceptional parameters quicker and supply higher consequences, the guide technique of qualifying water wi-fi in aquaculture may not give constant consequences, and it needs extra time and greater manpower. Because of the long period and complicated arrangement, the constituents of the water example might also alternate and, consequently, give less treasured information for monitoring water quality [23]. Hence, instigating greater sensors can improve the working of water excellent tracking system that in a roundabout way can assist authorities to put into effect short measures to increase the quality of water.

15.5 Water Quality Monitoring Methods

a. The Internet of Things (IoT) environment includes Intel Galileo Gen 2, which acts as a connection to attain statistics from various electronic systems [24].
b. Supervisory Control and Data Acquisition (SCADA) equipment combined with IoT generation to detect water pollution, channel leaks and automatically take reading of multiple water sources without going through Global System for Mobile Communications (GSM) modules [25].
c. The instantaneous microbial sensor detects four different specific water types. It demonstrates the advantages of microbial sensors over turbidity sensors in watching microorganisms in consumption water as an initial cautioning of microbial contamination [26].
d. Regular methods are based on criteria for organic and abiotic diseases. Then there are embedded sensor systems, microfluidic sensors, and spectroscopic techniques for online sensor analysis [27].
e. Near-infrared (NIR) reflectance spectroscopy estimates the water superiority of the Aceh river [28].

f. Toxicity tests focus on toxic substances, especially toxins, and adenosine triphosphate (ATP) level technology indicates the pathogenicity of bacteria [29].
g. IoT technology has developed ThingSpeak (TS) to watch the quality of water. Reading from various sensors interlinked to Arduino is transferred to the cloud with the help of TS [30].
h. Fast ultraviolet (UV)/and visible light (UV-Vis) spectroscopy technology provides excellent tracking of water from agricultural root vegetable care systems. This calculation depends on UV-Vis absorbance and uses statistical procedures regression, partial least squares, and principal component analysis [31].
i. Biochemical pollutants in island water, Polymerase Chain Reaction (PCR) is an appropriate technique for the recognition of microorganisms in river and is entirely depending on the abstraction and repetition of deoxyribonucleic acid (DNA) fragment structure [29].
j. The IoT is mainly based on systems that monitor water through LEDs. LED lights are based on changes in water quality sensed by various radars. The machine was linked to a Raspberry Pi and planned with a Java programme [32].
k. Surface-enhanced Raman scattering (SERS) as an innovative diagnostic technique is depending on ultrasensitive vibration spectroscopy on water surfaces to detect bacteria [33].
l. Lake uses Laser Induced Fluorescence (LIF) spectroscopy efficiency to find organisms depending on the fluorescence concentration ratio and fluorescence depth to obtain the concentration of organisms [34].
m. Multi Source Transmission Learning (MSTL) for water superiority forecast efficiently uses water superiority data from multiple nearby observations to recover water quality precision and decrease bias [35].
n. A virtual recognizing instrument uses corporal sensor technology to evaluate water anomalies and focuses on the possibility of agricultural water usage [36].
o. Many fish pond water quality monitoring systems (WQSN) use IoT and underwater recognizing sensor to continuously monitor critical standards at specific times using the C programming language using Arduino/Raspberry Pi boards [37].

15.6 Approaches of Controlling Water Class in Different Countries

Water class monitoring approaches have been recognized in different countries as shown in Figure 15.1.

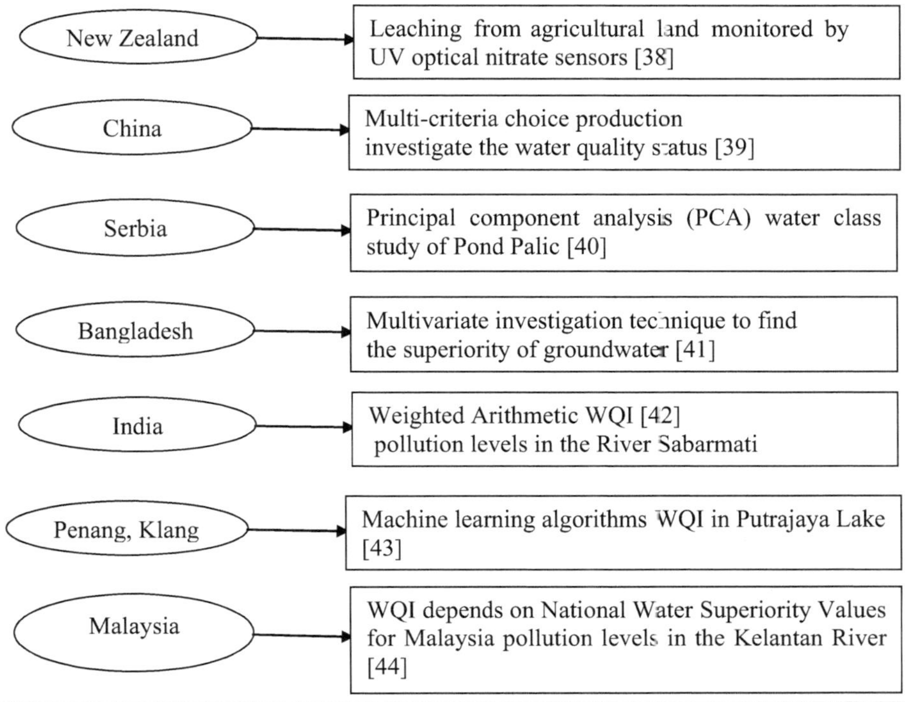

Figure 15.1 Water class monitoring approaches in different countries.

15.7 Water Quality Control System

15.7.1 Virtual Detection System

Virtual sensing systems outperform all physical systems by a large margin. In the evaluation of physical sensors, it uses the concept of existing auxiliary data models of digital or soft sensors and their ability to estimate constraints [36]. It translates various input data from inexpensive sensors and outputs them to extra complicated and luxurious sensors. This process has three important methods: understanding only according to law and ideas from truth or common sense [45]. The development of virtual sensors for monitoring the aquatic environment is generally divided into four steps: acquisition of reality, preliminary data, modelling, and updating.

15.7.2 IoT and Water Quality Monitoring with Actual Period Implementation

Water class monitoring facilities can use the IoT to send data over the Internet. The device uses a voice query to send a message to an online station (called the "cloud").

The present water resistance process is effectively improved in the planned structure, but the water resistance can be improved by the accumulation of additional constraints, including an oxygen reduced capacity sensor or oxygen sensor [28].

15.7.3 Cyber Physics Systems

Cyber-physical systems (CPS) were first proposed by Helen Gill of the American Plant Science Basement in 2006 [46]. CPS is a machine that can easily feed physical additives into computational algorithms. It is the destiny of entrenched systems. CPS provides additional benefits because it is a user-friendly option like fuzzy logic to handle the complicated measured variables called many sensor nodes or sensor arrays. CPS can be utilized to analyse water class. These are intelligent community structures that operate through entrenched sensors, mainframes, and actuators [47].

15.7.4 Optical Technology

Optical sensors and spectral services are other examples of emerging technologies. Recently, smart devices work with electronic and visual devices to expand and manage tracking devices. Electronic measurement is easily portable and solved, while current optical measurement does not affect the water sample and gives more accurate results [48].

15.8 ANFIS (Adaptive Neuro-Fuzzy Inference System)

ANFIS was the primary and was established by Jang in the year 1993. ANFIS can collect feedback using entry–exit numbers and human data in the method of a fuzzy if-then structure [49]. It is the first Fidelity National Information Services (FIS) used in the structure of ANN [50]. ANNs consist of many interconnected neural networks organized into hierarchical structures. Briefly, the ANN model usually has three special layers: entry, exit, and hidden layers. Each layer comprises many neurons, each of which has certain properties and can be fixed or tuned [50]. This method has proven its superiority over traditional methods due to its ability to handle uncertainty and misinformation in data entry as well as nonlinear relationships, which are problems in many applications around the world [51].

15.8.1 Study Area

The study location of the research project is Kanchipuram. The city of Kanchipuram is located on the southwestern road, 76 km from Chennai. The city is positioned in at a height between 2 m and 83 m, decreasing gently to the sea. The city covers an area of 36.14 sq. km. There are small rivers in the region, such as the Palaru river

Figure 15.2 Study area.

and Cheyyar river. Palaru stream is the core water foundation for Kanchipuram city. Surface water ponds will be used in the study area. The location of sample point is shown in Figure 15.2. The lake is square in shape and is fed by rain, and water falls from the surrounding area during rain. Total area 3,481 m^2 (59 m × 59 m) or 37,469.17 ft^2.

15.8.2 Collection of Water Samples

Lake water samples were collected in 1 l polyethylene bottles. Rinse the bottle three to four times with lake water and collect the water sample by immersing the bottle half a meter above the lake surface. Fill the top of the bottle with pond, water and close the bottle strongly with the lid. The collected pool water samples were sent to the laboratory for physical, chemical, and biological analysis and detailed data were collected for further analysis.

Samples were sent to a laboratory and the following values were measured: pH, oxygen level, total softened solids, chloride, sulphate, heavy metal iron, and coliform count. In the four years from 2017 to 2020, the same trend continues in March because the pollution in lake water is highest in this month. Water quality testing procedures are shown in Table 15.1.

15.9 Water Quality Index Determination Procedure

Pond water quality is measured numerically according to the WQI value. In this procedure seven parameters of water were used for calculation and water quality values were calculated according to Bureau of Indian Standards (BIS) surface water category C. Calculation of water superiority constraints is achieved by the weighted

Table 15.1 Water Class Testing Procedures

Sl. No	*Water Class Parameters*	*Methods for Finding the Water Quality Parameters*	*BIS Class – C Standards*
1.	pH	pH meter	6.5–8.5
2.	Dissolved oxygen	Winkler's method	4 mg/l (min)
3.	Total Dissolved Solids	TDS meter	1500 mg/l
4.	Chloride	Silver Nitrate method	600 mg/l
5.	Sulphate	UV spectrometer	400 mg/l
6.	Iron	Atomic absorption spectrophotometer	50 mg/l
7.	Total coliform count	Multiple tube fermentation test	5000 MPN

arithmetic method. The method used to calculate the water quality index is given in Figure 15.3 and represents the BIS category C superficial water standards as shown in Table 15.1.

The pond water class index was derived from the following equation:

$$WQI = \sum [(Qi\, W_i) / \sum W_i] \tag{15.1}$$

The Q_i formula is given below.
The Q_i of *n*th reading taken water class constraint is

$$Q_i = 100(V_n / S_n) \tag{15.2}$$

Q_i = 0 when water is impure, Q_i = 100 if the taken value of limit is equivalent to its acceptable value of pond water. Hence, the larger the value of Q_i, the more pollution is available in the water. The evaluation for pH and oxygen level has superior consideration. The acceptable limit for pH in water is 6.5–8.5. Thus, the superiority assessment for pH and oxygen level (DO) was calculated by the following formula:

$$Q_i = [(V_n - V_{id}) / (S_n - V_{id})]X100 \tag{15.3}$$

Q_i = Superiority rating for nth constraint of the given sample.
V_n = Assessed value of the nth constraint at a particular point.
S_n = Normal suitable worth of the nth constraint.
V_{id} = Ideal worth of nth constraint in clean water.

The unit weights

$$W_i = K / S_n \tag{15.4}$$

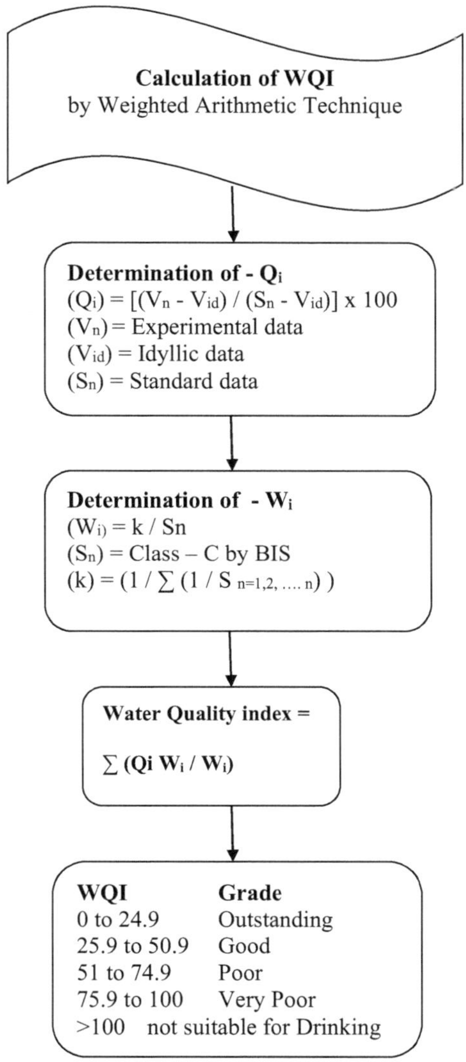

Figure 15.3 Procedure for the calculation of WQI.

K= proportionality constant, K= $[1 / \sum (1 / S_{n=1,2, \ldots n})]$

W_i readings are shown in Table 15.2. The importance of using weighting factors for every parameter when calculating the WQI is to weight the influence of each water quality parameter. Each factor has a dissimilar influence; to weigh the factor, the result of the WQI would be more accurate.

Table 15.2 Unit Weight WI Standards

BIS Standards for Class – C			
	(Sn)	*(1/Sn)*	*(Wi)*
pH	8.5	0.11765	0.2995
Dissolved oxygen	4	0.249	0.63665
Total dissolved solids	1500	0.00067	0.0017
Chloride	600	0.00167	0.004
Sulphate	400	0.00249	0.006
Iron	50	0.021	0.05
Total coliform count	5000	0.00019	0.000501
		0.39267	1

15.10 ANFIS Modelling Methodology

The WQI of temple pool water was predicted based on BIS surface water Class C standards. In the ANFIS modelling process, 30 data readings are necessary; due to the availability of a minimum dataset for the training stage of work, ANFIS created an artificial dataset as per the standard limits of BIS surface water Class C. This created reading has a variety of absent values, and the average interpolation method is implemented for all these absent readings [52]. The flow diagram for ANFIS modelling is represented in Figure 15.4.

ANFIS is a good estimator and one that can make equally good decisions in the neural network [53]. The ANFIS working method procedure has five types of coats, and each layer consists of numerous nodes. The graphic representation of the hybrid model is represented in Figure 15.5. The grid consists of m entries (A1,..., Am) and n member processes. Since the current system is an extension of the existing ANFIS R processes with fuzzy rule sets, it is a standard model. All processes (N) in the first layer can be viewed as the output of the amount of entry m and the number of cooperative operations n, i.e. (N = m × n). The number of nodes in the subsequent coat depends on the number of fuzzy guidelines (R). A description of every layer is provided.

Step 1: In the initial process, entry value is altered into the language form by utilizing then the membership procedure. The exit layer is represented in the equation as

$$O^1_{ij} = (X_i), i = 1,\cdots,m; j = 1,\cdots,n \tag{15.5}$$

μ_{ij} indicates the jth relationship function for input reading X_i. The membership function is in three types such as triangular, trapezoidal, and bell member functions.

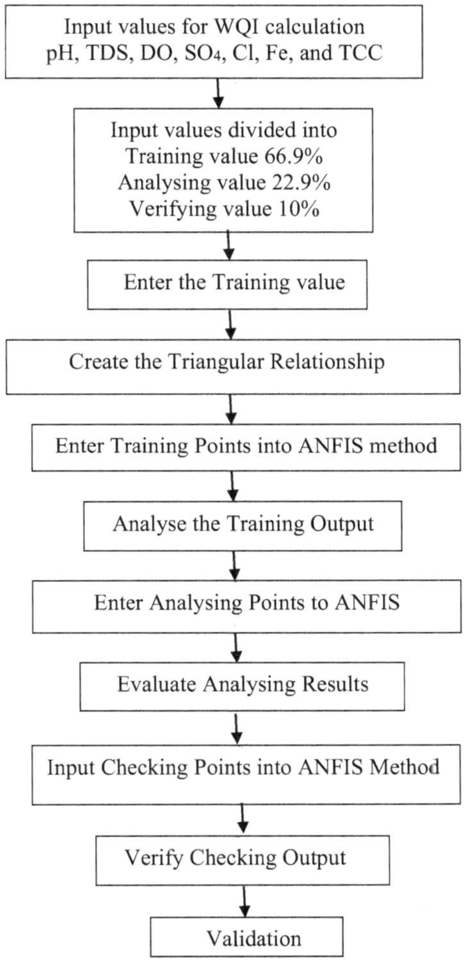

Figure 15.4 Methodology for ANFIS modelling.

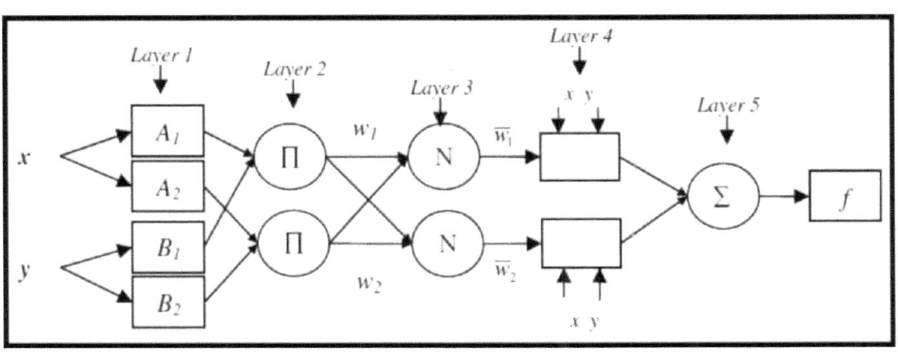

Figure 15.5 Graphic representation of ANFIS model.

Step 2: Second procedure, all standard nodes can be found by multiplying the dialectal values which were found in the prior layer:

$$O_k^2 = W_k = \mu_{1_{e_1}}(X_1)\mu_{2_{e2}}(X_2)\cdots\mu_{me_m}(X_m), \tag{15.6}$$

$$k = 1, \cdots, \text{R}; e_1, e_2, \cdots, e_m = 1, \cdots, \text{n}$$

Step 3: Third step, equivalent layer outcomes are central by the weighting feature for each node:

$$O_k^3 = \overline{W_k} = \frac{W_k}{W_1 + W_2 + \cdots W_R} \tag{15.7}$$

Step 4: Takagi-Sugeno prototypical if – then instructions were utilized for the calculation of exit of all nodes:

$$O_k^4 = \overline{W_k} f_k \tag{15.8}$$

Hence f_k- the exit of the kth method (TSK) of fuzzy, guidelines are specified under If $\left(X_1 \text{ is } A_{1_{e_1}}\right)$ and $\left(X_2 \; is \; A_{2_{e_2}}\right)$ and $\cdots$ and $\left(X_m \text{ is } A_{m_{e_m}}\right)$, then

$$f_k = \sum_{i=1}^{m} p_{ie_i} X_i + r_k \tag{15.9}$$

P_{ie}, r_k = consequent constraints, $e_1, e_2 \ldots . e_n = 1$, k = 1,

Step 5: the exit assembly of ANFIS model:

$$O^5 = Y = \sum_{k=1}^{n} \overline{W_k} f_k \tag{15.10}$$

RMSE value E

$$E = \sqrt{\frac{1}{M}\sum_{z=1}^{M}\left(S_z - y_z\right)^2} \tag{15.11}$$

$$E' = \sqrt{\frac{1}{M}\sum_{z=1}^{M}\left(S_z - y_z\right)} \tag{15.12}$$

$$R^2 = 1 - \frac{\sum_{i=1}^{N}\left(S_z - y_z\right)^2}{\sum_{i=1}^{N}\left(S_z - \bar{y}_z\right)^2} \tag{15.13}$$

M = overall number of training models; S_z = an investigational output; y_z = the forecast output; and $\bar{y}$ is the average output of work.

15.11 ANFIS Model for WQI Prediction

The ANFIS model obtains inputs from the data. The ANFIS model uses inputs from water superiority data. The readings of the current method include pH, TDS (mg/l), DO (mg/l), SO_4 (mg/l), and Cl (mg/l), heavy metal Fe (mg/l), biological parameter TCC (MPN), and the result is WQI. The mentioned procedure was followed for the calculation of WQI. The results of different studies show that triangle membership has the lowest training and analysing root mean square E values. For this reason, the membership triangle function and 2187 instructions were formulated to obtain the most suitable model for the ANFIS method system. The working principle of water superiority measurement in the ANFIS method is represented in Figure 15.6. In the

```
ANFIS info:
    Number of nodes: 4426
    Number of linear parameters: 2187
    Number of nonlinear parameters: 63
    Total number of parameters: 2250
    Number of training data pairs: 20
    Number of checking data pairs: 0
    Number of fuzzy rules: 2187

Warning: number of data is smaller than number of modifiable parameters

Start training ANFIS ...

   1      0.00144224
   2      0.00143149

Designated epoch number reached --> ANFIS training completed at epoch 2.

>>
```

Figure 15.6 ANFIS modelling working system.

Table 15.3 Water Quality Parameter Values

Sl.No	*Water Quality Parameters*	*Measured Range in the Year 2017–2020*
1.	pH	7.1255–8.9586
2.	Dissolved oxygen	6.21–6.71 (mg/l)
3.	Total dissolved solids	382–744 (mg/l)
4.	Chloride	84–136 (mg/l)
5.	Sulphate	49–74 (mg/l)
6.	Iron	0.091–0.842 (mg/l)
7.	Total coliform count	400–1600 (MPN/100 ml)

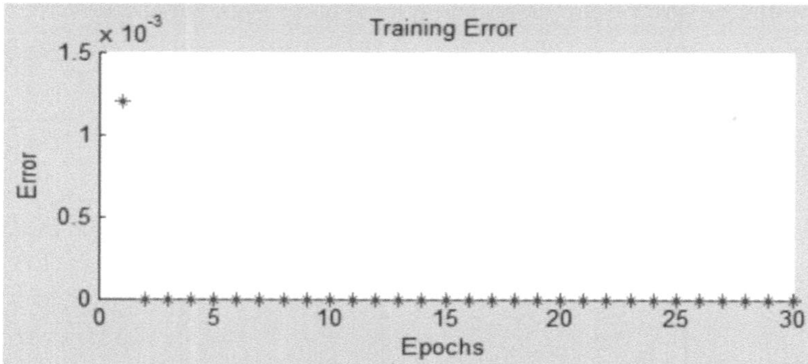

Figure 15.7 Training error (30 epochs).

30 datasets, 20 are training points, 7 are analysing points, and 3 are verifying points for ANFIS models.

The input water quality parameters given for 30 data points set the min and max values of the entry parameters, as given in Table 15.3.

The ANFIS modelling was trained, tested, and checked for 30 epochs, and the results were obtained. The output for training error for 30 epochs is shown in Figure15.7. The percentage of mistake amongst the expected numbers and experimental numbers is calculated by

$$\% \text{ of error} = ((\text{Expected WQI} - \text{ANFIS WQI})/\text{ANFIS WQI}) \times 100$$

The determined percentage of error between the experimental and ANFIS predictions of WQI is given in Table 15.4.

The ANFIS method was applied to the lake water class model [55]. The determined average error of the prediction of WQI in training, testing, and checking is 3.51, 0.961, and 3.21%, respectively. The tentative result has concluded that the strategic output can be applied for the programmed prediction of WQI accuracy of 96.80% [54].

The output forecast by the recreation ANFIS model signifies an outstanding correspondence with investigational outcomes for a wide range of working conditions.

15.12 Conclusion

The current work was done on Kanchipuram city temple tank water. Seven water class indicators were determined and checked for four consecutive years. The WQI value is determined according to BIS category C surface water values. The ANFIS model was developed and an automatic system to find WQI under uncertain

Table 15.4 The Percentage of Error between Experimental and ANFIS WQI

Calculated Output- WQI	*ANFIS Output – WQI*	*Error*
78.908	78.991	0.052
54.952	54.9535	0.104
84.957	84.951	0.022
72.937	72.938	0.162
71.921	71.922	0.060
87.980	87.981	0.064
61.920	61.921	0.041
70.911	70.9111	0.045
61.927	61.928	0.145
71.490	71.051	0.004
53.942	53.943	0.013
76.973	76.974	0.117
75.967	75.968	0.138
55.959	55.968	0.430
72.929	72.937	0.060
51.928	51.929	0.123
61.011	61.087	0.461
50.914	50.918	0.768
66.978	66.979	0.208
74.953	74.955	0.282
52.7901	52.8302	0.061
56.6303	56.8504	0.376
53.3000	53.6105	0.573
81.7202	81.9202	0.242
63.1901	63.5200	0.506
62.0102	62.5103	0.813
75.1602	75.5304	0.487
73.2100	73.6902	0.653
65.4101	65.6204	0.320
74.4300	74.8201	0.519

conditions was developed for the water department. The process is examined, tested, and verified. According to the statistical analysis, it was determined that the prediction of the ANFIS model in the training and analysing phases was very close to the test result of the ANFIS method. Since this method offers a simple process that would not require a lengthy WQI calculation, this study has important implications for designing and using this method to release information more quickly and analyse the negative impact of water on WQI.

References

[1] I. S. Rahat, et al. 'A step towards automated haematology: DL models for blood cell detection and classification', *EAI Endorsed Trans. Perv. Health Tech.*, vol. 10, p. 18, 2024.

[2] Kumar, K. Sunil Ratna, et al. "Iot and data mining techniques to detect and regulating of solar power system." *2023 International Conference on Inventive Computation Technologies (ICICT)*. IEEE (2023).

[3] S. Krishnasamy S. Bobbili Aravind, R. Gokulan, S. Balakumar, R. Muralikrishnan, B. Janakani Uday, R. Bonu, and N. Ali, 'Effective removal of reactive yellow 145 (RY145) using biochar derived from groundnut shell', *Adv. Mater. Sci. Eng.*, vol. 2022, pp. 1–7, Mar. 2022.

[4] R. Gokulan, S. Praveen, A. Avinash, and S. Saravanan, 'Soft computing-based models and decolorization of Reactive Yellow 81 using Ulva Prolifera biochar', *Chemosphere*, vol. 287, no. 132368, p. 132368, Jan. 2022.

[5] G. Kalyan, R. Gokulan, and S. Sujatha, 'Biosorption of zinc metal ion in aqueous solution using biowaste of Pithophora cleveana wittrock and Mimusops elengi', *Desalin. Water Treat.*, vol. 218, pp. 363–371, 2021.

[6] S. Praveen, J. Jegan, T. B. Pushpa, and R. Gokulan, 'Artificial neural network modelling for biodecolorization of basic violet 03 from aqueous solution by biochar derived from agro-bio waste of groundnut hull: Kinetics and Thermodynamics', *Chemosphere*, vol. 276, p. 130191, 2021.

[7] S. Şener, E. Şener, and A. Davraz, 'Evaluation of water quality using water quality index (WQI) method and GIS in Aksu River (SW-Turkey)', *Sci. Total Environ.*, vol. 584–585, pp. 131–144, Apr. 2017.

[8] M. Hameed, S. S. Sharqi, Z. M. Yaseen, H. A. Afan, A. Hussain, and A. Elshafie, 'Application of artificial intelligence (AI) techniques in water quality index prediction: A case study in tropical region, Malaysia', *Neural Comput. Appl.*, vol. 28, no. S1, pp. 893–905, Dec. 2017.

[9] R. Isaac Sajan and V. Bibinchristopherb, 'A Fuzzy Inference System for enhanced ground water quality assessment and index determination', *Water Quality Res. J.*, vol. 58, no. 3, p. 24, 2023.

[10] A. Shine and G. Madhu, 'Water quality modelling of river Periyar using artificial neural network and adaptive neuro-fuzzy inference system', *IOP Conf. Series: Earth and Environ. Sci.*, vol. 1125, p. 24, 2022.

[11] P. Banerjee, V. S Singh, K. Chatttopadhyay, P. C Chandra, and B. Singh, 'Artificial neural network model as a potential alternative for groundwater salinity forecasting', *J. Hydrol.*, vol. 398, pp. 212–220, 2011.

[12] E. Dogan, B. Sengorur, and R. Koklu, 'Modeling biological oxygen demand of the Melen River in Turkey using an artificial neural network technique', *J. Environ. Manage.*, vol. 90, no. 2, pp. 1229–1235, Feb. 2009.

[13] Q. Miao, H. Yuan, C. Shao, and Z. Liu, 'Water quality prediction of moshui river in China based on BP neural network', in 2009 International Conference on Computational Intelligence and Natural Computing, Wuhan, China, 2009.

[14] K. P Singh, A. Basant, A. Malik, and G. Jain, 'Artificial neural network modeling of the river water quality—A case study', *Ecol. Modell.*, vol. 220, pp 888–895, 2009.

[15] E. Dogan, A. Ates, E. C. Yilmaz, and B. Eren, 'Application of artificial neural networks to estimate wastewater treatment plant inlet biochemical oxygen demand', *Environ. Prog.*, vol. 27, no. 4, pp. 439–446, Dec. 2008.

[16] H. Elhatip and M. A. Kömür, 'Evaluation of water quality parameters for the Mamasin dam in Aksaray City in the central Anatolian part of Turkey by means of artificial neural networks', *Environ. Geol.*, vol. 53, no. 6, pp. 1157–1164, Jan. 2008.

[17] S. Palani, S. Y. Liong, and P. Tkalich, 'An ANN application for water quality forecasting', *Mar. Pollut. Bull.*, vol. 56, pp. 1586–1597, 2008.

[18] Y. Chen, L. Song, Y. Liu, L. Yang, and D. Li, 'A review of the artificial neural network models for water quality prediction', *Appl. Sci. (Basel)*, vol. 10, no. 17, p. 5776, Aug. 2020.

[19] I. S. Yeon, J. H. Kim, and K. W. Jun, 'Application of artificial intelligence models in water quality forecasting', *Environ. Technol.*, vol. 29, pp. 625–631, 2008.

[20] O. Kisi, K. S. Parmar, A. Mahdavi-Meymand, R. M. Adnan, S. Shahid, and M. Zounemat-Kermani, 'Water quality prediction of the Yamuna River in India using hybrid neuro-fuzzy models', *Water (Basel)*, vol. 15, no. 6, p. 1095, Mar. 2023.

[21] P. Khatri, K. K. Gupta, and R. K. Gupta, 'Raspberry Pi-based smart sensing platform for drinking-water quality monitoring system: A Python framework approach', *Drink. Water Eng. Sci.*, vol. 12, no. 1, pp. 31–37, Jun. 2019.

[22] M. Amrita, 'Analysing the water quality parameters from traditional to modern methods in aquaculture', *Int. J. Sci. Environ. Technol.*, vol. 7, pp. 1954–1961, 2018.

[23] S. Pasika and S. T. Gandla, 'Smart water quality monitoring system with cost-effective using IoT', *Heliyon*, vol. 6, no. 7, p. e04096, Jul. 2020.

[24] P. Salunke and J. Kate, 'Advanced smart sensor interface in internet of things for water quality monitoring', in 2017 International Conference on Data Management, Analytics and Innovation (ICDMAI), Pune, India, 2017.

[25] K. Saravanan, E. Anusuya, R. Kumar, and L. H. Son, 'Real-time water quality monitoring using Internet of Things in SCADA', *Environ. Monit. Assess.*, vol. 190, no. 9, p. 556, Aug. 2018.

[26] B. Hojris, S. N. Kornholt, S. C. B. Christensen, H.-J. Albrechtsen, and L. S. Olesen, 'Detection of drinking water contamination by an optical real-time bacteria sensor', *H2Open J.*, vol. 1, no. 2, pp. 160–168, Dec. 2018.

[27] S. N. Zulkifli, H. A. Rahim, and W.-J. Lau, 'Detection of contaminants in water supply: A review on state-of-the-art monitoring technologies and their applications', *Sens. Actuators B Chem.*, vol. 255, pp. 2657–2689, Feb. 2018.

[28] Y. Syahrul, P. Satriyo, and A. A. Munawar, 'Applying infrared reflectance spectroscopy to predict water quality in Aceh river', *Int. J. Sci. Technol. Res.*, vol. 8, pp. 969–972, 2019.

[29] M. F. Coimbra, M. R. Coelho, and M. B. Ribas, 'Review of Technologies for the Rapid Detection of Chemical and Biological Contaminants in Drinking Water', Publications Office of the European Union, 2020.

[30] A. Radzevičius, M. Dapkienė, N. Sabienė, and J. Dzięcioł, 'A rapid UV/Vis spectrophotometric method for the water quality monitoring at on-farm root vegetable pack houses', *Appl. Sci. (Basel)*, vol. 10, no. 24, p. 9072, Dec. 2020.

[31] A. Radzevicius, E. Dapkien, N. Sabien, and J. Dzi˛eciol, 'A rapid UV/Vis spectrophotometric method for the water quality monitoring at on-farm root vegetable pack houses', *Appl. Sci.*, vol 10, p. 9072, 2020.

[32] P. Mahajan and P. Shahane, 'An IoT based system for water quality monitoring'. *SSRN Electron. J.*, pp. 489–495, 2021.

[33] A. Canciu, M. Tertis, O. Hosu, A. Cernat, C. Cristea, and F. Graur, 'Modern analytical techniques for detection of bacteria in surface and wastewaters', *Sustainability*, vol. 13, no. 13, p. 7229, Jun. 2021.

[34] R. Du, D. Yang, and X. Yin, 'Rapid detection of three common bacteria based on fluorescence spectroscopy', *Sensors (Basel)*, vol. 22, no. 3, p. 1168, Feb. 2022.

[35] J. Zhou, J. Wang, Y. Chen, X. Li, and Y. Xie, 'Water quality prediction method based on multi-source transfer learning for water environmental IoT system', *Sensors (Basel)*, vol. 21, no. 21, p. 7271, Nov. 2021.

[36] T. Paepae, P. N. Bokoro, and K. Kyamakya, 'From fully physical to virtual sensing for water quality assessment: A comprehensive review of the relevant state-of-the-art', *Sensors (Basel)*, vol. 21, no. 21, p. 6971, Oct. 2021.

[37] M. Manoj, V. Dhilip Kumar, M. Arif, E.-R. Bulai, P. Bulai, and O. Geman, 'State of the art techniques for water quality monitoring systems for fish ponds using IoT and underwater sensors: A review', *Sensors (Basel)*, vol. 22, no. 6, p. 2088, Mar. 2022.

[38] L. Burbery, P. Abraham, D. Wood, and S. de Lima, 'Applications of a UV optical nitrate sensor in a surface water/groundwater quality field study', *Environ. Monit. Assess.*, vol. 193, p. 303, 2021.

[39] Y. Li, X. K. Wang, H. Y. Zhang, J. Q. Wang, and L. Li, 'An integrated regional water quality assessment method considering interrelationships among monitoring indicators', *Environ. Monit. Assess.*, vol. 193, p. 223, 2021.

[40] M. Horvat, Z. Horvat, and K. Pastor, 'Multivariate analysis of water quality parameters in Lake Palic, Serbia', *Environ. Monit. Assess.*, vol. 193, no. 7, p. 410, Jun. 2021.

[41] M. N. Hasan, K. Rahman, Tajmunnaher, and M. R. Bhuia, 'Assessment of ground water quality in the vicinity of Sylhet City, Bangladesh: A multivariate analysis', *Sustain. Water Resour. Manag.*, vol. 6, no. 5, p. 29, Oct. 2020.

[42] N. Khatri, S. Tyagi, D. Rawtani, and M. Tharmavaram, 'Assessment of river water quality through application of indices: A case study River Sabarmati, Gujarat, India', *Sustain. Water Resour. Manag.*, vol. 6, no. 6, p. 29, Dec. 2020.

[43] A. Najah, F. Y Teo, M. F. Chow, Y. F. Huang, S. D. Latif, S. Abdullah, M. Ismail, and A. El-Shafie, 'Surface water quality status and prediction during movement

control operation order under COVID-19 pandemic: Case studies in Malaysia', *Int. J. Environ. Sci. Technol. (Tehran)*, vol. 18, no. 4, pp. 1009–1018, Feb. 2021.

[44] K. N. A. Maulud, A. Fitri, W. H. M. W. Mohtar, W. S. W. M. Jaafar, N. Z. Zuhairi, and M. K. A. Kamarudin, 'A study of spatial and water quality index during dry and rainy seasons at Kelantan River Basin, Peninsular Malaysia', *Arab. J. Geosci.*, vol. 14, p. 29, 2021.

[45] A. S. Pattanayak, B. S. Pattnaik, S. K. Udgata, and A. K. Panda, 'Development of chemical oxygen on demand (COD) soft sensor using edge intelligence', *IEEE Sens. J.*, vol. 20, no. 24, pp. 14892–14902, Dec. 2020.

[46] E. A. Lee, 'The past, present and future of cyber-physical systems: A focus on models', *Sensors (Basel)*, vol. 15, no. 3, pp. 4837–4869, Feb. 2015.

[47] J. Bhardwaj, K. K. Gupta, and R. Gupta, 'Towards a cyber-physical era: Soft computing framework based multi-sensor array for water quality monitoring', *Drink. Water Eng. Sci.*, vol. 11, no. 1, pp. 9–17, Feb. 2018.

[48] N. A. Razman, W. Z. Wan Ismail, N. A. I. M. Kamil, S. N. Zainurin, I. Ismail, J. Jamaludin, M. Sahrim, K. N. Z. Ariffin, and S. R. Balakrishnan, 'A review on water quality monitoring methods based on electronics and optical sensing', *J. Adv. Res. Appl. Sci. Eng. Tech.*, vol. 26, no. 2, pp. 1–7, Apr. 2022.

[49] J.-S. R. Jang, 'ANFIS: adaptive-network-based fuzzy inference system', *IEEE Trans. Syst. Man Cybern.*, vol. 23, no. 3, pp. 665–685, 1993.

[50] V. Profillidis and G. Botzoris, Modeling of transport demand: analyzing, calculating, and forecasting transport demand. Elsevier Inc, 2018.

[51] M. Stojcic, 'Application of the ANFIS model in road traffic and transportation: A literature review from 1993 to 2018', *Oper. Res. Eng. Sci. Theor. Appl.*, vol. 1, no. 1, pp. 40–61, Dec. 2019.

[52] O. Jayalakhmidevi and S. L. Belagadi, 'Water quality assessment from different districts of southern Karnataka', *Nat. Environ. Pollut. Technol.*, vol. 4, pp. 589–596, 2005.

[53] I. Maher, M. E. H. Eltaib, A. A. D. Sarhan, and R. M. El-Zahry, 'Investigation of the effect of machining parameters on the surface quality of machined brass (60/40) in CNC end milling—ANFIS modeling', *Int. J. Adv. Manuf. Technol.*, vol. 74, no. 1–4, pp. 531–537, Sep. 2014.

[54] S. Emami, H. Dehghanisanij, M. Achite, N. Al -Ansari, and N. T. Thuy Linh, 'Application of ANFIS, ELM, and ANN models to assess water productivity indicators based on agronomic techniques in the Lake Urmia Basin', *Appl. Water Sci.*, vol. 13, no. 55, pp. 1–15, 2023.

Chapter 16

WSN-Based Optimal Crude Oil Storage Health Monitoring Framework

Ismail Keshta, Mukesh Soni, Sagar Dhanraj Pande, Roshan Rajendrakumar Karwa, and Haewon Byeon

16.1 Introduction

A long-standing topic of intense academic interest has been the control of WSNs' health monitoring [1]. Researchers [2] have already accomplished a great deal in the field of health monitoring [3] management through their work using WSNs. A neural network based on rough set optimization for probability was suggested by the author [4] as a method for diagnosing node faults in WSN. Using the Bayesian theory, the author [5] suggested a way to identify faults in WSN online. Author [6] summarized the advantages and disadvantages of WSN and provided methods to improve security. These studies can effectively improve the reliability and security of WSNs.

In practical engineering applications, a WSN consists of multiple sensor devices installed at different locations. Probabilistic decision-making framework is an important aspect of WSN health monitoring management [7]. Providing users with the optimal health monitoring timing of WSNs can effectively reduce the probability of WSN failures. Probabilistic decision-making frameworks for real systems mainly face two problems: lack of detection data and complex system mechanisms [8]. On

 DOI: 10.1201/9781003479062-16

the one hand, with the development of the equipment manufacturing industry, the reliability of sensors continues to improve, and the probability of sensor failure keeps decreasing, resulting in fewer available failure data [9].

According to the obtained small sample failure data, it is impossible to accurately provide health monitoring decisions. Providing more WSN data improves the accuracy of the probabilistic decision-making framework In contrast, when a monitored system's status is the responsibility of a WSN, its monitoring positions are scattered, and the monitored features are also different. Relying solely on sensor judgement for the optimal health monitoring timing may be too delayed, increasing the health monitoring cost of the WSN because experts cannot provide accurate network information for a probabilistic decision-making framework. Sensor knowledge is already difficult to use due to its inherent fuzziness and lack of clarity. In conclusion, the two issues highlighted earlier in the probabilistic decision-making framework of WSNs must be resolved to increase the dependability of these systems and decrease the expenses associated with their health monitoring [10]. Fuzzy logic, the IF-THEN probabilistic rule, and evidence reasoning form the basis of the author's suggested confidence rule base-probabilistic sensor system (PSS) [11]. By combining quantitative detection data with qualitative sensor information, the PSS framework [12] can handle the ambiguities, uncertainties, and incompleteness that come with sensor knowledge. The PSS framework's great performance has led to its extensive use in many real-world engineering domains, including health monitoring status prediction, defect detection, and safety evaluation. Authors [13] and [14] offered safety evaluation methods that combine the PSS framework and conditional generalized minimum variance to handle probabilistic rule combination explosion, while Author [15] proposed a BRB-based fault forecasting system that accounts for environmental interference factors. An effective direction approach-based optimization technique for PSS safety evaluation frameworks was also developed [16].

The PSS technique enhances the information sources of WSNs [17] by integrating detection data with sensor knowledge, hence addressing engineering application challenges such as limited detection data and complex system operations [18]. Conventional forecasting techniques, such as the Kalman process, estimate the future condition of a system based on its present condition. The Kalman process is utilized in several sectors, such as life prediction and health monitoring status prediction, due to its technical adaptability. This study presents a decision-making framework for WSNs that utilizes the PSS framework and Kalman process, based on probabilistic principles. The framework incorporates a health monitoring outcome prediction based on the Kalman process and a health monitoring status evaluation based on the BRB. The status of WSNs in health is evaluated using the BRB-based health monitoring status framework, and the future status is predicted using the Kalman process-based framework for forecasting health monitoring outcomes. The preliminary structure and coefficients for the health monitoring status framework are provided by domain specialists [19]. Because of the ambiguity in sensor knowledge, the initial approach to evaluating the health monitoring state could not effectively assess the health monitoring of the WSN. The P-CMA-ES optimization approach can simultaneously train the health monitoring status and forecasting systems [20].

In Section 16.2, we point out the difficulties and outcomes of upkeep. In Section 16.3, we outline the methodology and theoretical foundation of the approach, and in Section 16.4, we describe methods employed to enhance the framework's efficiency. In Section 16.5, we verify the framework by conducting experiments or running simulations, and in Section 16.6, we evaluate results and make recommendations for further study.

16.2 Problem Description

There are two main issues in the probabilistic decision-making framework for WSNs: insufficient detection data and WSN complexity.

16.2.1 Issues in the Probabilistic Decision-Making Framework

The health monitoring approach for WSNs encounters two primary challenges:

Problem 1: The accuracy of the data collected by sensors is essential for effectively managing the health monitoring of WSNs. Due to ongoing enhancements in electrical component quality, the reliability of sensor devices within WSNs is steadily growing. The likelihood of sensor failure is quite minimal. Nevertheless, it is unavoidable for sensors to malfunction after extended periods of operation. However, the existing data on sensor failures is restricted. The lack of sufficient failure data is the main obstacle that needs to be addressed in order to develop a reliable probabilistic decision-making framework.

Issue 2: WSNs are used to monitor the operation status of systems. For complex systems, the distribution of sensors is extensive. The detection information from sensors in the WSN exhibits strong non-linearity and coupling. To optimize the upkeep of WSNs, correct decision-making data cannot be supplied by domain specialists. Sensor knowledge is already difficult to use due to its fuzziness, incompleteness, and unpredictability. In addition to specialized expertise, various types of data must be considered. We must now attend to this second issue.

This probabilistic decision-making approach is built to tackle the two difficulties mentioned earlier:

$$U_{\text{opt}}(\alpha) = \psi\left(\Xi\left(x_1(\alpha), x_2(\alpha), \cdots, x_M(\alpha)\right), K_{\text{thre}}, S\right) \tag{16.1}$$

where $U_{\text{opt}}(\alpha)$ indicates the optimal health monitoring period of the sensing network in real-period α; $\Xi(\cdot)$ and $\Psi(\cdot)$ represent the nonlinear health monitoring prediction framework and the nonlinear framework for predicting health monitoring outcomes, respectively; K_{thre} is the health monitoring threshold provided by experts; s indicates the sensor knowledge used in the probabilistic decision-making framework; $x_1(\alpha), x_2(\alpha), \cdots, x_M(\alpha)$ represent the detection data of N features of the sensing network in real-period α.

16.2.2 Composition of the Optimal Health Monitoring Prediction Framework

The health monitoring prediction framework and the framework for forecasting health monitoring outcomes make up the suggested probabilistic decision-making framework for WSNs. As a sensor system, the PSS framework integrates detection data with sensor knowledge in real-period [21]. The health monitoring prediction framework has many probability rules, where the kth PSS rule is stated as

$$C_k(\alpha)\text{If } x_1(\alpha) \text{ is } B_1^k \wedge x_2(\alpha) \text{ is } B_2^k \wedge \cdots \wedge x_N(\alpha) \text{ is } B_N^k,$$

$$\text{Then } G(\alpha) \text{ is } \{(E_1, \beta_{1,k}), (E_2, \beta_{2,k}), \cdots, (E_N, \beta_{N,k})\}$$

With probabilistic rule weight θ_k, feature weight

$$\delta_1, \delta_2, \cdots, \delta_N \tag{16.2}$$

where $G(\alpha)$ indicates the health monitoring status evaluated by the sensing network in real-period u; N indicates the number of features of the WSN; $x_i(\alpha), i = 1, 2, \cdots, N$ indicates the detection detail of the ith network feature at period α; $B_1^k, B_2^k, \cdots, B_N^k$ represent the points of reference of the N feature attributes in the kth PSS rule, used to convert detection data of different formats into a unified data format; $\{(E_1, \beta_{1,k}), (E_2, \beta_{2,k}), \cdots, (E_N, \beta_{N,k})\}$ indicates the level output results of the health monitoring prediction framework, $(E_1, E_2, \cdots, E_N)$ are N reference levels of the health monitoring framework, $(\beta_{1,k}, \beta_{2,k}, \cdots, \beta_{N,k})$ are the corresponding confidence levels; θ_k indicates the weight of the kth rule; $\delta_1, \delta_2, \cdots, \delta_N$ are weights of N features [22].

Then, based on status of the sensing network's health monitoring, the WSN's future framework for predicting health monitoring outcomes based on the Kalman process can be represented as follows:

$$T = G(\alpha) + \varphi(\alpha)\alpha + \psi(\alpha)\Xi(\alpha) \tag{16.3}$$

where T indicates the predicted future health monitoring status; $\varphi(\alpha)$ and $\psi(\alpha)$, respectively, denote the parameters based on the Kalman process at time α; $\Xi(\alpha)$ is a Brownian motion following a standard normal distribution.

Together, the health monitoring threshold Kth and the anticipated health monitoring condition Ŧ decide when the WSN should be serviced optimally. Figure 16.1 shows the steps used to create the best health monitoring choice framework.

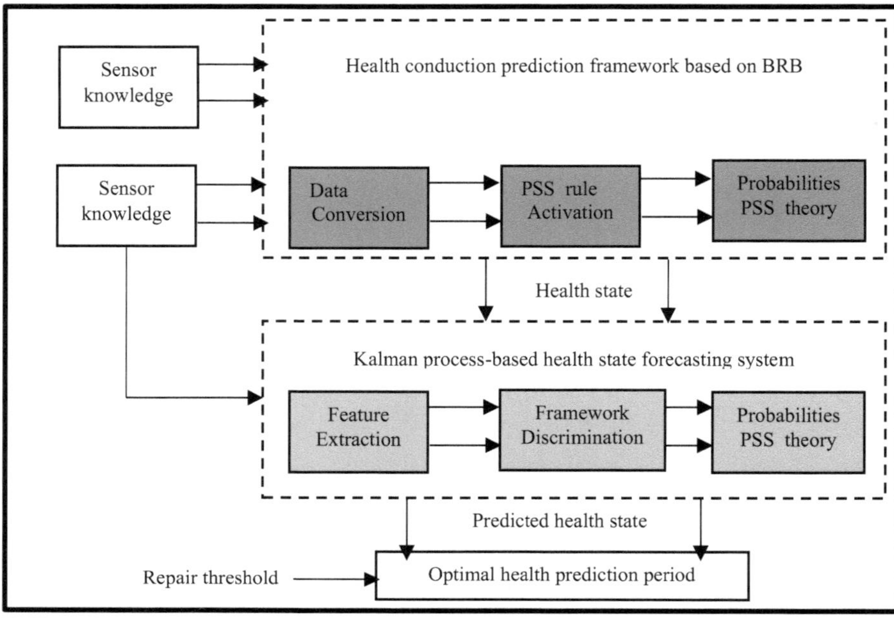

Figure 16.1 Frameworkling process of the optimal health monitoring prediction framework.

16.3 Reasoning of the Optimal Health Monitoring Prediction Framework

16.3.1 Health Monitoring Prediction Framework Based On BRB

The health monitoring prediction framework that is based on PSS goes through four phases in the frameworkling process: data transformation, PSS rule activation, PSS rule combination, and health monitoring status estimation.

Step 1: In engineering applications, different features have different data formats. The following equation is used to convert the detection data of network features into a unified data format:

$$n_j^i(\alpha) = \begin{cases} \dfrac{B_{i(k+1)} - x_i^*(\alpha)}{B_{i(k+1)} - B_{ik}}, j = k \text{ if } B_{ik} \leqslant x_i^*(\alpha) \leqslant B_{i(k+1)} \\ \dfrac{x_i^*(\alpha) - B_{ik}}{B_{i(k+1)} - B_{ik}}, j = k+1 \\ 0, j = 1, 2, \cdots, |x_i|, j \neq k, k+1 \end{cases} \tag{16.4}$$

where $n_j^i(\alpha)$ shows the accuracy of identifying within the ith feature and the j-th starting point at period u; B_{ik} and $B_{i(k+1)}$ represent the starting points of the i-th consist of in the kth and (k+1)th PSS rules, and $|x_i|$ represents the number of PSS rules containing the feature [23].

Step 2: The following equations are used to acquire the degree of match of the kth PSS rule and the input data of network features after computing the matched degree of each starting point:

$$\bar{\delta}_i = \frac{\breve{\delta}_i O}{\max_{i=1,2}\left\{U_k\left\{\delta_i\right\}\right\}}, 0 \le \bar{\delta}_i \le 1 \tag{16.5}$$

$$n_k = \prod_{i=1}^{U_k}\left(n_k^i\right)^{\delta_i} \tag{16.6}$$

where n_k indicates the matching degree between the input information [24] and the kth PSS rule; U_k indicates the number of features in the kth PSS rule; $\bar{\delta}_i$ is the relative weight of the i-th feature, indicating its relative importance among U_k features [25], $0 \le \delta i \le 1$.

The PSS rules in the rule base are then triggered to varying degrees depending on the determined match degree of the input information. The following equation determines the degree of matching for the kth PSS rule:

$$z_k = \frac{\theta_k n_k}{\sum_{l=1}^{M}\theta_l n_l}, \quad k = 1,2,\cdots,M \tag{16.7}$$

where z_k indicates the kth PSS rule's activation weight; M indicates the number of rules in the health monitoring framework based on BRB; θ_k indicates the probabilistic rule weight, highlighting how significant the kth rule is inside the rule base.

Step 3: Determine the rules' health monitoring status output using their activation weights. It is not possible to directly acquire the health monitoring status because of the various PSS rule output formats. The evidentiary reasoning (ER) algorithm's analytical technique may be used to determine the final output result, as indicated in the following equations:

$$\beta_n = \frac{\mu\left[\prod_{k=1}^{M}\left(z_k\beta_{n,k} + 1 - z_k\sum_{j=1}^{N}\beta_{j,k}\right) - \prod_{k=1}^{M}\left(1 - z_k\sum_{i=1}^{N}\beta_{j,k}\right)\right]}{1-\mu\left[\prod_{k=1}^{M}\left(1-z_k\right)\right]} \tag{16.8}$$

$$\mu = \left[\sum_{n=1}^{N} \prod_{k=1}^{M} \left(x_k \beta_{n,k} + 1 - x_k \sum_{j=1}^{N} \beta_{j,k} \right) - (N-1) \prod_{k=1}^{M} \left(1 - x_k \sum_{j=1}^{N} \beta_{j,k} \right) \right]^{-1} \tag{16.9}$$

where β_n indicates the combined output confidence of the n-th output reference level E_n [26]; M and N represent the number of PSS rules and the number of output levels [27], respectively (Note: In the frameworkling process [28], status of the sensing network's heal this continuously changing, and the optimal health monitoring timing is determined by the current health monitoring status [29]).

Step 4: The combined confidence of the output health monitoring reference levels indicates the probability of the health monitoring status at different reference levels. The final health monitoring status is calculated using

$$v(x(\alpha)) = \sum_{n=1}^{N} v(E_n)\beta_n \tag{16.10}$$

where $v(x(\alpha))$ shows the health monitoring state as of period b [30], assessed using the data obtained for detection purposes [31] of the network of sensor features $x(\alpha)$ [32]; $v(E_n)$ is the utility value of the reference level provided by experts [33], used to measure the impact of different reference levels in the final health monitoring status evaluation [34].

16.3.2 Health Monitoring Status Forecasting System Based on Kalman Process

A traditional technique for making predictions [35], the Kalman process has several applications in engineering. This research uses the Kalman process to forecast WSN health. Next, the BRB-based health monitoring system evaluates the WSN's health monitoring condition, and the following framework is employed to forecast the health monitoring state of the WSN in the future:

$$\mathrm{T}(\alpha_i) = \Theta(\alpha_{i-1}) + \varphi(\alpha_i)\Delta\alpha + \psi(\alpha_i)\Xi(\Delta\alpha) \tag{16.11}$$

where $\mathrm{T}(\alpha_i)$ and $\Theta(\alpha_{i-1})$ represent the condition of the monitoring system in real-period α_i and α_{i-1}, respectively; $\Delta\alpha = \alpha_i - \alpha_{i-1}$ indicates the period interval between two period points; $\Xi(\Delta\alpha)$ indicates Brownian motion, $\Xi(\Delta\alpha) \sim N(0, \Delta\alpha)$; $\varphi(\alpha_i)$ and $\psi(\alpha_i)$ represent the degradation coefficient and diffusion coefficient of medical condition forecasting system, respectively. The variation of health monitoring status can be obtained through the following theorem.

Theorem: The degradation coefficient and diffusion coefficient of medical condition forecasting system can be calculated by the following equations:

$$\breve{\varphi}(\alpha_i) = \frac{\sum_{j=1}^{U(\alpha_i)} \Delta T_j}{\sum_{j=1}^{U(\alpha_i)} \Delta \alpha_j} \tag{16.12}$$

$$\breve{\psi}(\alpha_i) = \sum_{j=1}^{U(\alpha_j)} \frac{\left(\Delta T_j - \varphi(\alpha_i)\Delta\alpha_j\right)^2}{\Delta\alpha_j} \tag{16.13}$$

where ΔT_j indicates the j-th change in status of the sensing network's e-health; $U(\alpha_i)$ indicates the availability of the condition of the monitoring system in real-time α_i.

Proof:

1. The change in state transition can be calculated by the following equation:

$$\Delta T = T(\alpha_i) - T(\alpha_{i-1}) = \varphi(\alpha_i)\Delta\alpha + \psi^2(\alpha_i)\Xi(\Delta\alpha) \tag{16.14}$$

2. The mean and variance of the state transition change can be calculated by the following equations:

$$F(\Delta T) = \varphi(\alpha_i)\Delta\alpha \tag{16.15}$$

$$E(\Delta T) = \psi^2(\alpha_i)\Delta\alpha \tag{16.16}$$

3. The change in state transition follows a normal distribution $N\left(\varphi(\alpha_i)\Delta\alpha, \psi^2(\alpha_i)\Delta\alpha\right)$, and the probability distribution of ΔT can be described as

$$f(\Delta\tau) = \frac{1}{\sqrt{2\pi\psi^2(\alpha_i)\Delta\alpha}} \exp\left\{-\frac{\left(\Delta\tau - \varphi(\alpha_i)\Delta\alpha\right)^2}{2\psi^2(\alpha_i)\Delta\alpha}\right\} \tag{16.17}$$

4. The following equations demonstrate the formulation of the greatest likelihood function:

$$M(\Delta\tau)=\prod_{j=1}^{U(\alpha\alpha_i)}\frac{1}{\sqrt{2\pi\psi^2(\alpha_i)\Delta\alpha_j}}\exp\left\{-\frac{\left(\Delta\tau_j-\varphi(\alpha_i)\Delta\alpha\alpha_j\right)^2}{2\psi^2(\alpha_i)\Delta\alpha_j}\right\} \tag{16.18}$$

$$\ln M(\Delta T\tau)=-\frac{1}{2}\ln 2\pi\psi^2(\alpha_i)+\sum_{j=1}^{T(\alpha_j)}\left(-\frac{1}{2}\ln\Delta\alpha_j\right)-\sum_{j=1}^{U(\alpha_i)}\frac{\left(\Delta\tau_j-\varphi(\alpha_i)\Delta\alpha_j\right)^2}{2\psi^2(\alpha_i)\Delta\alpha_j} \tag{16.19}$$

5. The degradation coefficient and diffusion coefficient of medical condition forecasting system can be calculated by the following equation:

$$\begin{gathered}\frac{\partial\ln M(\Delta\mathrm{T})}{\partial\varphi(\alpha_i)}=0\\ \frac{\partial\ln M(\Delta\mathrm{T})}{\partial\psi(\alpha_i)}=0\end{gathered} \tag{16.20}$$

Thus, through this theorem, estimated values of the degradation coefficient $\breve{\varphi}(\alpha_i)$ and the diffusion coefficient $\psi(\alpha_i)$ of medical condition forecasting system can be obtained.

Based on the prediction of the status of the sensing network's e-health, the optimal health monitoring timing can be given by the following equation:

$$U_{\text{optimal}}=\alpha_p,\ \text{if}\ \tau(\alpha_p)<\mathrm{T}_{\text{thre}} \tag{16.21}$$

where U_{pptimel} indicates the optimal health monitoring timing; T_{thre} indicates the health monitoring threshold given by domain experts considering the reliability and health monitoring cost of the WSN.

Note: If the anticipated condition of the WSN's health monitoring falls below the health monitoring threshold, it could be due to both external noise and system noise. To enhance the precision of determining the ideal timing for health monitoring. This chapter continuously monitors the predicted health monitoring status at multiple consecutive period points. For example, only when the predicted values of the health monitoring status at times $\alpha, \alpha+1$, and $\alpha+2$ are all below T_{thre}, will the timing at period α be considered as the optimal health monitoring timing.

16.4 Optimization of the Optimal Health Monitoring Prediction Framework

The first optimal health monitoring prediction framework lacks the ability to provide precise health monitoring period information due to the ambiguity of sensor knowledge. This chapter presents an optimization framework for making optimal health monitoring decisions and provides a summary of the framework methodology.

16.4.1 Optimization Framework Based on P-CMA-ES

The primary objective of the health monitoring prediction framework is to minimize the discrepancy between the real health monitoring condition and the predicted health monitoring condition. The accuracy of the health monitoring prediction mode is quantified by the mean square error (MSE) between the actual and estimated health monitoring states. This can be computed using the following equation:

$$MSE = \frac{1}{U}\sum_{t=1}^{U}\left(\text{output}_{\text{framework}} - \text{output}_{\text{actual}}\right)^2 \tag{16.22}$$

The output framework and output α_{actual} reflect the output value of the WSN health monitoring prediction framework and the actual health monitoring status value, respectively. α denotes the number of detection data in the health monitoring framework. The network health monitoring prediction framework, which incorporates the PSS framework, relies on domain specialists to offer the PSS parameters, which have precise physical interpretations. Thus, in order to guarantee the comprehensibility of the parameters in the framework procedure, it is necessary to meet the following conditions:

$$0 \leqslant {}^{-}{}_{k} \leqslant 1, k = 1, 2, \cdots, M \tag{16.23}$$

$$0 \leqslant \delta_i \leqslant 1, i = 1, 2, \cdots, \alpha - 1 \tag{16.24}$$

$$0 \leqslant \beta_{n,k} \leqslant 1, n = 1, 2, \cdots, N, k = 1, 2, \cdots, M \tag{16.25}$$

$$\sum_{n=1}^{N}\beta_{n,k} \leqslant 1, k = 1, 2, \cdots, M \tag{16.26}$$

The aforementioned parameters refer to the optimized parameters in the optimization framework. The limitations can be modified according to the condition of the WSN.

The optimization framework can be expressed as

$$\min \mathrm{MSE}\left(\tau_k, \beta_{n,k}, \delta_i\right) \tag{16.27}$$

$$\text{s.t. } (16.23) - (16.26)$$

Note 2: The parameters of medical condition forecasting system are not optimized in the optimization framework. Section 16.2.2 explains how to calculate them using maximum likelihood estimation.

16.4.2 Framework Process of the Optimal Health Monitoring Prediction Framework

The framework method for the optimal health monitoring choice framework for the WSN mentioned above can be stated as follows, based on the reasoning process:

Step 1: Partition the identified data set into a training set and a test set. The training set is utilized to train the optimal health monitoring framework, while the test set is used to evaluate its performance.

Step 2: Develop the most effective health monitoring framework by utilizing the training set data. The WSN's detection data is utilized to optimize the network health monitoring prediction framework. The resulting state of the WSN's health is then employed as input data to generate the parameters of the medical condition forecasting system.

Step 3: Evaluate the performance of the optimized framework by applying it to the test set data. The present condition of the WSN's health is determined by utilizing the network health monitoring prediction framework, which is based on BRB. Subsequently, the future state of the WSN's health is forecast based on the estimated health status at the present moment.

Step 4: Determine the ideal date for health monitoring by considering the health monitoring threshold provided by domain experts and the projected condition of the network of sensor health.

16.5 Experimental Verification

This chapter performs practical research to test the effectiveness of the developed optimal health monitoring framework. The research uses the optimal health monitoring scheduling of a wireless WSN that monitors the operational status of oil storage tanks near a port in Hainan Province as an example.

16.5.1 Problem Description of WSN Health Monitoring Prediction Framework

Optimizing health monitoring decisions for WSNs is a highly effective method for enhancing their reliability. A probabilistic decision-making framework must tackle two concerns:

1. The restricted availability of sensor failure data, caused by the high reliability and low failure rate (FR) of sensors, hinders the establishment of an accurate optimal health monitoring framework.
2. The sensor devices of the WSN are dispersed across various locations of the monitored objects, each with distinct monitoring capabilities. Furthermore, the presence of noise in the real working environment hampers the precision of the detection data, posing challenges for specialists in delivering precise knowledge. Moreover, the ambiguity around sensor knowledge amplifies the challenge of utilizing said knowledge. These two criteria significantly impact the precision of determining the ideal timing for health monitoring in the WSN.

The experiment utilizes a WSN to monitor various operating data of three crude oil storage tanks. This includes parameters, such as humidity, temperature, wind speed, wind direction, tilt, vibration, and settlement. The devices used for deploying WSNs include wireless gateways, data-collecting relay stations, on-site collection hosts, and monitoring centre hosts. The monitoring data is communicated to the gateway via a wireless unit operating at a frequency of 433 MHz. The communication technique utilizes a 433 MHz LORA low-power wireless communication scheme with a transmission power below 100 MW, a pulse width below 200 ms, and an average power consumption below 10 MW.

The FR and available range of the WSN are chosen as attributes to represent the features of the premise. The FR is determined by dividing the number of failed sensors in the WSN by the total number of sensors. The availability rate is determined by analysing many data points, including the deployment area of the wireless WSN (WSN), the number of nodes, the sensing radius of the nodes, and the communication radius of the nodes. A BRB-based health system is developed using these two aspects. Subsequently, a health status forecasting system is produced based on the Kalman process, utilizing reference health status levels and anticipated health status.

16.5.2 Construction of WSN Health Monitoring Prediction Framework

Domain experts create the optimal health monitoring prediction framework in this part. BS and HS are each defined by four and five reference points, respectively. In this context, M, G, TM, TG, and N correspond to low, high, slightly low, slightly high, and medium values. The experts have provided the reference values for the two attributes, which may be found in Tables 16.1–16.3. Table 16.3 displays the reference health status levels of the output health status from the health framework. The kth probabilistic rule of the WSN BRB-based health system is denoted as

$B_k(t)$: If $BS(t)$ is $B_1^k \wedge HS(t)$ is B_2^k,

$$\text{Then } G(t) \text{ is } \left\{(L,\beta_{1,k}),(TM,\beta_{2,k}),(N,\beta_{3,k}),(TG,\beta_{4,k}),(G,\beta_{5,k})\right\} \quad (16.28)$$

With ruMe weight τ_k, feature weight δ_1, δ_2

Among them, the initial rule weight and initial feature weight are both set to 1.

Based on the structure of the probabilistic rule, the health status framework based on PSS contains a total of 20 probabilistic rules. The initial health status framework provided by experts is shown in Table 16.4. The initial health status forecasting system based on the Kalman process is obtained from historical health states and can be represented by the following equation:

$$\tau(\alpha_i) = \tau(\alpha_{i-1}) + 0.0125 \cdot \Delta\tau + 0.0123 \cdot \Xi(\Delta\alpha) \quad (16.29)$$

16.5.3 Optimal Prediction of Health Conduction

During the construction of the optimal health monitoring framework, a total of 250 data sets were randomly selected from the detection dataset of the two features.

Table 16.1 Points of Reference and Reference Values of *BS*

Point of Reference	*M*	*N*	*TG*	*G*
Reference	5.3998	5.5	5.7	5.815

Table 16.2 Points of Reference and Reference Values of *HS*

Point of Reference	*M*	*TM*	*N*	*TG*	*G*
Reference	0.3696	0.4	0.42	0.44	0.45

Table 16.3 Points of Reference and Reference Values of Output Levels

Point of Reference	*M*	*TM*	*N*	*TG*	*G*
Reference	0	0.25	0.5	0.75	1

Table 16.4 Initial Prediction of Health Framework of WSN

No.	*Rule Weight*	*Feature*		*Output Distribution*
		BT	*GT*	*{T,TM,N,TG,G}*
1	1	M	M	(1 0 0 0 0)
2	1	M	TM	(0.7 0.3 0 0 0)
3	1	M	N	(0.6 0.4 0 0 0)
4	1	M	TG	(0.5 0.5 0 0 0)
5	1	M	G	(0.6 0.3 0 0 0)
6	1	N	M	(0.3 0.7 0 0 0)
7	1	N	TM	(0.2 0.8 0 0 0)
8	1	N	N	(0.1 0.9 0 0 0)
9	1	N	TG	(0 0.7 0 0 0.3)
10	1	N	G	(0 0.5 0 0 0.5)
11	1	TG	M	(0.2 0.8 0 0 0)
12	1	TG	TM	(0 1 0 0 0)
13	1	TG	N	(0 0.8 0 0 0.2)
14	1	TG	TG	(0 1 0 0 0)
15	1	TG	G	(0 0.8 0 0 0.2)
16	1	G	M	(0 0.9 0 0 0.1)
17	1	G	TM	(0 1 0 0 0)
18	1	G	N	(0 0.8 0 0 0.2)
19	1	G	TG	(0 0.6 0 0 0.4)
20	1	G	G	(0 0 0 0.4 0.6)

These data sets were used as training data, while the remaining data sets were used as testing data. The P-CMA-ES algorithm-based optimization framework was limited to a maximum of 300 optimization iterations.

By optimizing the framework for WSN health monitoring, it is possible to achieve optimal health framework and health status forecasting systems. Refer to Table 16.5 to get the health optimization framework. The output and error of the health monitoring prediction framework are shown in Figures 16.2 and 16.3, respectively. The health monitoring prediction framework effectively evaluates the health condition of the WSN, as demonstrated in Figures 16.2 and 16.3, together with Tables 16.7 and 16.8. The disparity between the framework's value and the actual health status is justifiable. The health state of the WSN falls below 0.92 at 217 sampling periods, as depicted in Figure 16.2 and Table 16.6, which further highlights the need for health monitoring. One can determine the health status of a WSN by employing a medical condition forecasting technique that relies on the Kalman

Table 16.5 Optimized Prediction of Health Framework of WSN

No.	*Rule Weight*	*Feature*		*Distribution*
		BT	*GT*	*{T,TM,N,TG,G}*
1	0.108 9	M	M	(0.244 2 0.060 6 0.098 7 0.290 7 0.305 9)
2	0.313 8	M	TM	(0.243 7 0.270 0 0.130 0 0.268 6 0.087 6)
3	0.538 7	M	N	(0.152 5 0.205 5 0.245 2 0.364 5 0.032 4)
4	0.249 4	M	TG	(0.051 5 0.116 8 0.319 8 0.108 8 0.403 1)
5	0.247 3	M	G	(0.450 2 0.248 7 0.009 1 0.240 0 0.052 0)
6	0.049 9	N	M	(0.130 9 0.132 9 0.154 4 0.211 5 0.370 3)
7	0.919 3	N	TM	(0.925 5 0.031 8 0.015 5 0.018 9 0.008 2)
8	0.095 1	N	N	(0.224 2 0.090 7 0.390 6 0.275 4 0.019 1)
9	0.370 8	N	TG	(0.147 6 0.234 3 0.103 5 0.502 3 0.012 3)
10	0.976 6	N	G	(0.299 6 0.084 1 0.091 8 0.181 8 0.342 8)
11	0.257 6	TG	M	(0.076 9 0.215 1 0.119 6 0.275 1 0.313 3)
12	0.118 5	TG	TM	(0.458 7 0.068 3 0.024 4 0.212 6 0.236 0)
13	0.603 5	TG	N	(0.857 6 0.070 7 0.026 1 0.005 3 0.040 4)
14	0.236 5	TG	TG	(0.492 7 0.197 9 0.231 0 0.036 5 0.041 9)
15	0.331 8	TG	G	(0.106 4 0.020 7 0.218 3 0.134 9 0.519 6)
16	0.552 1	G	M	(0.310 0 0.113 8 0.084 3 0.229 3 0.262 6)
17	0.471 6	G	TM	(0.378 1 0.166 5 0.100 0 0.210 8 0.144 6)
18	0.044 2	G	N	(0.391 8 0.005 9 0.033 0 0.414 0 0.155 3)
19	0.209 5	G	TG	(0.250 8 0.494 2 0.047 2 0.147 5 0.060 4)
20	0.307 9	G	G	(0.366 7 0.033 1 0.056 6 0.285 6 0.258 0)

process. Figure 16.4 demonstrates that the ideal timing for health monitoring varies about 217, suggesting that the projected optimal timing for health monitoring satisfies the condition of the WSN's health state being below 0.92. In summary, the experimental results indicate that the established health monitoring framework can accurately determine the appropriate timing for health monitoring.

16.6 Conclusion

A solution has been provided to tackle the problems of insufficient detection data and intricate system mechanisms in managing the WHSN. This solution involves the development of an optimal health monitoring prediction framework that incorporates the PSS framework and the Kalman process. The framework has two novel features: The objective of the optimal health monitoring prediction

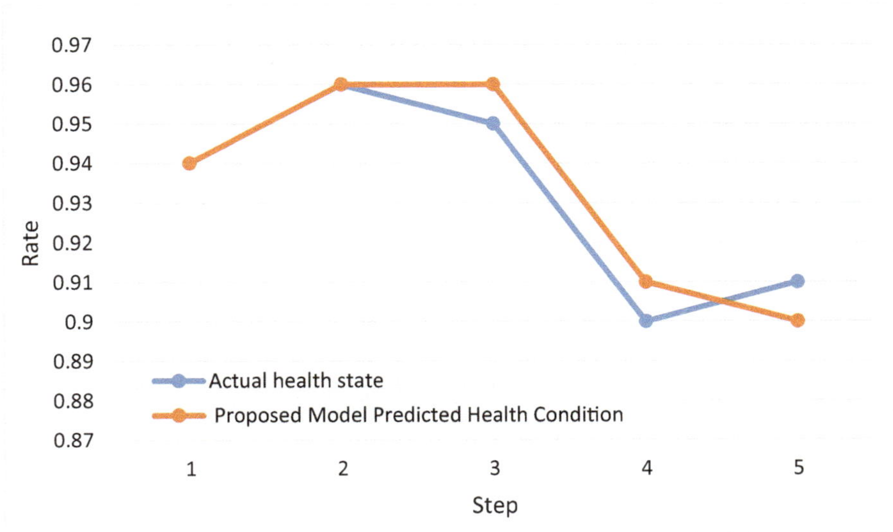

Figure 16.2 Prediction of health monitoring conduction.

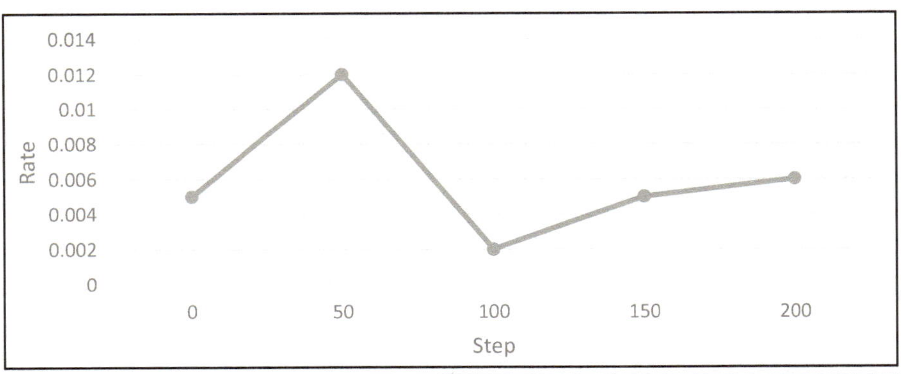

Figure 16.3 Error between evaluated health monitoring status and actual health monitoring status.

framework is to determine the most favourable scheduling for repair, taking into account the reliability of the WSN. The dependability and safety of a WSN rely on its precision. This work presents a health monitoring framework that is optimized using the PSS sensor system and the Kalman process. This chapter presents the derivation process of computing the degradation coefficient and diffusion coefficient using the maximum likelihood estimation approach in the WSN framework for predicting health monitoring outcomes based on the Kalman process. This study assumes that the detection data from the WSN is reliable, without

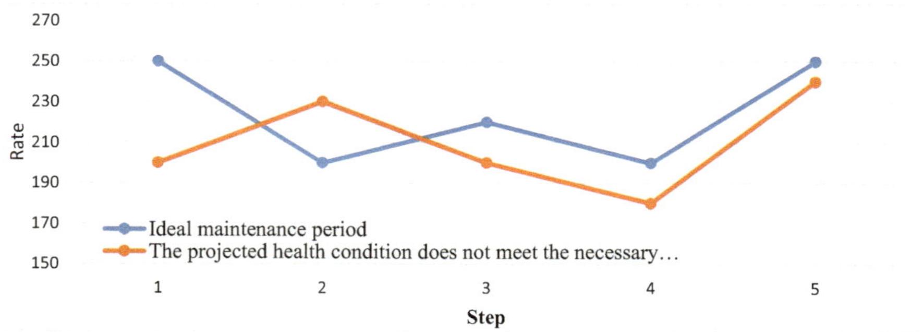

Figure 16.4 Ideal health monitoring period for the WSN.

considering any external noise interference in the WSN's working environment. Furthermore, the health monitoring prediction framework is unable to address the issue of various network properties due to the very volatile nature of the probabilistic rule combination. These problems present difficulties and areas of study for future endeavours.

References

[1] J. Yang, Z. Yue and Y. Yuan, "Noise-Aware Sparse Gaussian Processes and Application to Reliable Industrial Machinery Health Monitoring," *IEEE Transactions on Industrial Informatics*, vol. 19, no. 4, pp. 5995–6005, April 2023, doi: 10.1109/TII.2022.3200428.

[2] C. K. Wu *et al.*, "An IoT Tree Health Indexing Method Using Heterogeneous Neural Network," *IEEE Access*, vol. 7, pp. 66176–66184, 2019, doi: 10.1109/ACCESS.2019.2918060.

[3] X. Li, Y. Xu, N. Li, B. Yang and Y. Lei, "Remaining Useful Life Prediction With Partial Sensor Malfunctions Using Deep Adversarial Networks," *IEEE/CAA Journal of Automatica Sinica*, vol. 10, no. 1, pp. 121–134, January 2023, doi: 10.1109/JAS.2022.105935.

[4] X. Li *et al.*, "Using WSN for Tracing and Locating Air Pollution Sources," *IEEE Sensors Journal*, vol. 21, no. 10, pp. 12162–12170, May15, 2021, doi: 10.1109/JSEN.2021.3063815.

[5] C. Ingrao, R. Stella Evola, P. Cantore, P. De Bernardi, A. Del Borghi, E. Vesce and R. Beltramo, "The Contribution of Sensor-Based Equipment to Life Cycle through Improvement of Data Collection in the Industry," *Environmental Impact Review*, vol. 88, p. 106569, 2021, ISSN 0195-9255, doi: 10.1016/j.eiar.2021.106569.

[6] T. R. Wanasinghe, R. G. Gosine, L. A. James, G. K. I. Mann, O. de Silva and P. J. Warrian, "The Internet of Things in the Oil and Gas Industry: A Systematic Review," *IEEE Internet of Things Journal*, vol. 7, no. 9, pp. 8654–8673, September 2020, doi: 10.1109/JIOT.2020.2995617.

[7] Y. Duan, X. Cao, J. Zhao, M. Li and X. Yang, "A Spatiotemporal Fusion Autoencoder-Based Health Indicator Automatic Construction Method for Rotating Machinery Considering Vibration Signal Expression," *IEEE Sensors Journal*, vol. 23, no. 20, pp. 24822–24838, October 15, 2023, doi: 10.1109/JSEN.2023.3309013.

[8] A. E. Bondoc, M. Tayefeh, A. Barari, "LIVE Digital Twin: Developing a WSN to Monitor the Health of Belt Conveyor System," *IFAC-PapersOnLine*, vol. 55, no. 19, pp. 49–54, 2022, ISSN 2405-8963, doi: 10.1016/j.ifacol.2022.09.182.

[9] C. Liu, L. Zhang, Y. Liao, C. Wu and G. Peng, "Multiple Sensors Based Prognostics With Prediction Interval Optimization via Echo State Gaussian Process," *IEEE Access*, vol. 7, pp. 112397–112409, 2019, doi: 10.1109/ACCESS.2019.2925634.

[10] W. Zhang, D. Yang and H. Wang, "Data-Driven Methods for Predictive health monitoring of Industrial Equipment: A Survey," *IEEE Systems Journal*, vol. 13, no. 3, pp. 2213–2227, September 2019, doi: 10.1109/JSYST.2019.2905565.

[11] Y. Qamsane *et al.*, "A Methodology to Develop and Implement Digital Twin Solutions for Manufacturing Systems," *IEEE Access*, vol. 9, pp. 44247–44265, 2021, doi: 10.1109/ACCESS.2021.3065971.

[12] H. Karami, M. Kamruzzaman, J. A. Covington, M. Hassouna, Y. Darvishi, M. Ueland, S. Fuentes, M. Gancarz, "Advanced Evaluation Techniques: Gas WSN s, Machine Learning, and Chemometrics for Fraud Detection in Plant and Animal Products," *Sensors and Actuators A: Physical*, vol. 370, p. 115192, 2024, ISSN 0924-4247, doi: 10.1016/j.sna.2024.115192.

[13] M. Compare, P. Baraldi and E. Zio, "Challenges to IoT-Enabled Predictive Health Monitoring for Industry 4.0," *IEEE Internet of Things Journal*, vol. 7, no. 5, pp. 4585–4597, May 2020, doi: 10.1109/JIOT.2019.2957029.

[14] S. Al-Rbeawi, "A Review of Modern Approaches of Digitalization in Oil and Gas Industry," *Upstream Oil and Gas Technology*, vol. 11, p. 100098, 2023, ISSN 2666–2604, doi: 10.1016/j.upstre.2023.100098.

[15] A. Fuller, Z. Fan, C. Day and C. Barlow, "Digital Twin: Enabling Technologies, Challenges and Open Research," *IEEE Access*, vol. 8, pp. 108952–108971, 2020, doi: 10.1109/ACCESS.2020.2998358.

[16] S. I. Sezer, G. Elidolu, E. Akyuz, O. Arslan, "A Quantified Risk Analysis for Oil Spill during Crude Oil Loading Operation on Tanker Ship under Improved Z-Number based Bayesian Network Approach," *Marine Pollution Bulletin*, vol. 197, p. 115796, 2023, ISSN 0025-326X, doi: 10.1016/j.marpolbul.2023.115796.

[17] N. Krishnan, S. Raoux and D. Dornfeld, "Quantifying the Environmental Footprint of Semiconductor Equipment using the Environmental Value Systems Analysis (EnV-S)," *IEEE Transactions on Semiconductor Manufacturing*, vol. 17, no. 4, pp. 554–561, Nov. 2004, doi: 10.1109/TSM.2004.835705.

[18] Y. Yu, D. Ma, Y. Wang, "Structural Resilience Evolution and Vulnerability of Semiconductor Materials Supply Network in the Global Semiconductor Industry," *International Journal of Production Economics*, vol. 270, p. 109172, 2024, ISSN 0925-5273, doi: 10.1016/j.ijpe.2024.109172.

[19] S. Dehghani-Dehcheshmeh, M. Akhoondzadeh, S. Homayouni, "Oil Spills Detection from SAR Earth Observations Based on a Hybrid CNN Transformer Networks," *Marine Pollution Bulletin*, vol. 190, p. 114834, 2023, ISSN 0025-326X, doi: 10.1016/j.marpolbul.2023.114834.

[20] N. Ma, F. Yang, L. Tao and M. Suo, "State-of-Health for Aero-Engine Based on Density-Distance Clustering and Fuzzy Bayesian Risk," *IEEE Access*, vol. 9, pp. 9996–10011, 2021, doi: 10.1109/ACCESS.2021.3050792.

[21] N. Zhou, Y. Xu, S. Cho and C. T. Wee, "A Systematic Review for Switchgear Asset Management in Power Grids: Condition Monitoring, Health, and Health Monitoring Strategy," *IEEE Transactions on Power Delivery*, vol. 38, no. 5, pp. 3296–3311, Oct. 2023, doi: 10.1109/TPWRD.2023.3272883.

[22] X. Li, X. Jia, Y. Wang, S. Yang, H. Zhao and J. Lee, "Industrial Remaining Useful Life Prediction by Partial Observation Using Deep Learning With Supervised Attention," *IEEE/ASME Transactions on Mechatronics*, vol. 25, no. 5, pp. 2241–2251, Oct. 2020, doi: 10.1109/TMECH.2020.2992331.

[23] T. Lin, P. Wu, F. Gao, "Information Security of Flowmeter Communication Network Based on Multi-Sensor Data Fusion," *Energy Reports*, vol. 8, pp. 12643–12652, 2022, ISSN 2352-4847, doi: 10.1016/j.egyr.2022.09.072.

[24] M. Zhao and J. Lin, "Health of Rotating Machinery Using a Rotary Encoder," *IEEE Transactions on Industrial Electronics*, vol. 65, no. 3, pp. 2548–2556, March 2018, doi: 10.1109/TIE.2017.2739689.

[25] X. Jia, M. Zhao, Y. Di, Q. Yang and J. Lee, "Of Data Suitability for Machine Prognosis Using Maximum Mean Discrepancy," *IEEE Transactions on Industrial Electronics*, vol. 65, no. 7, pp. 5872–5881, July 2018, doi: 10.1109/TIE.2017.2777383.

[26] H. Yang, "A Practical Method for Connectivity and Coverage Reliability Analysis for Linear Wireless WSN s," *Ad Hoc Networks*, vol. 146, p. 103183, 2023, ISSN 1570-8705, doi: 10.1016/j.adhoc.2023.103183.

[27] M.-C. Li, X. Liu, K. Lv, J. Sun, C. Dai, B. Liao, C. Liu, C. Mei, Q. Wu, and M. Hubbe, "Cellulose Nanomaterials in Oil and Gas Industry: Current Status and Future Perspectives," *Progress in Materials Science*, vol. 139, p. 101187, 2023, ISSN 0079-6425, doi: 10.1016/j.pmatsci.2023.101187.

[28] S. Venkatraman and R. P. Sundarraj, "Health-Analytics Readiness: Elaborated Action Design Research and Nascent Theoretical Implications," *IEEE Transactions on Engineering Management*, vol. 71, pp. 4710–4726, 2024, doi: 10.1109/TEM.2022.3206270.

[29] V. Atamuradov, K. Medjaher, F. Camci, P. Dersin and N. Zerhouni, "Railway Point Machine Prognostics Based on Feature Fusion and Health State," *IEEE Transactions on Instrumentation and Measurement*, vol. 68, no. 8, pp. 2691–2704, August 2019, doi: 10.1109/TIM.2018.2869193.

[30] M. Shimizu, S. Perinpanayagam, B. Namoano and A. Starr, "Real-PeriodPrognostics and Health Management Without Run-to-Failure Data on Railway Assets," *IEEE Access*, vol. 11, pp. 28724–28734, 2023, doi: 10.1109/ACCESS.2023.3259221.

[31] S. N. Anbalagan, M. Schwarz, R. Bemthuis, and P. Havinga, "Assessing Factory's Industry 4.0 Readiness: A Practical Method for IIoT Sensor and Network Analysis," *Procedia Computer Science*, vol. 232, pp. 2730–2739, 2024, ISSN 1877-0509, doi: 10.1016/j.procs.2024.02.090.

[32] F. Wang, G. Chen, K. Liu, T. Zhang, Y. Li, and C. Ai, "Health of an Electro-Hydraulic Servo Pump Control System for Servomotor Based on LGA Deep Neural Network," *Heliyon*, vol. 10, no. 5, p. e26936, 2024, ISSN 2405-8440, doi: 10.1016/j.heliyon.2024.e26936.

[33] S. Sadeghi, N. Soltanmohammadlou, and F. Nasirzadeh, "Applications of Wireless WSN s to Improve Occupational Safety and Health in Underground Mines," *Journal of Safety Research*, vol. 83, pp. 8–25, 2022, ISSN 0022-4375, doi: 10.1016/j.jsr.2022.07.016.

[34] Y. Ke and X. Xing, "Construction of Rehabilitation Training Simulation Framework Based on Energy Monitoring and Transmission by Wireless WSN s," *Alexandria Engineering Journal*, vol. 85, pp. 19–28, 2023, ISSN 1110-0168, doi: 10.1016/j.aej.2023.11.031.

[35] F. Paiva Knebel, R. Trevisan, G. Santana do Nascimento, M. Abel, and J. A. Wickboldt, "A Study on Cloud and Edge Computing for the Implementation of Digital Twins in the Oil & Gas Industries," *Computers & Industrial Engineering*, vol. 182, p. 109363, 2023, ISSN 0360-8352, doi: 10.1016/j.cie.2023.109363.

Chapter 17

Cybersecurity Education Gamification: A Current Review and Research Agenda

Ramiz Salama and Fadi Al-Turjman

17.1 Introduction

Given the constantly changing nature of digital threats, learning cybersecurity is becoming increasingly important. At some point, it becomes necessary for people to recognize and minimize cyber dangers. Though successful, traditional teaching methods may fall short of maintaining students' attention long enough to impart the necessary practical abilities for real-world problem-solving. Gamification, which incorporates game elements into non-gaming activities to make learning more engaging and dynamic, has been a popular trend in recent years. The study on using gamification in cybersecurity education is surveyed in this report. We will make an effort to comprehend the state of the art in this area, the efficacy of gamified approaches, and offer fresh research concepts that can close gaps and create new avenues for investigation. The goal of the developing field of gamification is to enhance participation and education, particularly cybersecurity education. This chapter offers a cutting-edge analysis of the body of research on gamification in cybersecurity education, including the creation and application of gamified learning

 DOI: 10.1201/9781003479062-17

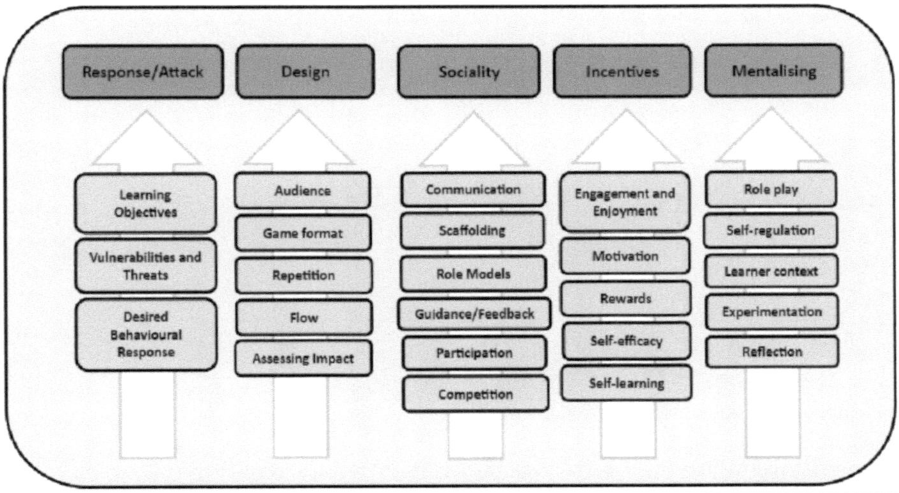

Figure 17.1 The RAD-SIM framework and associated sub-components.

environments. The review also demonstrates how gamification can enhance students' motivation, retention of information, and problem-solving abilities. The review serves as the basis for this report's research agenda, which highlights knowledge gaps and makes recommendations for future paths in the creation and assessment of gamification techniques in cybersecurity education [1–4]. The results indicate that although gamification has demonstrated promise, more investigation is required to comprehend the enduring effects and expandability of gamified methods in diverse educational settings (Figure 17.1).

17.2 Quantity of Prior Publications

17.2.1 An Overview of Educational Gamification

Gamification in education has also been studied in a variety of sectors, and the findings indicate that incorporating game-like features into instruction improves student motivation, interest, and academic performance [4–8]. Simulations, serious games, and competitive events like capture the flag (CTF) competitions are some examples of how gamification has been used in cybersecurity education.

17.2.2 Gamification in the Teaching of Cybersecurity

The potential of gamification to teach the fundamentals of cybersecurity has been examined in recent studies. These projects show how gamification may provide a

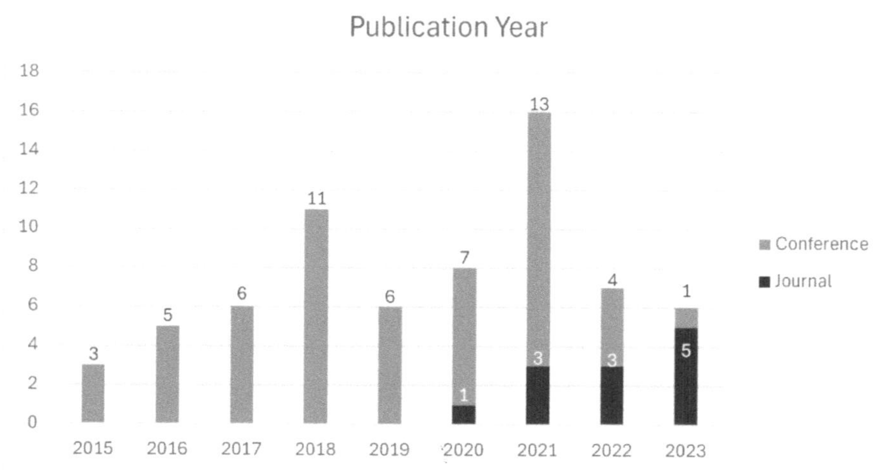

Figure 17.2 Publication trends over time for gamification in SEET.

series of practical exercises in a sandbox setting. The papers vary, though, with some merely presenting concepts and others presenting actual performance statistics demonstrating gamified tools or platforms.

17.2.3 Evaluation of Published Work Critically

There is a wide range of research on gamification in cybersecurity education. The majority of research found that gamification improved individuals' motivation or helped them retain the information they had acquired. Nevertheless, several studies also highlight the difficulties with excessive resource use and the possibility that gamification won't be successful if it doesn't align with learning objectives (Figure 17.2).

17.3 Gamification of Cybersecurity Education: A Current Assessment and Research Agenda

17.3.1 Method for Searching the Literature

A thorough search of pertinent databases from several libraries, including Google Scholar, Association for Computing Machinery (ACM) Digital Library, and IEEE Xplore, will be carried out. In order to incorporate the most recent studies that have been published in peer-reviewed journals, we concentrate on the last ten years [9–18]. Among the search terms used were "serious games," "gamification," "cybersecurity education," and "educational technology."

17.3.2 Criteria for Inclusion and Exclusion

Articles that were written in English, based on actual research or offered some theoretical concepts, and specifically addressed gamification in cybersecurity education were included. Many that were only theories with no real-world applicability or had nothing to do with cybersecurity had to be eliminated.

17.3.3 Extraction and Analysis of Data

In order to classify the gamification type, the educational environment (professional training or higher education), and the outcome type (engagement, knowledge retention, skill development, etc.), we organized the pertinent data. To identify recurring themes and research gaps, qualitative analysis was then conducted (Figure 17.3).

17.4 Results and Discussion

17.4.1 Effectiveness of Gamification in Cybersecurity Education

The analysis of findings demonstrated that, generally speaking, gamification improves practical skills and learning engagement in the area of cybersecurity education.

Figure 17.3 Gamification and why it is important for cybersecurity awareness training.

Numerous studies indicate that a gamified environment will enable students to use their theoretical knowledge in real-world simulations, which will help them get ready for real-world cybersecurity difficulties [19–23].

17.4.2 Implementing Gamification Presents Difficulties

Gamification has the potential to improve cybersecurity education, but there are still issues. These difficulties include the need for significant investment in the creation and upkeep of gamified platforms, the requirement that teachers receive cybersecurity and game design training, and the possibility that gamification could divert students from learning objectives if it is not properly integrated with the curriculum.

17.4.3 Research Agenda and Future Directions

The long-term effects of gamification on learning outcomes, how gamification approaches scale across various educational environments, the creation of standardized tools to gauge gamification effectiveness, the effects of various game design elements related to points, badges, and leaderboards on learning outcomes for cybersecurity education, and cultural variations in gamification effectiveness are just a few possible research topics that a systematic review can highlight [24–28] (Figure 17.4).

17.5 Conclusion

Gamification has been shown to be a useful technique for increasing cybersecurity education's efficacy by encouraging motivation, engagement, and the development of useful skills. This analysis demonstrates the wide variety of gamified methods being used to teach critical cybersecurity skills, ranging from interactive simulations to serious games. Notwithstanding the achievements, issues

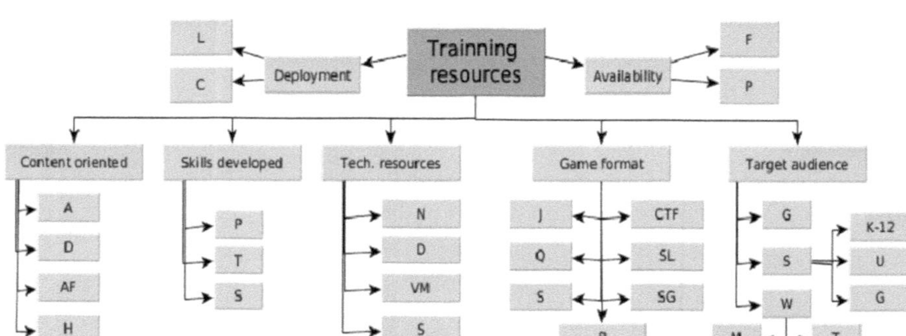

Figure 17.4 Cybersecurity training resources taxonomy.

like scalability, content adaptation, and long-term learner motivation still exist. A more sophisticated comprehension of learner demands, ongoing adaptive learning system development, and the incorporation of real-world scenarios to close the gap between theory and practice are all necessary to address these problems. In order to develop more immersive and customized learning experiences, the suggested research agenda places a strong emphasis on the necessity of interdisciplinary collaboration, the investigation of adaptive gamification strategies, and the integration of cutting-edge technologies like AI and VR. Stakeholders may create more reliable and scalable gamified learning resources that enhance learning outcomes and provide a workforce with the necessary skills to handle the always changing cybersecurity risks by addressing these future trends. In the end, developing gamification in cybersecurity education will help create a digital environment that is safer and more robust.

References

[1] Mouhssine, Y., Boutracheh, H., & Moumen, A. (2024, May). Teaching Cybersecurity to Engineering Students: A New Perspective through Literature Review. In 2024 4th International Conference on Innovative Research in Applied Science, Engineering and Technology (IRASET) (pp. 1–9). IEEE.

[2] Gwenhure, A. K., & Rahayu, F. S. (2024). Gamification of Cybersecurity Awareness for Non-IT Professionals: A Systematic Literature Review. *International Journal of Serious Games*, 11(1), 83–99.

[3] Razack, A. K. A., & Saad, M. F. M. (2024). Enhancing Cybersecurity Awareness through Gamification: Design an Interactive Cybersecurity Learning Platform for Multimedia University Students. *Journal of Informatics and Web Engineering*, 3(3), 21–40.

[4] Pahlavanpour, O., & Gao, S. (2024). A Systematic Mapping Study on Gamification within Information Security Awareness Programs. *Heliyon*, 10(19), e3874

[5] Tay, A., Hayes, S. M., Wilson, D., Hall, E., & Kaufman, D. (2024). Gamified Cybersecurity Education through the Lens of the Information Search Process: An Exploratory Study of Capture-the-Flag Competitions [Research-in-Progress]. *Issues in Informing Science and Information Technology*, 21, 001.

[6] Mason, O. J., Collman, S., Kazamia, S., & Boureanu, I. (2024). Preparing UK Students for the Workplace: The Acceptability of a Gamified Cybersecurity Training. *Journal of Cybersecurity Education, Research and Practice*, 2024(1), 1–20.

[7] Alothman, B. Y. (2024, April). Cyber Gamification: Implementing Gamified Adaptive Learning Environments for Effective Cyber Security Teams Education. In Proceedings of the 2024 5th International Conference on Education Development and Studies (pp. 33–40).

[8] Fatokun, F., Awang, Z., Hamid, S., Fatokun, J. O., & Norman, A. (2025). Cybersecurity Knowledge Deterioration and the Role of Gamification Intervention. *Journal of Advanced Research in Applied Sciences and Engineering Technology*, 43(1), 66–94.

[9] Mukherjee, M., Le, N. T., Chow, Y. W., & Susilo, W. (2024). Strategic Approaches to Cybersecurity Learning: A Study of Educational Models and Outcomes. *Information*, 15(2), 117.

[10] Qawasmeh, S. A. D., AlQahtani, A. A. S., & Khan, M. K. (2024). Navigating Cybersecurity Training: A Comprehensive Review. arXiv preprint arXiv:2401.11326.

[11] Bitrián, P., Buil, I., Catalán, S., & Merli, D. (2024). Gamification in Workforce Training: Improving Employees' Self-Efficacy and Information Security and Data Protection Behaviours. *Journal of Business Research*, 179, 114685.

[12] Ibrahim, A., McKee, M., Sikos, L. F., & Johnson, N. F. (2024). A Systematic Review of K-12 Cybersecurity Education around the World. *IEEE Access*, 12, 59726–59738.

[13] Valente, J., & Reith, M. (2024, June). Cyber Game-Based Learning for DoD CEs. In *European Conference on Cyber Warfare and Security* (Vol. 23, No. 1, pp. 795–802). Academic Conferences and Publishing International Limited.

[14] Dillon, R., & Tan, K. L. (2024). Cybersecurity Workforce Landscape, Education, and Industry Growth Prospects in Southeast Asia. *Journal of Tropical Futures*, 1(2), 172–181.

[15] Barletta, V. S., Caruso, F., Di Mascio, T., Greco, F., Islam, T., Rossano, V., & Xiao, H. (2024, June). CyberSecurity Education for Industry and Academia (CSE4IA 2024). In Proceedings of the 2024 International Conference on Advanced Visual Interfaces (pp. 1–4).

[16] Criollo-C, S., Guerrero-Arias, A., Buenaño-Fernández, D., & Luján-Mora, S. (2024). Usability Evaluation and Workload Assessment of a Cybersecurity Educational Game Application. *IEEE Access*,12, 12771–12784.

[17] Francis, S. P., Kolil, V. K., Pavithran, V., Ray, I., & Achuthan, K. (2024). Exploring Gender Dynamics in Cybersecurity Education: A Self-Determination Theory and Social Cognitive Theory Perspective. *Computers & Security*, 144, 103968.

[18] Namukasa, M., Chaparro Osman, M., Ficke, C., Piasecki, I., OConnor, T. J., & Carroll, M. (2024). Evaluation of an Internet of Things Device-Based Educational Approach to Engage a More Diverse Cybersecurity Workforce. *International Journal of Human–Computer Interaction*, 1–17.

[19] Ahmed, Y., Ezealor, M., Mahmoud, H., Azad, M., BenFarah, M., & Yousefi, M. (2024). Enhancing Security Awareness through Gamified Approaches. arXiv preprint arXiv:2404.09052.

[20] Pirta-Dreimane, R., Romanovs, A., Bikovska, J., Pekša, J., Vartiainen, T., Valliou, M., ... & Eltahawy, B. (2024). Enhancing Smart Grid Resilience: An Educational Approach to Smart Grid Cybersecurity Skill Gap Mitigation. *Energies*, 17(8), 1876.

[21] Köhler, D., & Meinel, C. (2024). The Right Tool for the Job: Contextualization of Cybersecurity Education and Assessment Methods. In ICISSP (pp. 234–241).

[22] Varghese, B., Murthy, R., & Hansen, S. W. (2024). Game On: Theorizing the Impacts of Gamification on Organizational Cybersecurity Training Efficacy. In *Thirtieth Americas Conference on Information Systems (AMCIS 2024)*, Salt Lake City.

[23] Pirta-Dreimane, R., Brilingaitė, A., Roponena, E., Parish, K., Grabis, J., Lugo, R. G., & Bonders, M. (2024). Try to Escape from Cybersecurity Incidents! A Technology-Enhanced Educational Approach. *Technology, Knowledge and Learning*, 1–30.

[24] Manganello, F., Earp, J., Fante, C., Bassi, G., Fabbri, S., Matteucci, I., ... & Gentile, M. (2024, June). Shaping the foundation of the SuperCyberKids Learning Framework: A comprehensive analysis of cybersecurity education initiatives. In *Frontiers in Education* (Vol. 9, p. 1375853). Frontiers Media SA.

[25] Douglas, M., & Reith, M. (2024, June). A Survey of Learning Technology Integration in Information Warfare Education. In *European Conference on Cyber Warfare and Security* (Vol. 23, No. 1, pp. 148–156). Academic Conferences and Publishing International Limited.

[26] Williams, J. A., Zafar, H., & Gupta, S. (2024). Fortifying Healthcare: An Action Research Approach to Developing an Effective SETA Program. *Computers & Security*, 138, 103655.

[27] Yasin, A., Fatima, R., Wen, L., JiangBin, Z., & Niazi, M. (2025). What Goes Wrong during Phishing Education? A Probe into a Game-Based Assessment with Unfavorable Results. *Entertainment Computing*, 52, 100815.

[28] Rooholamini, A., & Salajegheh, M. (2024). Health Profession Education Hackathons: A Scoping Review of Current Trends and Best Practices. *BMC Medical Education*, 24(1), 554.

Chapter 18

Artificial Intelligence and Cybersecurity in 6G Wireless Networks

Ramiz Salama and Fadi Al-Turjman

18.1 Introduction

The majority of popular online activities take place on social media, which is the most popular platform in the world. The younger generation is completely enamored with social media. The use of electrical devices has grown significantly over the past few years, and social media will become essential to both the present and future generations. Social media is growing more diverse as a result of increased expectations. Social media has significantly impacted learning and education.

Social media platforms like Facebook, LinkedIn, Twitter, and Instagram are used to network, connect, and communicate with individuals worldwide, including family and friends. By exchanging content, stories, images, video messages, and other materials on various social media platforms, networking entails making both personal and professional connections. Using social media to provide fresh viewpoints on well-established technologies is an essential part of the innovation process. It describes social media's function. The phenomenon of social media has existed for some time. There is no need to introduce well-known social media sites like Facebook, Instagram, Snapchat, YouTube, and Twitter. The early 2000s saw the development of numerous social media platforms that transformed communication

DOI: 10.1201/9781003479062-18

and information sharing between individuals and corporations. These platforms made it easier to create and distribute a variety of media with target audiences. By offering a platform for hiring, information sharing, advertising, and maintaining a brand presence, social media has also emerged as the fastest means of expanding businesses and brands. Organizations now include social media in their competitive strategies due to its significance and wide range of applications. Social media is used by businesses for marketing, customer interaction, customer relationship management, recruiting, and reaching customers in a variety of geographic locations. Because employees utilize both personal and organizational devices to interact with social media as people in both professional and personal ways, the lines between private and corporate social media use have grown hazy. Even while social media is widely used and has many advantages, there are risks and hazards associated with using it for both personal and professional purposes [1–9]. To assist in preventing cybercrime on social media, awareness is essential, as is the case with all cybersecurity and cybercrime issues. On social media, workers and organizations run the risk of becoming victims of cybercrime. To prevent unintentional cybercrime on social media and involvement in it, awareness is also necessary. Cybercrime awareness training and basic cybersecurity training must include instruction on cybercrime on social media. To determine the success of their awareness training initiatives, organizations must also assess the amount of cybercrime awareness among their staff members on social media. Cybercrime awareness on social media (CASM) is not well studied, and the majority of measures of cybercrime awareness do not take social media crimes into account. Moreover, no research has examined how hearing impairments affect social media awareness of cybercrime. Employees in the security-critical education, finance, government, information technology, legal, medical, military, policing, and STEM fields were asked to rate the impact of hearing impairments on their CASM. Information security managers in security-critical industries can benefit from the study's findings, which will highlight the need for more inclusive and successful CASM initiatives. Several well-known social media sites, such as Instagram, Twitter, and Facebook, are used by various stockholders worldwide to interact with one another. Numerous advantages have resulted from this change, including improved information accessibility and community involvement. Social media is a collection of online communication tools built on cooperation, content sharing, and community involvement. Social media has completely changed how we connect with one another, communicate, and exchange information. Numerous advantages have resulted from this, but there have also been numerous difficulties, particularly in the area of cybersecurity. This study explores the diverse impacts of social media, looking at both the positive and bad effects it has on various businesses. It highlights the hazards that people and businesses face by concentrating on the cybersecurity issues raised by the increasing use of social media. Common cybercrimes such as phishing, social engineering, virus attacks, cyberstalking, identity theft, cybercasting, and burglary through social networking are all examined in this examination. This study highlights the significance of a comprehensive and focused cybersecurity strategy that incorporates safeguards including user education, advanced email filtering, strong authentication, encryption technology, and privacy improvements. By addressing these issues and taking preventative measures, people and organizations may navigate the changing social media landscape with increased cyber resilience. Since the early 2000s, social media has maintained its high level of popularity. There are billions of active users on social media sites like Facebook, YouTube, Twitter, and Snapchat worldwide. Due to its ease of sharing material and information, as well as its capacity to reach and interact with particular audiences of interest, social media continues to be a vital tool for both individuals and organizations. Social media is a great place for cybercrime because of its large user base, ease of contact, and data exchange. Cybercriminals actively target social media users, advertise their illegal operations on social media, and utilize social media to facilitate their cybercrime activities. Organizations must implement cybercrime on social media policies to protect against the potential hazards posed by cybercrime on social media and teach staff CASM. As the lines between personal and business social media use become hazier, CASM is crucial. The purpose of this study was to assess the CASM scores of workers in security-critical industries and ascertain whether hearing impairments affected the scores in any way. Americans who worked in the fields of education, finance, government, information technology, law, medicine, the military, and policing were polled. The CASM score was average across all industries, according to the results. Although CASM scores were somewhat lower for employees with hearing impairments, there was no statistically significant difference between the groups with and without hearing difficulties. The findings indicated that employees in the assessed group require additional CASM training (see Figure 18.1).

18.2 Quantity of Prior Publications

18.2.1 Social Media's Benefits for Businesses

Social media is popular right now because it brings companies, brands, and organizations together to create news, influence others, build relationships, and grow communities. Businesses use online social networking sites to build their brand. One benefit of using social media on the Internet is that it facilitates contact between a company and its investors. The company may be promoted through a number of remote, unapproved locations where people can interact [10–14]. An organization uses a large portion of the promotion as a tool to grow its social media networks in order to attract the best customers.

Here are a few benefits:

- Social media makes it simpler to understand customer wants.
- Businesses may advertise globally with the aid of social media.

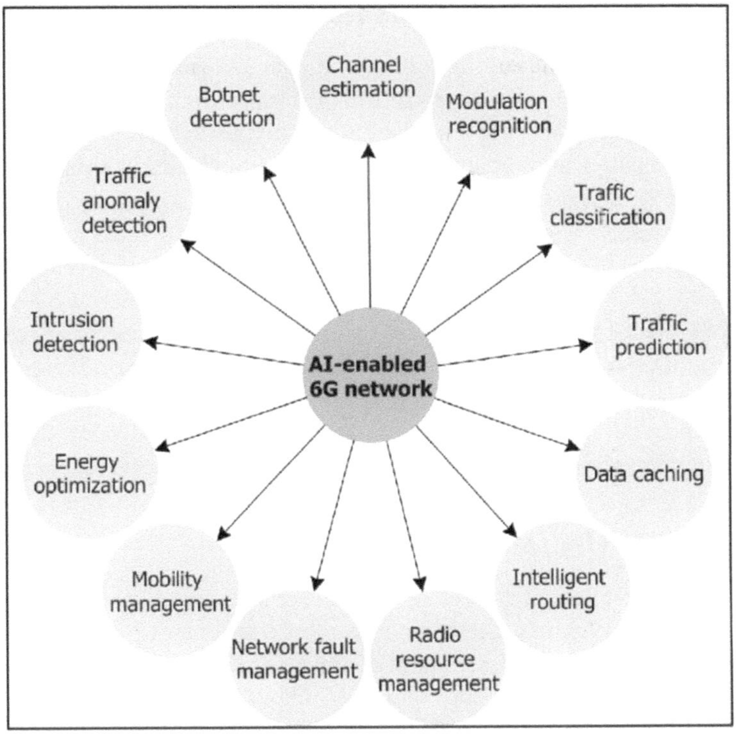

Figure 18.1 Artificial intelligence (AI)-enabled 6G networks.

- A social networking site will help to draw in new clients by providing a variety of services.
- Internet networking can help to gain insight into the market and expand further in comparison to its competitors.
- Social media builds deals and retains clients through a standard connection and a lucky client benefit.

The following are some drawbacks:

- Since many followers and supporters are permitted to leave comments on certain affiliations, social media poses no risk to businesses; but unfavorable remarks may create an association with disappointment.
- Customer reviews that are negative can be detrimental.
- It takes a long time.
- The amount of content on the Internet is increasing.
- Errors on social media are particularly hard to reverse.

18.2.2 Social Media's Benefits for Health

Through participation in various forms of digital networking, social media is a common movement that enhances social interaction, communication, and even abilities. Through social networking sites, there are numerous chances to get in touch with classmates, companies, and others who have gone through similar things. According to a survey, 16% of teenagers said they use YouTube constantly, and 71% said they use it at least once a day. Compared to younger kids, older teens are more likely to utilize social media sites like YouTube and Instagram. For example, just 45% of respondents say they use Instagram among youths 13 and 14, whereas 68% say they do so among those 15 and 17. As a result, the Internet and mobile devices account for a large amount of today's social and technological development. Some specialists have recently connected digital networking sites to mental health conditions like sadness, loneliness, and anxiety. Since digital networking sites are relatively new, there are still a number of unanswered questions regarding their potential effects on mental health.

18.2.3 Social Media's Benefit for Marketing

Social media marketing places an emphasis on user interaction, peer-to-peer contact, and user-generated content rather than bombarding consumers with information from a business. As a result, it increases brand confidence. There are numerous marketing platforms, including shopping apps (Amazon, Flipkart, etc.), promotion platforms (Instagram, Pinterest, etc.), and company marketing (HubSpot, ActiveCampaign). Social media use makes it simple for users to communicate with one another and gives advertisers multiple ways to reach consumers. By building direct relationships with their audience, Social Science and Medicine (SSM) companies can potentially increase the number of consumers for their products and services, which has several benefits.

18.2.4 With Social Media's Explosive Expansion

One benefit of using it in politics is its capacity to impact political outcomes and processes in a variety of ways. Social networking sites are used by politicians for campaign development and communication. It is often assumed that social media is essential for spreading information and political party claims [4]. The Internet allows people to debate or share political claims together. Due to differing views and viewpoints, social media is utilized to sway people's perspectives. The Internet has been characterized as a tool that gives citizens more access to the political process, as well as a chance to develop new avenues for political participation.

18.2.5 Social Media's Benefits for Finance

Social media accounts can be used to build strong relationships with clients and find out more about their opinions. When it comes to social networking, the financial services industry has lagged behind other sectors. By promoting openness through knowledge sharing and transfer, financial institutions can reduce their operating risks [8]. Social media can affect laws, regulations, and compliance and presents numerous risks to the financial sector.

Services related to finance: Social media marketing is the practice of promoting and advertising financial services, goods, and brands through social media platforms.

In the financial services sector, this entails producing content for those social media channels, interacting with users, and conducting research to identify and concentrate on advertising to particular target populations. The main goals of social media marketing are to raise brand awareness, create leads, improve website traffic, and acquire new clients.

18.2.6 Social Media's Benefit for Education

Nowadays, a sizable percentage of students have everyday access to smart devices and social media, which they browse in their free time. As a result, it's a fantastic method to make use of that time, gain knowledge, and accomplish something worthwhile. Social media websites and applications have changed to accommodate and collaborate with educational institutions' needs. The resources and online study materials are very beneficial to educators, parents, and students.

1. LinkedIn: This well-known website allows users to create a professional profile, search for jobs, and get updates on a range of subjects and industries. However, did you know that it also offers courses to improve your learning and upskill? To assist students in learning more, it provides both free and paid course materials.
2. YouTube: Since then, students have used YouTube to access entertaining and instructional videos. However, pupils were worried about the app's narrow selection of study resources. Many educators and educational organizations posted movies on a variety of subjects following the lockdown. As a result, students can now use YouTube as an educational tool more easily.
3. Instagram: Instagram took a while to become as well-known as it deserved to be. Students can now access instructional videos, share their projects and innovations, and use hashtags to reach a worldwide audience. It is a powerful tool for establishing connections with communities and individuals who have similar interests to your own: youngster.
4. Quora: Students mostly use Quora to post questions and receive answers, in addition to a number of other features. On Quora, students can also search for similar topics and view previously published articles or answers. It can be used to establish new online relationships because it is a social networking site.
5. Reddit, Telegram, and Facebook: Since millions of people use these social media platforms nationwide, they have a variety of groups that foster diversity.

With the help of these social media platforms, students can broaden their knowledge and make connections with new people, experts, and communities. It's also a good way to share and improve your knowledge. Teachers have created a number of Telegram channels to share study materials and information simultaneously.

6. These are video-conferencing apps rather than social media websites: Zoom, G Meet, and MS Teams. But because they could connect hundreds of individuals anywhere in the world with a single link, their appeal among parents, educators, and students surged significantly. It is actively used by educational institutions for programming, tests, and classes (see Figure 18.2).

18.3 Social Media Challenges

Social media, which connects people worldwide and facilitates the sharing of information, has emerged as a crucial component of modern communication in the digital age. However, social media's wide reach also presents a number of issues that need careful consideration. Some of the most common issues with social networking are listed below [15–19].

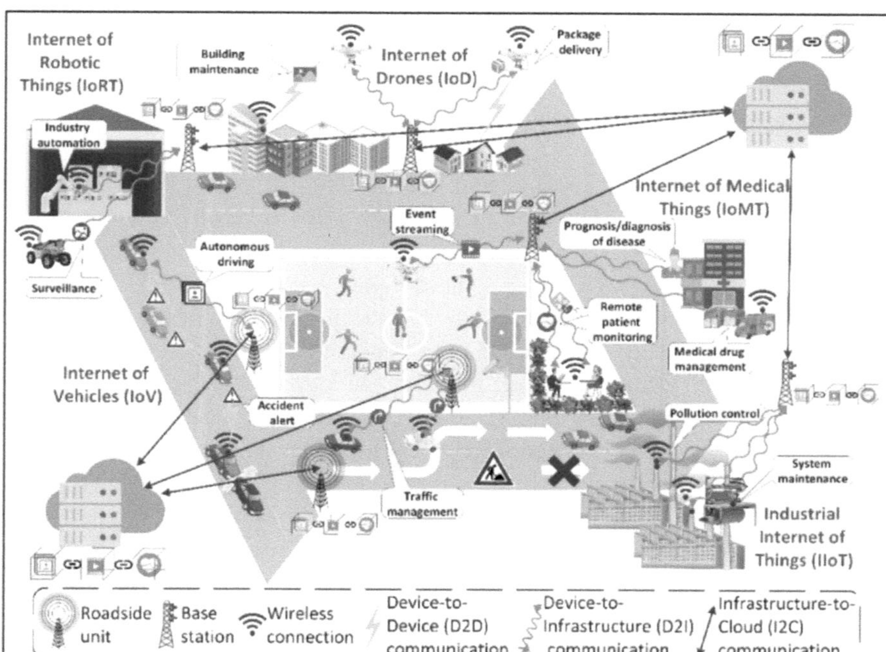

Figure 18.2 Artificial intelligence (AI)-enabled 6G networks.

Proliferation of fake news and disinformation: Because social media platforms allow for rapid information dissemination, fake news and disinformation are frighteningly more common in the digital age. Recent research by Vosoughi et al. emphasizes how erroneous information is spread quickly on social media sites like Twitter. This emphasizes the need for effective strategies to overcome this obstacle. Because disinformation poses a serious danger to public discourse and decision-making, widespread incursions are required to protect the credibility of online information.

Concerns about data exploitation and privacy are brought up by the way social media platforms collect and use a lot of users' personal information. The Cambridge Analytica scandal offers an evaluation of the probability of political misuse of personal data. It can be difficult for social media networks to provide individualized information while protecting user privacy. To maintain consumer confidence and trust, the digital world needs to undergo a constant revolution in privacy protections.

Mental health issues and psychological impacts: Twenge and Campbell claim that teens' increased use of social media is contributing to an alarming rise in depressive symptoms. Because these platforms are addicting, they exacerbate mental health difficulties by extending user interaction through algorithms. To successfully lessen the detrimental psychological effects of social media, initiatives must be taken at the platform and individual levels to create a better online environment.

Cybersecurity flaws: Because social media platforms are interconnected, users are vulnerable to various types of cybercrime, such as phishing attempts and identity theft. Current events, such as the 2020 Twitter Bitcoin hoax [19], draw attention to the flaws in even the most widely used websites. Persistent efforts are required to enhance security protocols, inform users about online dangers, and create cooperation between law enforcement and platforms in order to effectively manage cybersecurity issues (Figure 18.3).

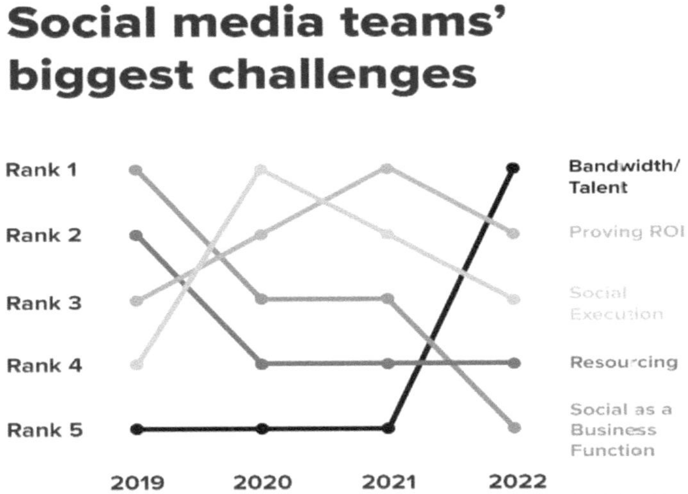

Figure 18.3 The main issues facing the social media team.

18.4 Approaches

Finding trends in qualitative and mixed methods approaches utilized in the developing area of social studies or the emerging field of social media research was the main goal of this literature study. In order to "attempt to summarize the contents of multiple studies and minimize any interpretation on the part of the reviewer," the review employs an integrative synthesis technique and is descriptive. A peer-reviewed journal paper summarizing the findings of a qualitative or mixed methods study in which social media was a key component served as the unit of analysis [20–29]. The literature review's purview was restricted to works released between 2007 and 2013. Because social media is a relatively new phenomenon, there was little literature published before 2007, which is why the first cutoff was made. Facebook was created in 2004 (Facebook, 2015), YouTube in 2005 (YouTube, 2015), and Twitter in 2006 (Twitter, 2015), according to business websites. Publications started to appear in 2006, according to a previous literature analysis on YouTube scholarship, although no research studies were available for a certain period. Since the first publications on Twitter and microblogging started to appear in 2007, Williams, Terras, and Warwick (2013) chose that year as the beginning of their literature study. Research on Facebook was first published in 2005 but appears to have gained traction in 2007. As a result, it was decided to set the first cutoff at 2007 and the latest cutoff at 2013, the year before the review was carried out in 2014 (Figure 18.4).

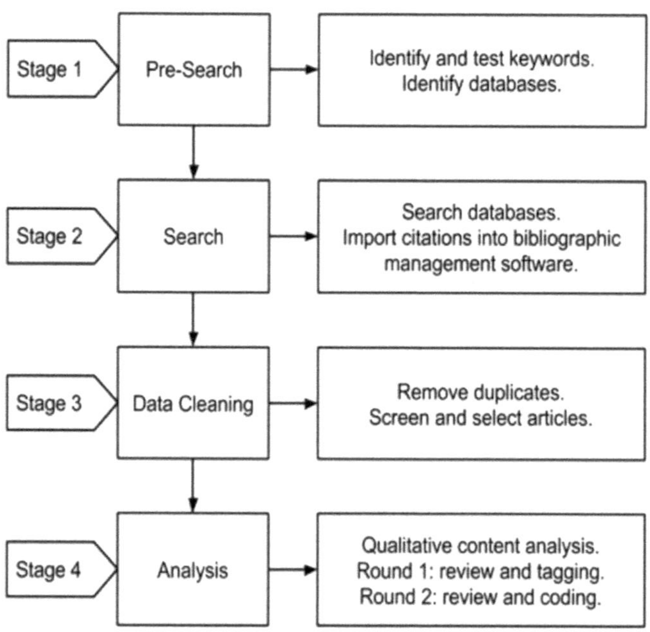

Figure 18.4 Trends in qualitative and mixed methods approaches utilized in the developing area of social studies.

18.4.1 Step 1: Preliminary Investigation

Keywords and databases were chosen during the research phase using a mix of (a) techniques from previous literature studies and (b) test searches using potential key terms, filters, and databases. Prior literature reviews have examined various facets of social media usage or content, independent of platform, while others have concentrated on particular social media platforms (e.g., Facebook, Twitter, and YouTube). The current study examines a variety of studies involving one or more social media platforms by combining platform-specific and general search terms. The following keywords were used: social media, social networking, YouTube, Facebook, and Twitter. In at least one previous literature review, each of these search terms has been utilized. A lot of the earlier assessments of the social media literature were based on specific research areas. Combinations of databases were searched, occasionally including databases that indexed field-specific literature (e.g., PubMed for medical-related literature reviews). The current review of the literature is multidisciplinary and focuses on research methodology trends across all fields. As a result, searches were limited to the following multidisciplinary databases, all of which have been used in previous literature reviews: Google Scholar, Web of Science, and Academic Search Premier. When combined, they provide significant and complimentary access to scholarly material across several fields.

18.4.2 Step 2: Look for

In January 2014, a search was conducted for publications published from 2007 to 2013 that contained bibliographic entries in the databases that were chosen. To make them reproducible for other researchers, the precise method for searching each of the databases (Google Scholar, Web of Science, and Academic Search Premier) is given here.

18.4.3 Step 3: Cleaning the Data

The papers' full-text versions and abstracts were examined to see if they qualified for analysis. The following criteria were used to select articles: (a) the study used qualitative research methodology or mixed methods research with a qualitative research component; (b) the study focused on online social media; (c) the article was published in a peer-reviewed journal; and (d) the article was available in full-text English. A subset of 55 of the 229 studies that satisfied the requirements used both qualitative and quantitative (i.e., mixed) approaches.

18.4.4 Stage 4: Evaluation

The review and analysis of the literature were organized using a qualitative content analysis methodology that was based on methodology. A descriptive research technique called qualitative content analysis entails creating a coding frame and qualitatively coding the data. According to Schreier, the coding frame was both concept-driven (predetermined) and data-driven (formed from data during coding). In essence, the concept-driven portion of the coding frame was created to categorize studies based on the social media emphasis and research design (qualitative and mixed methodologies). As will be briefly explained, the data-driven part of the coding frame was mostly derived by tagging and coding articles according to the research methodologies employed in the study.

18.5 Cybersecurity and Artificial Intelligence in 6G Wireless Networks

Social media platforms are online resources that allow users to produce, share, and communicate material and information with other people. Social media networks include Twitter, Facebook, Instagram, LinkedIn, and TikTok, to name a few. Together, they have grown to be a crucial component of social media marketing, enabling or assisting companies in reaching their target markets, increasing brand recognition, and boosting engagement and conversions. There are several ways that data analytics may help small businesses. Businesses can decide where to invest their resources and how much to charge for their goods and services by having a better understanding of data analytics. Additionally, data analytics can assist companies in understanding their clientele and identifying trends. By tracking and analyzing data, data analytics software enables you to produce dashboards and reports that are useful.

18.5.1 Cybercrime and Social Media

The "Criminal Use of Social Media" white paper from the National White Collar Crime Center claims that the recent rise of social media has fundamentally altered the nature of communication. Social media sites like Instagram, Facebook, LinkedIn, YouTube, and Twitter are used daily by millions of individuals. These platforms facilitate communication between people. These channels are used for hiring new employees, advertising, and regular communication within the public sector. According to the number of monthly active members, Figure 18.5 presents Statista's data on the most popular social media networks as of October 2023 [20].

The most common Internet activity, according to the research, is social media browsing. On a PC, social media networking takes up more than 12 minutes every hour, while on a mobile device, social media browsing takes up more than 18 minutes. Malefactors are taking advantage of social media platforms for malicious intent as a result of the evolving nature of communication. Five categories of

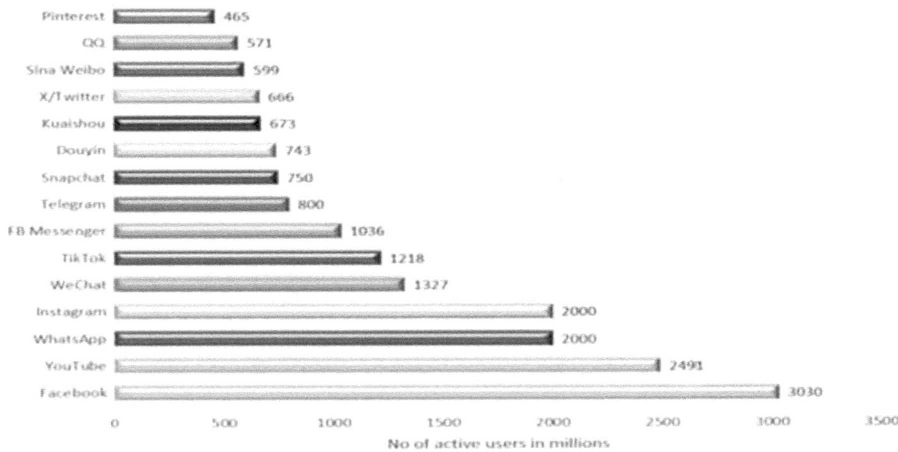

Figure 18.5 Social media users as of October 2023.

crimes involving social media were recognized in the National White Collar Crime Center (NW3C) report.

18.5.2 Social Network-Based Burglaries

Social media burglaries entail thieves searching social media for potential victims. Social media users frequently share details of their lives, such as going out to dinner or on a trip. Criminals can organize burglaries over a long period of time thanks to the information these actions give them about easy targets.

18.5.3 Social Engineering

Coercive control is used in social engineering to obtain private information. People frequently receive messages on social networking sites from friends requesting urgent financial support.

In order to deter criminality, these communications are typically transmitted without any password protection. Because bank account information is so easily evaluated, hackers employ a variety of strategies and techniques to obtain personal information about anyone through social engineering. Phishing emails, including tax return scams, phony iCloud update alerts, human resource survey scams, and owners requesting login credentials or bank information from staff members, appear extremely official. Cybercriminals attempt to intimidate their victims in order to prevent them from informing any cybercrime-related organizations. Millions of emails were sent by criminals to obtain information. Deceptive phishing is the most prevalent kind of phishing.

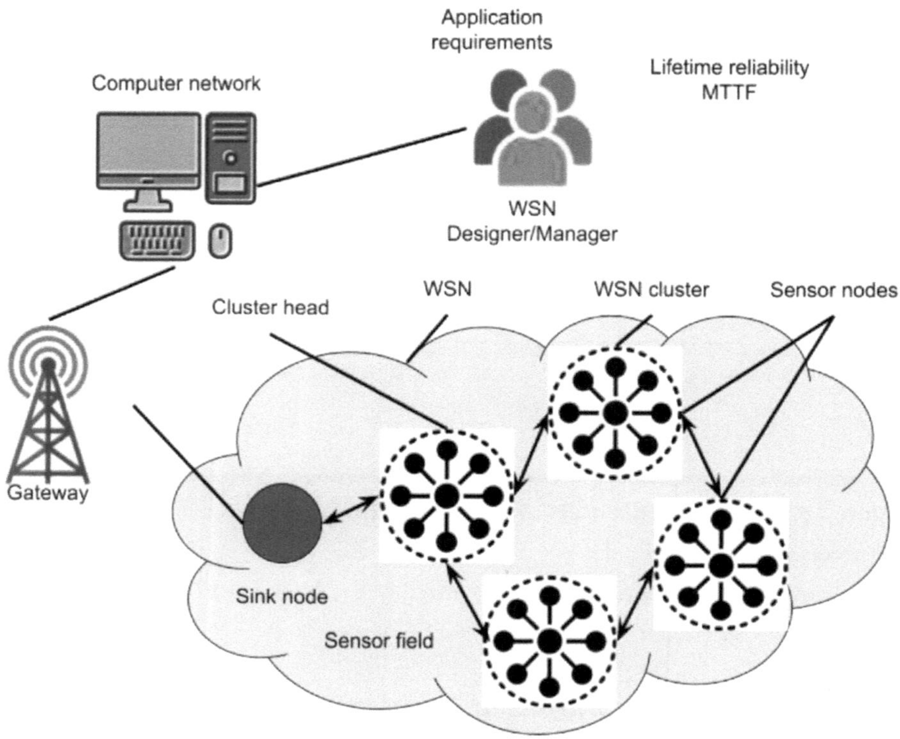

Figure 18.6 The proposed 6G wireless sensor network security analysis.

18.5.4 Dangerous Software

One important channel for the spread of malicious software and viruses is social media. Programs developed or constructed with the malicious purpose to hurt and damage users of our systems are referred to as malicious software. After the victims respond, the malware infects their computers without their knowledge or consent. According to the Symantec ISTR 2019, Emotet, a type of banking Trojan virus, is one of the most expensive and destructive malware that impact both the public and private sectors, despite a discernible decline in new malware varieties in 2018 and the quantity of malware cases within the last decade (Figure 18.6).

18.6 Conclusions

Through increased engagement, interactivity, and effectiveness, gamification has demonstrated significant promise for enhancing learning outcomes in cybersecurity education. Simulations, game-based learning, and competitive components can all help educational programs better prepare students to handle challenging cybersecurity problems. In addition to highlighting numerous successful implementations, this research has brought to light typical problems with gamified cybersecurity education, including scalability, long-term engagement, and content adaptation. In order to address these issues, future studies should concentrate on creating adaptive learning models that can tailor information according to each learner's requirements and development. Incorporating real-world situations and ongoing assessment techniques can also assist in closing the knowledge gap between theory and practice, guaranteeing that students are ready for real-world cybersecurity dangers. To create more comprehensive teaching resources, multidisciplinary cooperation is also required, bringing together knowledge from the fields of game design, education, and cybersecurity. The suggested study agenda identifies important topics for additional investigation, such as the use of cutting-edge technology like artificial intelligence (AI), machine learning, and virtual reality to improve gamification tactics. By tackling these issues, educators and developers may produce gamified solutions that are more efficient, scalable, and long-lasting, ultimately helping to build a workforce of cybersecurity professionals who are more resilient and skilled.

References

[1] Alhammadi, A., Shayea, I., El-Saleh, A. A., Azmi, M. H., Ismail, Z. H., Kouhalvandi, L., & Saad, S. A. (2024). Artificial Intelligence in 6G Wireless Networks: Opportunities, Applications, and Challenges. *International Journal of Intelligent Systems*, 2024(1), 8845070.

[2] Zhang, S., Zhu, D., & Liu, Y. (2024). Artificial Intelligence Empowered Physical Layer Security for 6G: State-of-the-Art, Challenges, and Opportunities. *Computer Networks*, 242, 110255.

[3] Talwar, R., Sharma, M., Anand, S., & Pandey, D. (2024). A Review on the Role of Artificial Intelligence in Security Challenges for 6G Communications. In *Security Issues and Solutions in 6G Communications and Beyond*, 12–25. IGI Global.

[4] Kaur, N., Kshetri, N., & Pandey, P. S. (2024). 6AInets: Harnessing artificial intelligence for the 6G network security: Impacts and challenges. arXiv preprint arXiv:2404.08643.

[5] Damaraju, A. (2024). The Future of Cybersecurity: 5G and 6G Networks and Their Implications. *International Journal of Advanced Engineering Technologies and Innovations*, 1(3), 359–386.

[6] Blika, A., Palmos, S., Doukas, G., Lamprou, V., Pelekis, S., Kontoulis, M., ... & Askounis, D. (2024). Federated Learning For Enhanced Cybersecurity and Trustworthiness In 5G and 6G Networks: A Comprehensive Survey. *IEEE Open Journal of the Communications Society*, 99, 23.

[7] Li, Y., Xiao, Y., Liang, W., Cai, J., Zhang, R., Li, K. C., & Khan, M. K. (2024). The Security and Privacy Challenges toward Cybersecurity of 6G Networks: A Comprehensive Review. *Computer Science and Information Systems*, 21, 16.

[8] Gupta, M., Jha, R. K., & Sabraj, M. (2024). Security Challenges in 6G Networks: A Game Theory-based Intrusion Detection with Network Slicing and AI/ML Perspectives.

[9] Nair, M. M., Deshmukh, A., & Tyagi, A. K. (2024). Artificial Intelligence for Cyber Security: Current Trends and Future Challenges. In *Automated Secure Computing for Next-Generation Systems*, edited by Amit Kumar Tyagi, 83–114: Wiley.

[10] Ganesh, S. S., Rajaram, G., Anusuya, V., Thirumalaikumari, T., Navaz, K., & Sobia, M. C. (2024). Next-Generation Threat Detection and Mitigation in 6G Wireless Networks Using IAM, ZTNA and Advanced Security Mechanisms. *Journal of Electrical Systems*, 20(5s), 2034–2041.

[11] Thanganadar, H., Yaseen, S. M., Shukla, S. K., Bist, A. S., Shavkatovich, S. N., & Vijayakumar, P. (2024). 6G Wireless Communication Cyber Physical System Based Smart Healthcare Using Quantum Optimization with Machine Learning. *Wireless Personal Communications*, 1–20.

[12] Awadallah, A., Eledlebi, K., Zemerly, J., Puthal, D., Damiani, E., Taha, K., ... & Yeun, C. Y. (2024). Artificial Intelligence-Based Cybersecurity for the Metaverse: Research Challenges and Opportunities. *IEEE Communications Surveys & Tutorials*, 2, 1008–1052.

[13] Aslam, A. B., Iqbal, F., Talpur, U., Syed, Z. S., & Shaikh, F. K. (2024). Artificial Intelligence-Enabled 6G Mobile Systems. In *Intelligent Technologies for Healthcare Business Applications*, 49–79. Cham: Springer Nature Switzerland.

[14] Javeed, D., Saeed, M. S., Ahmad, I., Adil, M., Kumar, P., & Islam, A. N. (2024). Quantum-Empowered Federated Learning and 6G Wireless Networks for IoT Security: Concept, Challenges and Future Directions. *Future Generation Computer Systems*.

[15] Dasika, S. R. C. M., Kakulapati, V., & Saligrama, S. (2024). Network Security and Data Privacy in 6G Environment: Impacts and Challenges. *Journal of Emerging Technologies and Innovative Research (JETIR)*, 11(2), D452–D461.

[16] Selvarajan, S., Manoharan, H., Khadidos, A. O., Khadidos, A. O., Alshareef, A. M., & Alsobhi, A. (2024). Secured 6G Communication for Consumer Electronics With Advanced Artificial Intelligence Algorithms. *IEEE Transactions on Consumer Electronics*, 70(99), 1.

[17] Rojek, I., Kotlarz, P., Dorożyński, J., & Mikołajewski, D. (2024). Sixth-Generation (6G) Networks for Improved Machine-to-Machine (M2M) Communication in Industry 4.0. *Electronics*, 13(10), 1832.

[18] Ara, I., & Kelley, B. (2024). Physical Layer Security for 6G: Toward Achieving Intelligent Native Security at Layer-1. *IEEE Access*, 12, 82800–82824.

[19] Haider, S. A., Ramesh, J. V. N., Raina, V., Maaliw, R. R., Soni, M., Nasurova, K., ... & Singh, P. P. (2024). Secure Artificial Intelligence for Precise Vehicle Behavior Prediction in 6G Consumer Electronics. *IEEE Transactions on Consumer Electronics*, 25.

[20] Mahmood, K., Obaidat, M. S., Shamshad, S., Alenazi, M. J., Kumar, G., Anisi, M. H., & Conti, M. (2024). Cost-Effective Authenticated Solution (CAS) for 6G-Enabled Artificial Intelligence of Medical Things (AIoMT). *IEEE Internet of Things Journal*, 11(13), 23977–23984.

[21] Liu, Y., Su, Z., Peng, H., Xiang, Y., Wang, W., & Li, R. (2024). Zero Trust-Based Mobile Network Security Architecture. *IEEE Wireless Communications*, 31(2), 82–88.

[22] Zhao, Y., Hu, C., & Wang, R. (2024). Support Vector Machine (SVM) to Predict Risk Factors in the 6G Cyber Digital Transformation Process of Enterprises. *Wireless Personal Communications*, 1–18.

[23] Saoud, B., Shayea, I., Yahya, A. E., Shamsan, Z. A., Alhammadi, A., Alawad, M. A., & Alkhrijah, Y. (2024). Artificial Intelligence, Internet of things and 6G methodologies in the context of Vehicular Ad-hoc Networks (VANETs): Survey. *ICT Express*, 20, 959–980.

[24] Shukur, H. M., Askar, S., & Zeebaree, S. R. (2024). The Utilization of 6G in Industry 4.0. *Applied Computer Science*, 20(2), 75–89.

[25] Aloqaily, M., Paik, H., Lunardi, W. T., Tunc, C., & He, F. (2024). Guest Editorial: Zero Trust Security Methods for Wireless Networks. *IEEE Wireless Communications*, 31(2), 12–13.

[26] Siddiky, M. N. A., Rahman, M. E., & Uzzal, M. S. (2025). Beyond 5G: A Comprehensive Exploration of 6G Wireless Communication Technologies, *Computers*, 14(1), Article 15. https://doi.org/10.3390/computers14010015

[27] Muniyandi, A. P., Balusamy, B., Dhanaraj, R. K., Sumathi, D., Nandakumar, S., Preetha, K. S., ... & Paramasivam, A. (2024). Intelligent Security System for Preventing DDoS Attacks for 6G Enabled WBSN Using Improve Grey Wolf Optimization. *IEEE Transactions on Consumer Electronics*, 70, 1–20.

[28] Nguyen, T., Nguyen, H., Ijaz, A., Sheikhi, S., Vasilakos, A. V., & Kostakos, P. (2024). Large language models in 6G security: Challenges and opportunities. arXiv preprint arXiv:2403.12239.

[29] Rzym, G., Masny, A., & Chołda, P. (2024). Dynamic Telemetry and Deep Neural Networks for Anomaly Detection in 6G Software-Defined Networks. *Electronics*, 13(2), 382.

Index

A
Accelerator, 2, 104
Accuracy, 8, 14, 31, 53, 59, 63, 70, 71, 74, 78, 86, 88, 90–92, 99, 104, 107–109, 111–115, 122, 134, 136, 139, 142, 170, 216, 217, 229, 230, 231, 243, 270, 277, 278, 281, 285
Attributes, 5, 9, 19, 29, 54, 76, 149, 230, 279, 287, 288
Automation, 1, 13, 15, 116, 136, 138, 139, 179, 186, 196, 200, 205, 230
Autonomy, 2, 154, 185

B
Balanced dataset, 86, 87, 94, 98
Big data, 1, 15, 18, 32, 66, 149, 194, 195, 218, 235, 238, 239

C
Cloud computing, 2, 203, 230
CNNs, 53–3, 64, 86, 87, 89, 97, 118
Communications, 9, 31, 66, 150, 181, 200, 259, 315, 317–319
Composite beam, 39, 42
Computational efficiency, 6, 146, 147
Conductivity, 35, 257, 258
Cyber-attack, 54, 56, 65, 67

D
Data analysis, 2, 55, 104, 108, 202, 203, 217, 225
Data collection, 104, 108, 121, 189–191, 208, 219, 292
Diagnostics, 2, 13, 50, 86, 104, 198, 230

E
Edge clouds, 54, 64
Electric motor, 17–20, 22, 29–31
Encoding, 53, 54
Energy losses, 153, 155
Energy management system, 68, 83, 84, 154, 179–181
ESPs, 58, 65

F
F1 score, 62, 92, 99, 100, 142, 145, 146

G
Green energy, 68, 73, 83, 84, 179, 181

H
Hidden layers, 62, 89, 111, 120, 124, 127, 129, 130, 134, 135, 257, 262
Human operators, 2, 13

L
Latency, 4, 11–13, 59, 61, 138

M
Machinery, 1, 32, 85, 102, 103, 116, 149, 199, 200, 222, 233–235, 237, 240, 292–294, 298
Mechanistic models, 118, 121
Model training, 20, 29, 134

N
Network architecture, 62, 97, 98, 120, 138, 149, 152, 153, 156, 158, 159, 168
NLP, 54, 149
Nutrients, 117, 192, 258

P
Parameter, 6, 35, 36, 45, 46, 48, 76, 94, 95, 124, 138, 234, 256, 258, 265, 269

PG, 54–56, 58, 65, 162
Power systems, 15, 52, 55–57, 62, 63, 66, 67, 72, 82, 83, 102, 179, 181, 192

R

Random forests, 79, 214, 222, 225
Real-time analysis, 9, 13
Resources, 4, 12, 13, 19, 52, 59, 65, 68, 70, 72, 83, 117, 118, 134, 136, 138, 151, 157, 177, 185–188, 192, 206, 300, 301, 309, 314, 317
Robotics, 2, 15, 182–186, 196, 197, 208

S

Sensitivity, 36, 43, 45, 48, 49, 257
Sensor fusion, 2, 8, 50
Smart grids, 67, 69, 70, 180
SODA, 140–148
Software, 2–5, 7, 105, 136, 150, 200, 202, 236, 238, 239, 256, 314, 316, 319
Support vector machine, 34, 83, 90, 106–108, 116, 214, 222, 226, 319
Swarm intelligence, 184, 185, 187–189, 192, 194, 196

T

Training loss, 55, 61, 62, 130